Daily Poison

Johann G. Zaller

Daily Poison

Pesticides - an Underestimated Danger

Copernicus Books is a brand of Springer

Johann G. Zaller
Institute of Zoology
University of Natural Resources and Life Sciences Vienna
Vienna, Austria

ISBN 978-3-030-50529-5   ISBN 978-3-030-50530-1   (eBook)
https://doi.org/10.1007/978-3-030-50530-1

English translation from the German language edition: Unser Täglich Gift - Pestizide - Die Unterschätzte Gefahr by Johann G. Zaller Copyright © Deuticke im Paul Zsolnay Verlag Wien 2018

This Copernicus imprint is published by the registered company Springer Nature Switzerland AG.
The registered company address is: Gewerbestrasse 11, 6330 Cham, Switzerland

The truth is reasonable for people.

Ingeborg Bachmann

Foreword

Why writing a book on pesticides? According to the advocates of pesticide-based agriculture, pesticides are among best studied substances in the world that specifically target weeds, pests or diseases and after fulfilling their duties, are broken down into harmless substances. Should traces of pesticides nevertheless be detected in food, the environment or the human body, this is said to be primarily due to the refined analytical methods. Usually, we consumers are comforted that pesticide residue levels in food are always below legal limits. In addition, everyone knows the century-old principle according to which only the dosage makes the poison, and one can even die from excessive water consumption! Consequently, anyone who speaks out against the use of pesticides is considered a dreamy environmentalist who denies the reality of modern food production and in doing so risking the starving of millions of people. This is the rough outline of the public presentation on the subject of pesticides. With this book I would like to shed some light on these and other statements about pesticides and challenge the truthfulness of the above-mentioned statements. This seems more important than ever, especially in our times of ubiquitous influence of politics through various interest groups, fake news and alternative facts.

You are only marginally interested in this topic, since you are neither a farmer nor a hobby gardener, and as a city-dweller you feel that you are not confronted with pesticides whatsoever? Well, let's consider you treat yourself with a pizza with mixed salad, a glass of wine or apple juice and a banana for dessert. Perhaps you will be surprised that for the production of the main ingredients, wheat for the pizza dough, tomatoes, corn, peppers, herbs, salad, wine and apple production, more than 1000 different pesticides are permitted in Europe, the USA or other countries (EPA 2019b; European Commission

2019)! Of course, all these pesticides will not be used at the same time, but they are theoretically available in the pesticide arsenal of conventional, pesticide-intensive agriculture. This pesticide number is conservative and does not include chemical substances that are allowed for further food processing, preservation, storage, taste improvement, cellar technology and so on. I hope that this example has convinced you that you are most likely also affected by pesticides, whether you like it or not!

I started with the research on the effects of pesticides on non-target organisms several years ago. In the hindsight I was quite naive when I started with this. Before that I spent almost 15 years studying the effects of elevated-CO_2, ultraviolet-B radiation and other environmental and climatic factors on ecological interactions between plants and animals. As many ecologists I preferred to investigate native ecosystems that are close to nature, because they usually inhabit interesting and rare species in great diversity. By contrast agricultural or other human-influenced ecosystems are often treated somewhat pejoratively by ecologists because the factors driving ecological interactions are simpler than in native ecosystems.

My curiosity on the pesticide topic rose after moving into a very nice region in Austria dominated by viticulture. Raised in a mountainous area with dominating grassland farming and no pesticide use I found it bizarre to spray pesticides on crops we would later consume or feed to livestock animals. Back then, of course, also the general media turmoil around glyphosate caught my attention. A first orientation on the topic in the scientific literature was eye-opening, because despite thousands of scientific studies published there were still many gaps in our knowledge. Just as most non-scientists and also many fellow scientists do, I believed that the pesticides that are used every day were, of course, rigorously tested before they could be applied to the environment in order to control pests and weeds. Of course, I remembered reading the famous *Silent spring* of Rachel Carson (Carson 1962) dealing with the devastating effects of pesticides on our environment but this was almost 60 years ago and things surely have changed. Well, after just a brief immersion into the matter, I concluded that many of these pesticide-related standard phrases were myths spread by pesticide manufacturers and various interest groups (Leu 2014). This eye-opener and what I experienced in the course of our relevant work at the University of Natural Resources and Life Sciences in Vienna inspired me to write this book. Most of what I compiled here is already published in various scientific papers, but who other than fellow scientists or a few science journalists reads and understands these articles often presented in very technical language?

The following aspects prompted me to write this book:

- The contamination even of newborn babies with traces of pesticides in their bodies.
- Parkinson's disease is a recognized occupational disease for winegrowers in France.
- The steady increase in legal limits for pesticide exposure and residue levels in recent years.
- The inadequacy for many modern synthetic pesticides of the much-used quotation from Paracelsus, according to which the dosage makes the poison.
- The many pesticide counterfeits with unknown ingredients which make up to 25% in some countries (UNEP 2018).
- Agrochemical companies pay several million Euros on compensation to Italian winegrowers because treatment with a recommended pesticide product led to complete crop failures.
- Toxic waste landfills of agrochemical industries are ticking time bombs especially in the event of natural disasters.
- Scientists who take a critical look at pesticides are quickly denounced in internet forums aiming to undermine their integrity?

Nobody can seriously say how the well over 100,000 chemicals that are currently in use affect our health and nature, as their side effects are insufficiently investigated. Theoretically, in Europe the precautionary principle is stipulated in the European Treaties. However, this principle is widely ignored also when it comes to free trade agreements (TTIP, Mercosur etc.). In my assessment of the situation I am strongly guided by this precautionary principle, which calls for restrictions in marketing a substance when there is a suspected risk to human health or the environment even when scientific evidence is not completely clear.

The first part of this book begins by outlining the problems regarding our use of pesticides, the quantities applied and where pesticides are mainly used. The second part gives an insight into the everyday scientific research of pesticide effects and their results. It also discusses how scientific results are disseminated and received in the public. If you now ask yourself how we can feed the growing world population without modern pesticide-intensive agriculture, then you have apparently been taken in by the marketing machinery of the agricultural lobbyists! The third part shows that pesticide-intensive agriculture is actually a quite unsustainable business model that causes tremendous costs for our economies and societies. The supposed benefits in crop yields by no means outweighs this. Fortunately, there are many practical alternative concepts that work without synthetic pesticides and even representatives of conventional agriculture admit that it would be possible to save half of the pesticides without causing drops in yields. An outlook chapter summarizes what is urgently needed for a transformation of pesticide-intensive agriculture and what policymakers need to contribute.

This book is based on a German popular science book (Zaller 2018). Although the German book already addresses international topics, I further expanded the scope of this English version in order to interest more international readers and updated the references with relevant studies. There remains a bias towards the pesticide situation in Europe, but that simply reflects my main sphere of activity. All statements in the text are supported by scientific studies that can easily be found in the internet. However, because of the broadness of the scope of the book I could only include a small part of the available studies—apologies if I forgot some important contributions. If the reader gets the impression that pesticides exclusively have detrimental effects on humans and the environment it is important to emphasize that there are of course studies out there that show only little effects. However, for my look at this matter both the precautionary principle and also some pragmatism is decisive. Why should we take unnecessary risks when there is scientific evidence on the harmfulness of certain substances and beyond, many alternative methods successfully demonstrate that there are pesticide-free measures to deal with pests, diseases and weeds? Additional information is knitted-in from numerous discussions with pesticide experts and practitioners.

Good scientific practice is characterized by the fact that studies are reviewed by other scientists (usually anonymously) and then published in international journals. In many modern journals, the studies, the underlying raw data and the reports of the reviewers are now made freely accessible for everybody. If studies on the approval of pesticides are still kept confidential by manufacturers or approval authorities, then this is scientifically suspicious and gives room for speculations that not everything might have been performed correctly. This secrecy is justified by manufacturers and authorities by protecting business secrets.

Readers who have not yet dealt with the pesticide topic will find some aspects simply unbelievable. However, it can be safely assumed that the real situation may well be even more serious than described in this book.

What I certainly don't want to do with this book is denouncing farmers that apply pesticides. My aim is rather to sensitize the farmers, other pesticide applicators and the wider public for this topic and perhaps ultimately also make it clear to politicians that there is an acute need for action for the benefit of our environment and our health. Moreover, often the farmers are ill-advised by people with unilateral economic interests. The mechanisms and entanglements between the agrochemical industry and spokespersons of the farm industry that lead to recommendations for such excessive use of pesticides must also be addressed in this context.

When I talk about pesticides in this book, I am referring to the substances—herbicides, insecticides, fungicides and others—that are sometimes called plant protection products. They are used in agriculture, forestry, by road-keepers, municipalities, railway companies and private persons in their gardens or at home. I personally refuse to use the term plant protection product because it is misleading and euphemistic. The absurdity of the term becomes clear when herbicides, i.e. substances that are made to kill plants, are also called plant protection products.

This book will not deal with biocides which are chemical substances that are used to control organisms that are harmful to human or animal health or that cause damage to natural or manufactured products. It is a very diverse group of poisonous substances including wood preservatives, insecticides, disinfectants, and rodenticides mainly used in non-agricultural applications. The terms "biocides", "pesticides" and "plant protection products" are often used interchangeably, the split up in different categories has mainly legal reasons. In Germany alone about 25,000 biocide products are known (UBA 2019), they are used outside of agricultural land in many areas of private or professional life such as in the building industry. Effects of these biocides on the environment and humans are particularly poorly studied and would fill another book. Biocides are approved throughout the EU, and are recorded in a positive list. Substances that are categorized both as a pesticide and a biocide are termed "Dual-Use" substances. Maximum residue levels for food also apply to these substances.

Already in this preface I mentioned the terms pests or weeds. Generally, both words are taboo words in an ecological perspective. In an ecosystem there are actually no unwanted organisms, since each species plays a specific role in the entire system. If an organism becomes harmful, then only if a certain population level is present. A few potato beetles, or thistles, have important functions in ecosystems; only if they occur in masses that harm our crops we see them as problematic and worth controlling. Since these terms for animals and plants, which occur at the wrong place in too large quantities, are so widely used in general language, I will use them also throughout the book.

Contents

1

What Is the Problem? Pesticides in Our Everyday Life

Literally, pesticides are substances that kill pestiferous organisms. In practice, they include insecticides against insect pests, herbicides against weeds, fungicides against fungal pathogens, acaricides against spiders and mites, molluscicides against pest slugs, and so on. The widespread use of these toxins in agriculture is being justified by securing higher yields that are considered more important to human society than potential side effects of these pesticides. We will later see that there is actually quite little scientific evidence that more pesticides secure higher yields. Pesticide manufacturers and regulators are reassuring critics that everything is alright as long as pesticides are used correctly with proper application equipment and protective clothing. In this context, the term "application according to good agricultural practice" is often used. What sounds like a standard only means the farming practice common in a given region. It is actually a rather empty phrase without clear rules of applications, quantitative requirements, or any legal significance. Nevertheless, the term is often used to reassure consumers that food production is well-controlled and environmentally friendly. One wonders, however, why reports are mounting that, despite this good practice, pesticide residues are so widespread in our environment and food.

For about 60 years, conventional agriculture has relied on the intensive use of synthetic chemical pesticides. In order to safeguard agricultural yields, around 466 active substances are currently approved in the European Union; among them are also about 25% microorganisms, insect pheromones, and plant extracts which are not synthetic chemicals (EC 2019c). These active ingredients are mixed or formulated with so-called adjuvants or co-formulants to a great variety of pesticide products. This sums up to more than 1700

© Springer Nature Switzerland AG 2020
J. G. Zaller, *Daily Poison*, https://doi.org/10.1007/978-3-030-50530-1_1

pesticides approved alone in Germany (BVL 2019). It is not easy to get official numbers of approved pesticides for other important agricultural countries
such as the USA or China, but numbers mentioned in studies indicate that it
is substantially higher. There is still much to be said in later chapters about the
role of these adjuvants, which are commonly considered chemically inert and
ineffective.

In Europe, around 374,000 tons of active substances of pesticides are sold
annually on average between 2011 and 2016 (Eurostat 2019). The demand
for pesticides is increasing worldwide, and the amount of pesticide use has
risen 50-fold since 1950. Figure 1.1 provides an overview of pesticide use in
different world regions based on the United Nations Food and Agricultural
Organization (FAO, www.fao.org/faostats). The database contains national
data collected from 205 countries and territories via a questionnaire filled out
by national statistical offices, ministries of agriculture, or other relevant agencies. Since 1990 the amount of total pesticides used in the world has increased
by 80% until 2017. The only region with decreasing pesticide amount during
this time was Europe, while all other world regions showed enormous increases
by 218% in Oceania, 116% in the Americas, and 97% in Asia.

Many pesticides are produced and sold by European, US, and Chinese
chemical companies. The whole pesticide business is gigantic and accounts for

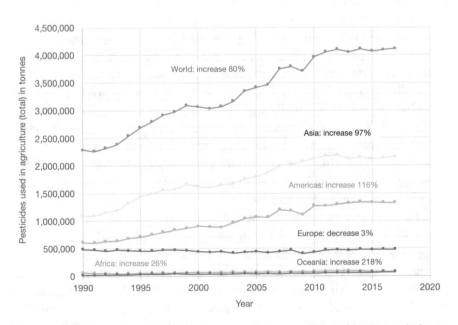

Fig. 1.1 Pesticide use in different world regions between 1990 and 2017. Graph drawn based on data from FAOSTATS 2019

an estimated 49 billion € worldwide (EPA 2017). Only four agrochemical companies control around 75% of the global pesticides business: two of them have their headquarters in Europe (BASF, Bayer/Monsanto), one in China (ChemChina/Syngenta), and one in the USA (Dow/DuPont now called Corteva Agriscience). The Chinese-Swiss company ChemChina/Syngenta is the world's largest pesticide manufacturer with a share of approximately 23% (MultiWatch 2016). The fact that the majority of research into the effects and side effects of pesticides is carried out or commissioned by these companies themselves is also notable.

It sounds unbelievable, but cautious estimates indicate that at most 10% of pesticides used are actually effective against the pests or diseases they are intended to control (Pimentel et al. 1998). Over 90% of the pesticides used affect areas not meant to be treated or impact so-called nontarget organisms, i.e., organisms that were not supposed to be controlled. Anyone who has ever observed pesticide sprayers on land, or worse, pesticide spraying via planes (so-called crop duster), can easily acknowledge this estimate of pesticide wastage. The consequences are inevitably a worldwide loss of biodiversity which is increasingly linked to pesticide use (Brühl and Zaller 2019). Furthermore, the pesticides, their degradation products, and co-formulants accumulate in the soil and impair nutrient cycles and the natural interaction between beneficial organisms and pests in nature. Sooner or later, these pesticides will also be found in drinking water or in our food and impair our health. Pesticides are now held responsible for neurological and hormonal dysfunctions, miscarriages, cancer, and other chronic diseases. These topics will also be elaborated further in later chapters.

Even though direct links between pesticide exposure and our health are difficult to prove, some health-related side effects are now legally recognized. Parkinson's disease caused by pesticides is officially accepted as an occupational disease for French winegrowers (agrarheute 2012; Gunnarsson and Bodin 2017). In order to be entitled to a retirement pension, winegrowers or farm worker must prove to have come into contact with the pesticides for at least 10 years and the disease must have broken out no later than 1 year after use. In Germany, too, several farmers and gardeners have already been granted pensions for occupational disease after developing Parkinson's. When I mention this aspect in public talks, it regularly leaves the audience in disbelief, consternation, and anger. Indeed, it is unbelievable that our farmers deal with such products. Sometimes it feels as if farmers are meant to sacrifice their health for the sake of cheap food. But even the agricultural chambers and extension services are silent about this and leave their members out in the pesticide rain. This is a taboo subject in many countries. Politicians responsible for agriculture, the environment, or health, or even representatives of farmers, seem to have more important issues on their agendas. But perhaps they are simply in a dilemma, since many of them sit in governing boards of agrochemical companies and are therefore in a conflict of interest if they would have to advocate for stricter regulations. This is at least documented for Germany (Balser et al. 2019; Nischwitz et al. 2019) and the USA (UCS 2018).

Agriculture, along with mining and construction, is one of the three most dangerous professions in the world. Of the many millions of accidents at work every year, at least 170,000 are fatal. The main causes for these fatalities are accidents involving machinery, poisoning with pesticides and other agrochemicals, physical overload, noise, dust, allergies, and animal-borne diseases (IAASTD 2009).

Pragmatics may now reply that pests and diseases need to be combated, because food production is at stake and farmers have to make a living from yields. It is assured that pesticides are only sprayed when there is acute danger and the harvest is threatened by pests, weeds, or fungal diseases. Unfortunately, this is not really the case. In many situations, pesticides are not only used when diseases or pests occur, but also as a preventive measure. This inevitably leads to some agricultural crops being treated with pesticides several times in the course of a vegetation period, for example because certain weather conditions increase the probability of fungal diseases according to computer models. Modern farmers are nowadays informed with text messages on their smartphones about the necessary pesticide treatments. This handy spraying advice is often provided free of charge by pesticide manufacturers. One might see a clear conflict of interest here as these advisors after all of course aim at selling their pesticides. I have also been told that unnecessary pesticide applications are also recommended by agricultural warehouses because salespersons do not want to risk any recourse claims from farmers in the event of crop failures (Zaller 2018).

Agricultural crops are differently affected by pests and diseases and treated with pesticides at different intensities. Generally, it is difficult to get hard data on the number, times, and frequency of applied pesticides. However, fortunately, we have published information from Germany (JKI 2019). These data include around 100 farms for each of the following crops: covering winter wheat, winter rye, winter oilseed rape, sugar beet, maize, potatoes, apples, and grapes, and 80 farms cultivating hops (Fig. 1.2).

Of course, these farms applied pesticides according to good agricultural practice. The most doubtful leader in the use of pesticides is apple cultivation with an average of 31 pesticide treatments per growing season, followed by viticulture with 18 and potatoes with 12 applications per season. Considering that the apple growing season lasts only 24 weeks (6 months), this is more

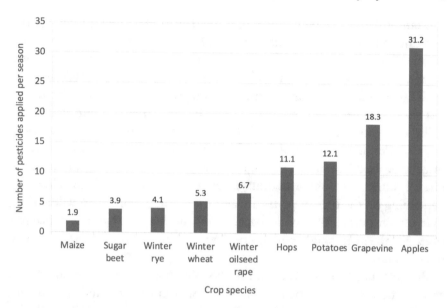

Fig. 1.2 Average number of pesticides applied in different crops assessed between 2011 and 2018 throughout Germany. Note that pesticides used for seed dressings are not included. Graph drawn based on data from JKI (2019)

than one pesticide treatment per week. In contrast, hops with an average of 11, wheat with five, or maize with two pesticide applications per season appear like near-natural forms of cultivation. Especially for the arable crops these numbers are too low as pesticides used in seed treatments are not considered. Also, pesticides used to store fruits are not included.

Comparisons with other countries are difficult because very little data is available for the public. However, it appears that the pesticide usage might be even heavier in other countries. As an indication the average number of pesticide treatments applied to crops in the UK increased dramatically (PAN UK 2017): for wheat from averaged 1.7 applications in 1974 to 20.7 in 2014, for potatoes from 5.3 applications in 1975 to 32 in 2016, and for onions and leeks from 1.8 applications in 1966 to 32.6 in 2015. Different time spans are mentioned due to the availability of data. An illustration of the available pesticides for different crops in the UK is also quite astonishing: there are 62 approved active ingredients with 633 approved products for potato production, and 93 active ingredients and 1432 products approved for wheat production (FERA 2020).

The figure makes clear that fruit and wine growing is very pesticide intensive. So, parental advice to children to wash fruits before eating was obviously well meant. However, unfortunately, many pesticides contain active substances that act systemically within the treated plant that means that they not only adhere to the outside of the plants or fruits but are distributed throughout the entire plant. Researchers actually tested a few different ways (water, bleach, baking soda) to get rid of pesticide residues from apples (Yang et al. 2017). Holding an apple under running water for a couple of seconds is not enough to get rid of the pesticides on its skin. However, almost all pesticides were gone from the apple surface after soaking the apples in a baking soda solution for 12–15 min! Still, about 20% of the fungicide and 4% of the insecticide used had soaked through the apple's skin and could not be washed off. Of course, the residue amounts found on fruits are usually low; however, depending on the type of pesticides and the amount you eat we will later see that certain pesticides may do harm to our health in the tiniest concentrations. So, the safest way is perhaps peeling your non-organic apples or buying organic ones.

But fortunately, as we read in the specialist books, modern pesticides are among the best tested chemical substances, comparable only to medicines (Hallmann et al. 2009). It is also constantly claimed that modern pesticides, in contrast to the older ones, are easily biodegradable and they quickly decompose into harmless components. Surprisingly, we regularly find pesticide residues in our food and every year pesticides are being banned because of their burden on the environment and human health. I do not want to sound polemical, but these are often the arguments in debates about pesticides.

The excessive use of pesticides is leading to another problem, which is now spreading around the globe. More and more animal pests or plant diseases become resistant against pesticides and weeds develop to so-called superweeds. It would be unfair to suggest that there is a business model behind this, since resistances ultimately lead to the development and sale of ever new pesticides. In any case, basic evolution in school biology already teaches us that organisms react to chronic stressors—and regular pesticide administration is nothing else—with evolutionary adaptations. Experts also warned very early on of the danger of resistance formation, but were ignored by the agrochemical industry. The situation is now so serious that even pesticide manufacturers are advising their farmers to switch to products of their competitors or apply more traditional methods of mechanical weed control (UCS 2013). With the logic of agrochemical companies, conventional research is now moving toward developing pesticides with several active ingredients or the development of genetically modified organisms (GMOs) that are resistant to several herbicides.

So far, we also contoured the problem area of this book. The following questions will be addressed further:

- How well do we actually know the substances that are applied in these large quantities?
- How rigorously are these products tested before they are marketed?
- What side effects do these pesticides have on our environment and on humans?
- Are the risks of pesticides outweighed by their benefits?
- What role does and should critical science and policy-makers play?
- Can we feed the growing world population without using pesticides?

But first we will briefly take a look on how the so-called conventional agriculture became so dependent on agrochemicals.

1.1   Agriculture in the Pesticide Treadmill

The history of pesticide use is almost as old as the history of agriculture itself because pestiferous organisms and plant diseases were always attacking crops. In Mesopotamia, elemental sulfur was used to combat pests on agricultural crops as early as 2500 BC. In the fifteenth century, the use of arsenic, mercury, or lead was widespread. In China during the Ming Dynasty (1368–1644), a number of plants and minerals were used as pesticides such as veratridine, flavescens, arsenolite, realgar, orpiment, and lime (Zhang et al. 2011). The voyages of discovery in the eighteenth century then brought to light the fact that plant-active substances can also be used against pests—nicotine from tobacco plants or pyrethrin from chrysanthemums plants. The widespread use of pesticides in the field began in the second half of the nineteenth century. The decisive trigger was probably the introduction of various pests around the globe with the upcoming of international trade, leading to catastrophic crop failures.

Back in 1845 Irish people based their diet largely on potatoes imported from the New World. Consequently, it was devastating when the potato blight (*Phytophthora infestans*), a fungal pathogen that is favored by moist, cool environments infected potato crops. In Ireland about one million people died and at least as many emigrated to the USA and other places; also, thousands left the Highlands of Scotland. The rest of Europe had more diverse crops and were not hit as hard. The disease is still around in our days and damaging potato crops and other crops of the Solanaceae family such as tomatoes. Around 1878, the highly aggressive fungus-like organism downy mildew of the family Peronosporaceae was introduced with seed potatoes from America to French vineyards. These are just a few examples of diseases that today's agriculture is still fighting against particularly when growing potatoes, grapes, tobacco, and cucurbits. In addition, worldwide trade flows constantly bring new potential pests or diseases to regions where they have no natural counterparts.

The first synthetic insecticide was developed as early as 1892 by the German agrochemical company Bayer. But the era of modern pesticides did really begin in the 1930s with organochlorine insecticides, including active ingredients such as the notorious DDT (dichlorodiphenyltrichloroethane) and lindane (gamma-hexachlorocyclohexane). It was in the 1940s that the first synthetic herbicide called 2,4-D (2,4-dichlorophenoxyacetic acid) was developed. All these substances are still around in our environment. We will see later that 2,4-D even celebrates a renaissance nowadays in the fight against glyphosate-resistant superweeds. Many pesticide-active substances, especially organophosphates, were used as combat gases during World War II or even in Iraq War in 2003. They were easy and cheap to produce and shown to be effective against a wide range of insect pests.

In many parts of the world, agricultural yields increased after the application of pesticides; therefore, their use was increasingly popular. In 1957, the first factory producing organophosphorus pesticides was built in China. Meanwhile, China has become the largest producer and user of pesticides in the world (Zhang et al. 2011).

Soon it turned out that many pesticides also have side effects on humans and the environment and that they were persistent and accumulated in the food chain.

In contrast to small-scale, traditional, and subsistence agriculture, modern conventional agriculture is characterized by specialization, monocultures, and maximum efficiency. At least that is how it is communicated to the public. Monocultures, i.e., the cultivation of an agricultural crop over huge areas, promote the development and spread of plant diseases and pests. The trend toward specialization is mainly found in conventional farming and to a lesser degree in organic farming.

Agriculture has undergone enormous changes in recent decades which heavily affected rural populations. In Europe in the 1950s more than half of the population was still employed in agriculture, whereas nowadays only 2.8% of Europe's working population is employed in agriculture (Eurostat 2018). Still we have about 10.5 million agricultural holdings in the EU in 2016, two-thirds of which were less than 5 ha in size. EU farms cultivated 173 million hectares of land in 2016, 39% of the total land area of the EU. The number of farms in the EU has been in steep decline, but the amount of land used for production has remained steady. Similar declines can be seen in all industrial nations.

Perhaps because of this decline in the labor force, the importance of agriculture compared with other branches of the economy is also notoriously underestimated by politics. However, a representative survey across Germany reveals that the importance of agriculture is still appreciated (TNS Emnid 2012). According to this survey most Germans think that a well-functioning agriculture is a basic prerequisite for the quality of life and viability of the country and that rural life is an important component of German culture. The survey also showed that ecological aspects are becoming increasingly important for people and that the production of renewable energies and climate protection are increasingly linked to agriculture. Compared with a similar survey 5 years earlier, the interest of the public in agricultural issues has increased significantly, but changed from a focus on the production of sufficient food to product quality and compliance with animal welfare standards. Now, almost 10 years later we are in the midst of the discussion how a transition of our societies toward more sustainability could be achieved. The keyword for agriculture is multifunctionality, i.e., not only the pure production of food, but also the creation of an attractive landscape, a healthy environment, and other agroecosystem services. I will look at this aspect in more detail later.

The basic principle of agriculture is actually to use solar energy as efficiently as possible through the cultivation of crops in order to produce food. It uses one of the oldest biological processes on earth: photosynthesis. Every plant utilizes solar energy, takes up the greenhouse gas CO_2, water, and nutrients, and builds up biomass, wood or fruits, which we can then use in a variety of ways. Fortunately for us, the waste product of photosynthesis is oxygen, which in turn is vital for most living beings (there are actually some viruses, single-celled microbes, and even multicellular animals that live without oxygen). The challenge for humans is to use this fascinating cycle in the most sustainable way possible.

Usually, we think that modern agriculture, with all its great machinery and sophisticated production facilities, is much better and more efficient at taking advantage of this solar energy link than old-fashioned, traditional agriculture. But this is not necessarily true. Around 1830, every unit of energy input in agriculture resulted in an output of five units in the form of food and harvested biomass. By 1910, this ratio even increased to 1:9, for energy input *versus* energy output. This was possible by better seed material and better equipment for field cultivation and smarter cultivation techniques. However, in the year 2000, the ratio between energy input and output in agriculture was only about 1:1, and some production methods now require more energy than can be extracted of the system (Krausmann 2001). Similar ratios apply for individual agricultural crops: the input/output ratio for maize was 1:10 in 1700 and 1:2.5 in 1980; traditional rice cultivation in the Philippines has a ratio of 1:108, but only 1:5 in the USA (Nentwig 2005). This means that modern agriculture, as the central supplier of human nutrition, is getting further and further away from being sustainable (Pimentel et al. 2005).

What are the reasons for this high energy consumption in agriculture? For US farmers, it has been calculated that almost 29% of energy costs are used for fertilizers, almost 6% for pesticides, and the remaining 65% for electricity and fuels (Schnepf 2004). If only fertilizers and pesticides are considered, then the fertilizer has an energy share of 77% of the crop product, followed by 23% for pesticides and seeds (FAO 2000). Synthetically produced fertilizers and pesticides used in agriculture have increased food production, but both fertilizers and pesticides also demand a lot of energy in production. Especially pesticide use and its energy demand are hardly addressed in the climate change debate. Nitrogen is the most important plant nutrient contained in fertilizers in terms of quantity. The production, transport, and application of one ton of nitrogen fertilizer corresponds to the energy content of about two tons of crude oil. Nitrogen fertilizer production uses large amounts of natural gas and coal and can account for more than 50% of total energy use in some sectors in conventional agriculture. Petrol accounts for between 30% and 75% of energy inputs of UK agriculture, depending on the cropping system (Woods et al. 2010).

Also, synthetic pesticides are made of fossil fuels and their production is very energy-intensive. For UK farming, it is estimated that pesticide manufacture accounts for less than 10% of energy input (Woods et al. 2010). Due to the great variety of pesticides and different production methods, there is a great variety on estimates on the energy intensity of its production. One can only estimate that the demand must be huge when seeing the great oil depots of agrochemical companies; also, many of those agrochemical companies are located near oil refineries. Estimates of carbon emissions, a measure for energy consumption, are in the range between 0.9–1.8 kg carbon per kg of nitrogen fertilizer. In comparison, average emission of carbon per kg of active ingredient is 6.3 kg for herbicides, 5.1 kg for insecticides, and 3.9 kg for fungicides (Lal 2004). The energy input devoted to pesticides of course depends on the agricultural sector. Fruit producers who treat their crops with pesticides 30 times per season have a higher proportion of their energy balance devoted to pesticides than cereal farmers who apply pesticides only four times per season. Despite the high energy costs of fertilizers and pesticides, farm machines are great diesel guzzlers. For example, in viticulture with about 25 pesticide applications during the season a diesel consumption of more than 250 liters/ha was estimated. Converted to 100 km, this corresponds to an incredible 300 l of diesel, or 0.8 miles per gallon (Moitzi 2005). Perhaps, this explains better why many farmers have their own filling station and why farmers fiercely fight for tax cuts on agricultural diesel.

Compared with that in 1950, the amount of pesticides used has increased about 50-fold (Tilman et al. 2002). If one projects pesticide production, which in the past has been constantly increasing, into the future, then pesticide production would almost double again by 2020, and almost quintuple by 2050. The exposure of nature and humans to pesticides will therefore also increase in the future.

There is little official information on the most commonly used pesticides used in different agricultural sectors in different countries. Of course, this would also be dependent on the agricultural sectors, actually be prevalent in different countries. As an example, the situation in the USA is shown in Table 1.1.

Of the 10 most commonly used active ingredients half of them are herbicides, with glyphosate, atrazine, and metolachlor-S being in the top three, and five are fumigants mainly used as fungicides or nematicides. Insecticides do not make it under the top ten of most commonly used pesticides in the USA.

It would be unfair to deny at this point the enormous achievements of agriculture in recent decades. During the first 35 years of the so-called *Green Revolution*, grain production doubled, thus satisfying the growing world population's demand for grain. The *Green Revolution* was proclaimed by the US Development Aid Organization as a countermovement to the violent Red Revolution in the Soviet Union. The American agronomist Norman E. Borlaug really pushed this concept forward in several regions of the world and even received the Nobel Peace Prize in 1970 for his work. The result was what we

Table 1.1 Most commonly used active ingredients of synthetic pesticides in the agricultural market sector in the USA in 2012. Data according to EPA (2017)

Active ingredient	Type	Amount applied (million kgs act. ingred.)	Rank
Glyphosate	Herbicide	122.5–131.5	1
Atrazine	Herbicide	29.0–33.6	2
Metolachlor-S	Herbicide	15.4–20.0	3
Dichloropropene	Fumigant	14.5–19.1	4
2,4-D	Herbicide	13.6–18.1	5
Metam	Fumigant	13.6–18.1	6
Acetochlor	Herbicide	12.7–17.2	7
Metam potassium	Fumigant	7.3–11.8	8
Chloropicrin	Fumigant	3.6–8.2	9
Chlorothalonil	Fungicide	2.7–7.3	10

call conventional, industrial-style agriculture that is characterized by large monocultures, large machines, chemical inputs in the form of fertilizers and pesticides, specialization, and a focus on a few high-performance crops. Since we have heard before how energy-intensive and machinery-intensive this type of agriculture is, it was probably not just a coincidence that it was mainly the oil giant Rockefeller and the agricultural machinery manufacturer Ford that supported this type of agriculture all over the world (Brown 2016).

Despite these achievements, almost 821 million people are still starving on our planet. Accordingly, the food industry and big agrocorporations agree unequivocally that we must press ahead with intensification in order to combat hunger in the world (WHO 2019). The fact that about 1.1 billion adults and children suffer from overweight and pathological obesity at the same time is deliberately overlooked in this discussion. The conclusion to be drawn is that enough food is actually produced worldwide, but that it is just unfairly distributed. In any case, never before have so much cereals been produced as today, but less than half of this amount actually serves as a direct food source. The rest is used as animal feed for meat production and as so-called agro-fuel mixed with fossil fuel or processed as industrial raw material.

Another example is palm oil. The EU is the second largest importer of crude palm oil in the world and more than half of it (around four million tons) is currently used to make "green" fuel. A study revealed that biodiesel made from palm oil is three times worse for the climate than regular diesel while soy oil diesel is two times worse (EC 2019a). This is because growing demand for these biofuels increases pressure on agricultural land leading to deforestation in tropical regions. A rapidly growing share of global agricultural areas is devoted to the production of biomass for nonfood purposes; yet, this sector has attained little critical attention in midst the type of bioeconomy. The European Union is a major processor and the biggest consumer of cropland-based nonfood products, while at the same time relying heavily on imports (Bruckner et al. 2019). Two-thirds of the cropland required to satisfy the EU's nonfood biomass consumption are located in other world regions, particularly in China, the USA, and Indonesia, giving rise to potential impacts on distant ecosystems (Bruckner et al. 2019). This example also raises the question of whether the agro-industry, with their plea for pesticide-intensive agriculture, is really primarily concerned with food security or rather devoted to nonfood products.

With this form of industrial agriculture, however, we are now in a conflicting situation. Agriculture is the source of more than a third of the world's population's income and livelihood, and food industry is still the world's most important economic sector. Yet, the agricultural sector is itself becoming one of the most important contributors to climate change, species extinction, environmental poisoning, and water scarcity. It is estimated that up to 40% of all greenhouse gas emissions are caused directly or indirectly by our agriculture and food production, its processing, transport, consumption, and disposal (Tubiello 2019).

Excessive pesticide use usually goes hand in hand with huge monocultures and agriculture on an industrial scale. Less well acknowledged is, however, that many small farms and part-time farmers often also use pesticides. Even quite aesthetical, diverse, and richly structured landscapes where rice, tea, apples, or grapevine is cultivated could receive a heavy load of pesticides. In Europe alone, several hundred pesticides are permitted for treating grapevines; many of them are applied before a pest or disease is making problems. Incidences in Italy and other regions suggest that winegrowers are also unwittingly serving as guinea pigs for agrochemical industry. The German agrochemical giant Bayer paid Italian winegrowers two million € in compensation because the treatment with a recommended fungicide against the *Botrytis* fungus resulted in a total loss of yields (ORF 2016b). Around 800 Italian winegrowers treated their grapes with this fungicide, and noticed enormous damage to the vine blossoms and many crop failures. The producer then published an official recommendation not to use the fungicide in viticulture "for precautionary reasons." So much for now regarding the myth that pesticides are the best investigated chemical substances.

In earlier years, pesticide application was perhaps more ruthless. Back then many pesticides were persistent and accumulated in the fatty tissue of animals and humans. Whether the newer pesticides are really better cannot be answered seriously, since we simply lack experience with them. A scientific textbook from the USA in 1974, for example, states that a mixture of 10–20 gallons of diesel oil, two to three pints of the herbicide dinitrophenol, and 100 gallons of water is recommended as a good herbicide for weed control in vineyards (Winkler et al. 1974). Note the exact specifications for the dosages of the various ingredients. This mixture should then be used four times a season. This pesticide was patented as an insecticide at the end of the nineteenth century, but also as a fungicide and in the 1970s as an herbicide. Sounds more like a trial and error approach than a targeted development of special preparations. The same book also refers to a new very promising and environmentally friendly active ingredient that is completely degraded into harmless components: glyphosate!

Worth reporting in this respect is also an episode with a student, whose father is a jute farmer in Bangladesh. Asked what pesticides they use on their jute farms, the agricultural science student told me that everything was organic and no synthetic pesticides were used. At my skeptical request as to what would be used against pests if necessary, the student explained that they only use kerosene (Zaller 2018). Kerosene, also known as aviation fuel, is not produced by agrochemical companies and therefore not considered a synthetic pesticide in his definition.

1.2   Pesticides Are Also Used Beyond Agriculture

Farmers, who feel wrongly criticized because of their pesticide use, often reply that railway companies are actually much bigger users of pesticides. The argue that nobody is talking about these users because they have a better lobby and are often state-owned. Let us look into the matter.

Fact is that herbicides are used by railway companies to keep the tracks free of weeds, as otherwise there could be derailments or an increased risk of fire. A parliamentary questionnaire about the use of herbicides on the rail network of the German Railways (Deutsche Bahn) showed that around 70 tons of glyphosate were used per year on a rail network of around 34,000 km (Bundestag 2009). This makes the German Railway not the largest, but by far the second largest consumer of herbicides in Germany after agriculture, which uses about 4000 tons annually. The herbicide is applied with a special spray train that travels at night, recognizing and selectively spraying the vegetation instead of applying the herbicide over a large area. In Austria, this has allegedly saved 75% of the original amount of herbicide applied. Herbicides are also used on track lines that lead through nature reserves. Public relations people have been careful in the wording calling the former "weed killer" an "anti-growth agent" and the applied glyphosate a biologically effective agent.

Pesticides are also used for so-called game deterrence, to ward off wildlife browsing in forests or birds feeding on agricultural crops. The substances used there are not harmless either and can be toxic to earthworms and aquatic organisms, toxic when inhaled, and suspected of triggering Parkinson's disease in the doses used (Wang et al. 2011).

Pesticides are used in a wide range of fields outside agriculture such as aquacultures, i.e., farms in which fish and seafood are produced. Random tests of salmon, trout, gilthead, and sea bass from aquacultures also revealed that a pesticide (ethoxy quinine) was found to be above the permitted limit (Weiland 2016). Due to possible carcinogenesis, genotoxicity, and alterations of the liver metabolism, this pesticide has been off the market in the European Union since 2011; however, hardly understandable, the chemical is still allowed as a feed additive. The problem has disappeared from European markets, but not from European plates. In a consumer market study, the chemical was found in 44–54 fish products; a salmon sample exceeded the maximum permissible quantity by the factor 17. Fish meal producers from all over the world use ethoxyquin to preserve their product for transport.

Pesticides are also used, for example, to decimate naturally occurring shrimps in commercial shrimp farms (Pisa et al. 2015). It sounds strange, but pesticides are used in shrimp farms to combat fish, crabs, snails, fungi, algae, and climbing plants (Gräslund and Bengtsson 2001). Furthermore, large quantities of disinfectants are used to prevent the formation of pathogens at the bottom of the sea, which could endanger farming, due to shrimp excrements. As in all intensive animal breeding operations, also antibiotics are used in large quantities in aquacultures, but this will not be further elaborated here.

Following the devastating floods caused by Hurricane Harvey in Texas in the fall of 2017, insecticides against mosquitoes were extensively applied over several weeks via C-130 Hercules military aircrafts (Kumar 2017). Within a few days, around 7500 km² of flood area was treated aiming at preventing the outbreak of diseases transmitted by mosquitoes such as West Nile fever or the Zika virus. Although authorities applying insecticides admitted that most mosquito species, which occur after flooding, do not act as disease vectors, but it is feared that mosquitoes will primarily molest inhabitants and helpers. Such actions have already been performed after the hurricanes Katrina, Rita, and Gustav the years before. Mainly naled, an insecticide manufactured by a partner of Monsanto, is used for this purpose (Webb 2017). In Europe, this insecticide is banned because of "unacceptable risks" to humans. Pesticide flights during hurricane season take place day and night; at least residents are warned to be cautious and beekeepers are asked to cover their hives during insecticide applications.

Insect repellents against mosquitoes, ticks, fleas, chiggers, and leeches often contain the active ingredients diethyltoluamide (DEET) and icaridin. The substances were also used by the military in World War II, Vietnam, and Southeast Asia. Besides its high mortality to salamanders (Almeida et al. 2018), it can also be found in wild mushrooms from Russia, Belarus, Poland, and Bulgaria or teas and tea-like products from Tanzania (Scherbaum and Marks 2019). But also in Europe the control of mosquitoes from helicopters along the great rivers such as the Rhine, the Danube, or also the Lake Chiemsee in Bavaria has been in use for decades (Nazarewska 2013). Insecticides from the group of pyrethroids or, more recently, a bacterium (*Bacillus thuringiensis israelensis*, Bti) is used. Pyrethroids are used against adult insects but are also toxic for many nontarget insects and are also classified as hormonally active substances. Mosquito control based on Bti is regarded as an environmentally friendly alternative. However, it is only effective against mosquito larvae that still live in water and not against adult insects; moreover, it can also affect nontarget chironomid midges that are recognized as a central resource in wetland food webs (Kästel et al. 2017). This shows the dilemma when humans intervene in nature even without the use of synthetic pesticides.

Insecticides are also used on a large scale in landscape gardening. A dramatic incident is reported from the USA where a landscaping company decided to treat some lime trees with an insecticide (Statesman Journal 2014). The trees stood on a parking lot and had infestation with aphids, which occur regularly in lime trees. The "danger" was that cars parked below the trees would be covered by sticky honeydew secreted by the aphids. A few days after spraying the trees with insecticides (neonicotinoid dinotefuran) the parking lot was littered with at least 50,000 dead bumblebees, the biggest reported bumblebee death in history. Probably many more died unnoticed elsewhere. Pesticide use in agriculture could at least be justified by the fact that it is intended to ensure food production, because after all we all need something to eat. But do we really have to spray highly toxic substances into the environment only to meet an exaggerated requirement for tidiness?

It might sound strange, but pesticides are also applied in nature conservation areas around the world (SNH 2017). Some plants and animals, so-called invasive alien species, neophytes (plants), neozoa (animals), or generally neobiota (all organisms), might pose a threat to the conservation interest of protected areas. Where the protected habitats and species may be threatened by these fewer desirable species, it is often accepted even among the conservation community that pesticides could be used as a component of management. Unfortunately, these pesticide applications are little monitored regarding their effects on nontarget species and are rarely made public either.

Pesticides are also used where it is not suspected at first: in fine arts museums, for example, where wooden picture frames or canvases of invaluable art works are threatened by wood-eating insects or fungi. In natural history museums all over the world, the exhibitions containing organic material are or have been heavily treated with pesticides. The use of pesticides in museums began in the 18th century (Ornstein 2010). Decades of pesticide use have resulted in a great deal of toxic cultural property being stored in the museum depots. The pesticides used in the twentieth century comprised arsenic, the insecticides DDT, lindane, and PCP (pentachlorophenol). Many of them are banned nowadays, but the art objects are still contaminated with these very persistent substances. It is estimated that around two-thirds of the collection in the well-known Ethnological Museum Berlin-Dahlem, Germany, is contaminated with pesticide residues. The topic is regarded as delicate, one speaks only carefully about the contamination of the archives and possible health problems of the people working in museums. Restorers and depot staff told me about dizziness, extreme fatigue, respiratory problems, and skin rashes after working with contaminated objects for a couple of hours (Zaller 2018). The danger for museum visitors is considered to be low, as affected objects are often presented in glass showcases in public exhibitions. In the meantime, even scientific conferences debate this subject (Wetzenkircher and Llubic Tobisch 2014).

Objects contaminated with pesticides have been returned to Native American tribes raising concerns about the risks posed to human health (Ornstein 2010). A survey of the American Association of Museums revealed that some of the most common pesticides found in collections were arsenic especially prevalent in taxidermy preservation, mercury on botanical specimens, naphthalene, and paradichlorobenzene (PDB) commonly known as "mothballs," DDT until the 1970s applied as a insecticide or disinfectant to biological and animal specimens as well as library materials. Many museums are aware of this problem and have guidelines for the handling of pesticides. The situation in tropical regions is even more serious. A colleague working on pest control in museums showed me pictures from museums in Southeast Asia, where termites not only eat away the picture frames of art objects, but also all the furniture, including the museum door frames.

Instead of pesticides, museum staff in modern museums increasingly rely on pest monitoring and ideal climate conditions to prevent pest development. As alternatives to poison, for example, insect traps are used in the Vienna Wagenburg, a museum with lots of wooden carriages, that emit nontoxic sexual attractants for wood pest species. This attracts the animals and keep them away from the valuable objects. If, despite all prevention, objects are affected by pest infestation, the objects are fumigated in chambers with a high nitrogen content for a couple of weeks. During this time both the animals and their eggs are killed. This procedure is efficient and not hazardous for the workers dealing with it.

An environmental physician told me that residues of the insecticide DDT are still detected in blood samples taken from actors (Zaller 2018). Although DDT has been banned for decades, it is assumed that the former treatment of historical costumes and wigs with DDT-containing moth powder was responsible for this contamination of the actors.

This is how people come into contact with pesticides at their working place, as farmers and gardeners, in museums or theaters. However, it is also very likely that you get in contact with pesticides during vacation, for example, in intercontinental flights. Especially on flights to Australia, New Zealand, India, the Seychelles, Mauritius, or South Africa, chances are good of getting in contact with pesticides (ORF 2010). In order to comply with the regulations of destination countries, on-board personnel is required to spray insecticides in the cabin. The purpose of these measures is to prevent agricultural insect pests and carriers of dangerous diseases such as malaria from entering holiday destinations and *vice versa*. Insect pests introduced into foreign ecosystems by plane or ship can cause considerable damage or even transmit dangerous diseases. In the vicinity of airports, malaria cases can occasionally occur in Central Europe, which can be traced back to accidentally introduced mosquitoes. Another example is the Western corn rootworm (*Diabrotica virgifera*), the most serious insect pest of corn in North America was introduced by US military aircraft during the Balkan War in the 1990s (Nentwig 2005). The Asian tiger mosquito reached the USA via bamboo imports and led to the spread of the dangerous dengue fever there. However, the effectiveness of pesticide application in airplanes is being challenged. In recent years, there has been an increase in complaints because of health problems of the flight crew that is constantly exposed to these toxins. The use of these pesticides with long-term effects is particularly controversial as they accumulate in human fatty tissue. A life-threatening asthma attack by an Irish businessman on a plane caused by an on-board pesticide application is an extreme example of adverse effects (Müller 2011). Because of the asthma attack, the plane had to make an emergency landing; the passenger barely survived the incident and was finally awarded 50,000 € in damages by the airline.

Arrived in the holiday resort, there is again a high probability of coming into contact with pesticides because pesticides are often used in hotel parks. Complaints about this can easily be found on various booking platforms on the Internet from hotels in Italy, Turkey, Egypt, or elsewhere. Additionally, pesticides are sprayed on camping sites, nowadays even using drones. Especially in warmer areas, insecticides are usually applied overnight to protect hotel guests from unpleasant crawling or molesting insects. The question is what will have the greatest long-term effects: the nuisance caused by insects or the chemical club? A very drastic case is reported from Thailand where bed bug pesticide poisoning killed a tourist guide and seven tourists in only 3 months (DMR 2011). All seven stayed at or used facilities at the hotel. Police initially dismissed the mystery deaths as a terrible case of food poisoning caused by eating toxic seaweed from a bazaar. Later, test results found small traces of an insecticide called chlorpyrifos inside the room—a chemical that is often used to get rid of bed bugs. Fears that some Thai hotels could be using unsafe chemicals first came to light 2 years before when two other tourists died at a different resort. Their symptoms, beginning with severe chest pain, and progressing to vomiting and fainting, were almost identical to the seven other tourists who died later. These are just a few examples of many very serious cases of pesticide effects in tourist locations around the world.

Another area for which there are few reliable figures available is the use of pesticides in local communities, kindergartens, playgrounds, or private gardens (Table 1.2). According to the US Environmental Protection Agency, approximately 88 million US households use about 6% of the total pesticides used each year (EPA 2017).

It is scary to see that private people without any training in the use of pesticides apply these substances in their private gardens or homes. Moreover, we know that only few will wear protective clothes or even read the user manual of the applied products.

Aware of the problem that protective clothing and equipment are rarely used, the Food and Agriculture Organization of the United Nations (FAO) has therefore also issued a Code of Conduct on the Use of Pesticides. It states that pesticides whose handling requires the use of personal protective equipment that is uncomfortable to wear, expensive, or not readily available should be avoided, especially in the case of small consumers in tropical climates (FAO 2013). However, this is only a voluntary code of conduct. As soon as health problems with pesticide exposure become public, manufacturers respond that the risk is acceptable if the safety regulations are followed.

The situation is probably not substantially different in more industrialized countries. Perhaps there is a slight trend for reduced pesticide use in private gardens. For instance, in Canada, 19% of garden-owning households used chemical pesticides in 2013; in 1994 this was up to 31% (ECCC 2016). For Canada it is believed that a ban on so-called cosmetic pesticides, those that only improve the appearance of fruits or lawns, has contributed significantly to this reduction. In Austria, a survey of over a thousand home and garden

Table 1.2 Most commonly used active ingredients of synthetic pesticides in the home and garden market sector in the USA in 2012. Data according to EPA (2017)

Active ingredient	Type	Amount applied (million kgs act. ingred.)	Rank
2,4-D	Herbicide	3.18–4.08	1
Glyphosate	Herbicide	1.81–2.71	2
MCPP	Herbicide	0.91–1.81	3
Pendimethalin	Herbicide	0.91–1.81	4
Carbaryl	Insecticide	0.91–1.81	5
Acephate	Insecticide	0.91–1.36	6
Permethrin and other pyrethroids	Insecticide	0.91–1.36	7
Dicamba	Herbicide	0.91–1.36	8
MCPA	Herbicide	0.91–1.36	9
Malathion	Insecticide	0.91–1.36	10

MCPP methylchlorophenoxypropionic acid, *MCPA* 2-methyl-4-chlorphenoxyacetic acid owners showed that 35% use pesticides (Sattelberger 2001). According to this survey, pesticides are used more often by women, by people over 50 years of age, and by farmers. Eleven percent of pesticides are used against flour moths, clothing moths, and other insects. Only 17% of the respondents stated that they adhere to the recommended dosage, but only 12% said they use protective measures such as gloves.

An important indicator for pesticide uses in households can be house dust. When pesticides are used indoors, the persistence of the active substances can be considerably longer than in the field. A study in Germany collected 336 dust samples from vacuum cleaners over four weeks and found pesticide residues in almost all households (Walker et al. 1999). Strikingly, residues of DDT were found in over two-thirds and pentachlorophenol (PCP) in almost all samples, although these persistent active substances have been banned for decades. Other insecticidal active ingredients can be detected in the indoor air samples even 14 months (permethrin) to 7 years (lindane) after their application (Butte 1999).

On a walk through the neat neighborhood in early summer with all the burned looking margins or paving scratches and the monotonous laws without any herbs, it gets obvious that pesticide applications are pretty common among private people too. I would guess there are only a few households in which not at least an insect spray, mosquito repellent, rat poison, ant powder, slug pellets, or the like can be found.

1.3   What Substances Are We Talking About?

A few active ingredients, such as the insecticides DDT and lindane or the herbicide glyphosate, have already been mentioned. I guess, it makes little sense to elaborate here on the hundreds and hundreds of active ingredients that are used in pesticides. These are usually considered in specialist books on plant protection. Rather, I would like to pick out those product groups that are used the most and with which we are confronted the most: herbicides, insecticides, and fungicides. All these pesticides are applied in a variety of ways: manually with a sprayer carried like a rucksack called backpack sprayer; by tractors, helicopters, or airplanes; scattered as dusting agents or granulates or mixed in irrigation water. Pesticides are also often used for seed dressings by covering seeds with a layer of active substances to protect them from fungal diseases and insect attacks.

Pesticides are usually a mixture of several chemicals, so-called formulations. These include the active substances, which are the actual pesticide active ingredients (e.g., glyphosate), and a large number of so-called co-formulants or adjuvants (FAO WHO 2014). The adjuvants improve the handling, absorption, and efficacy of the active substances. They enable the active ingredient to adhere better to weeds or pests so that they are not easily washed away by a rain shower. Adjuvants also improve the mixing of the active substance in the spray liquid or the penetration of the active substance into the organism to be controlled. In Germany alone, around 1600 adjuvants are registered (BVL 2018). Many adjuvants are harmless such as talcum, bentonite, kaolin, diatomaceous soil, or lime. Less harmless are organic solvents, mineral oils, surfactants, emulsifiers, stabilizers, PVC, or organosilicons.

Adjuvants or co-formulants are usually not declared by the manufacturers, as the recipe of the pesticide is considered a business secret. This is very puzzling as the adjuvants often make up the predominant amount of the pesticide formulation. The very popular herbicide Roundup, for instance, only contains 7–37% of the active ingredient glyphosate, while the remaining ingredients are unknown. Per convention, adjuvants are considered chemically inert, which means that they do not react with other substances or only to a negligible extent and have thus no harmful effect on other organisms. The term inert is actually misleading, as many active substances can only develop their full efficacy in the presence of the adjuvants. Studies comparing the effect of the pure active ingredients with that of the ready-to-use pesticide formulation have also shown that the pesticide formulation is much more toxic than the pure active substance (Mullin et al. 2016). When the effects of nine different pesticides were tested on human cell cultures, it could be shown that eight of them were up to a 1000 times more toxic than the pure active ingredients (Mesnage et al. 2014). Among the pesticides tested, fungicides were the most toxic substances, followed by herbicides and insecticides, even at concentrations 300–600 times lower than the recommended dosages. Thus, describing these adjuvants as inert and chemically inactive is quite euphemistic. We will get back to this issue later when dealing with the approval procedures of pesticides where in most cases only the toxicity of the active ingredients is considered.

1.3.1 Herbicides Destroy Plants

Herbicides are agents used to destroy unwanted grasses, herbs, or trees. Soil herbicides are taken up by plants via the roots, while leaf herbicides enter the plants via the leaves and green stem parts. Weeds are by far the most important "pests" in agriculture worldwide. This is remarkable because weeds do not cause direct damage or diseases to cultivated plants. Their harmful effects are caused indirectly by competition with the crop plant for water, nutrients, or light. Weeds can also impede the harvesting process for instance when they block the pickup reel of combine harvesters. The German word for weeds is "Unkräuter" meaning unwanted herbs and therefore a rather improper term for an ecologist. Mark Twain more pragmatically defines weeds as all plants that grow again after weeding.

Weeds are simply plants that grow in the wrong place in the wrong time. This can lead to such grotesque situations that the sunflower crop or rapeseed plants become weeds when they grow in subsequent cereal crops in the following year, for example because seeds have been spread during the harvesting process. It is estimated that the weed damage potential is around 30% and thus about twice as high as that of fungal diseases and a third higher than that of pests—in relation to the world's eight most important crops (Hallmann et al. 2009). The importance of herbicides as the most important pesticides is also derived from these numbers.

Active chemical herbicides have only been available since the 1940s. The oldest act like plant growth hormones and kill plants by triggering uncontrolled growth. The systemic herbicide 2,4-D (2,4-dichlorophenoxyacetic acid), which selectively kills most broadleaf weeds, is one of them. This active ingredient is on the market since 1945 and achieved its devastating fame in the Vietnam War of 1967/68 as a component of Agent Orange which was used by the US army in millions of liters to defoliate the rainforest. It still causes unspeakable suffering in the form of genetic defects in newborns. It is estimated that about 2.5 kg per capita and year of this herbicide was released of the exposed rural population during the Vietnam War (Cribb 2014). This active ingredient can still be found in over 1500 commercial lawn herbicide mixtures, and is widely used as a weed killer on cereal crops, pastures, and orchards. Other known herbicidal agents such as atrazine or paraquat interfere with plant photosynthesis, while agents such as glyphosate interfere with metabolic pathways that build amino acids and other biomolecules.

Before chemical herbicides were so popular and widely used, unwanted plants were kept in check mechanically using plows, harrows, or hoes or by controlling weeds by smart crop rotations or variations in crop seeding densities. These nonchemical methods of weed control are still successfully used in organic farming.

Usually, herbicides cannot distinguish between cultivated plants and weeds—they kill all plants alike. These are called broad-spectrum or broad-band herbicides. However, there are also herbicides that specifically act on monocotyledonous plant species such as grasses; others only kill dicots, i.e., herbs. The latter herbicides are for example frequently used when the aim is to create lawns without dandelions or other unwanted herbs. The special case of genetically modified plants that are tolerant against specific herbicides will be considered in a later chapter.

Broadband herbicides are popular since 1980; among them are several sulfonylureas (e.g., flazasulfuron or metsulfuron-methyl) that can be taken up from the soil via the roots and kill plants by interfering with plant biosynthesis of certain amino acids in the plant physiology. These sulfonylureas are still one of the most important herbicide groups. For example in 2010, more than 30% of the German cereal cultivation areas were treated with these herbicides (Drobny et al. 2012). The introduction of sulfonylureas led to a reduction in application rates from previously about 2000 g/ha to 3–60 g/ha. However, continuous application of sulfonylurea herbicides over many years led to the development of resistant weed populations in many countries. Interestingly, these sulfonylureas are also used in medicine as antidiabetic drugs to treat diabetes mellitus type 2 and increase insulin release from the pancreas.

When a broadband herbicide is sprayed on a head of lettuce or another crop species, it would die within a few hours or days. Nevertheless, they are also used in perennial cropping systems such as apple orchards, olive orchards, or vineyards. If used improperly or at higher wind speed, herbicides can also kill the cultivated plants that were actually meant to be protected from weeds. For example, if herbicides are used in order to control the undergrowth of grapevines, it is important to make sure that the grapevine leaves are not sprayed with herbicides; otherwise the vine would die. Therefore, some herbicides used in viticulture may only be used under at least 4-year-old grapevines because the risk of damaging the younger vine would be too great. One might think that it is not a big deal if a couple of leaves of a grapevine die from herbicide drift. However, many herbicides act systemically meaning that they are distributed across the whole plant which would ultimately lead to a complete die-off of the grapevine. In arable farming, herbicides are usually applied before the crop is sown in order to kill all weeds and create a tidy seed bed for the following crop.

A rather strange aspect of herbicide use is directed against the cultivated crop itself, when used before harvesting as a so-called ripening spraying or crop desiccation. Why is this done at all? Well, if you see a cereal field shortly before the harvest, you can often notice greenish areas where cereals are not yet fully ripe or where nests of weeds are still green but the crop already ripe. A spraying of herbicides onto the crops makes sure that every plant on the field is killed evenly which can facilitate the harvesting process. In many regions with unfavorable weather conditions for instance in northern Europe or Canada, ripening spraying is frequently performed. However, also in many other countries around the world, it is often used in the cultivation of various cereals, potatoes, hops, rape, field beans, fodder beets, sugar beets, or lentils. With sugarcane, desiccation can increase sucrose concentration before harvest. Desiccation spraying is particularly controversial because it increases the risk for pesticide residues in food products despite required waiting periods between spraying and harvesting. Herbicide application on the crop plant is also performed in nonfood crops. For instance, in cotton, reliance on natural frost performing natural defoliation may be too late to be effective in some regions. Thus, leaves that remain on the cotton plant will interfere with mechanical harvesters and stain the white cotton resulting in a lower quality grade.

Worldwide there are more than 300 herbicidal active ingredients in use. However, for the sake of clarity, I will mainly elaborate on glyphosate and a few other active ingredients that are widely used by professional and private users.

1.3.1.1 Herbicidal Active Ingredient Glyphosate

Glyphosate is probably the best-known active ingredient worldwide, besides the infamous insecticide DDT. In any case, it is the active ingredient in the world's most widely used broadband herbicides. The chemical was discovered in the 1940s and was originally patented in 1964 as a chelating agent used to clean out calcium and other mineral deposits in pipes and boilers of residential and commercial hot water systems (USPA 1964). Later in 1974, it was patented as a herbicidal agent which kills plants by disrupting the shikimate pathway; it was considered to have no effect on humans because the shikimate pathway is not present in mammals (USPA 1974). The last patent for glyphosate was granted in 2010 as an antimicrobial agent for the prevention and therapy of pathogenic infections caused by protozoan parasites such as malaria (USPA 2010). Knowing this wide action spectrum of glyphosate one can stop wondering about the numerous nontarget effects of this substance. Again, doubts arise as to whether the development of pesticides is actually as scientifically planned as often claimed.

Glyphosate is a systemic pesticide, which means that the active ingredient is absorbed by leaves and then distributed across the treated plant. In contrast, contact poisons are not distributed in the plant, which means that only the treated plant parts are killed. Glyphosate belongs to the group of organophosphorus chemicals that block the synthesis of aromatic amino acids in all green parts of the plant and makes the plant to die after a few hours. Glyphosate was marketed widely in the early 1970s by the US agrochemical company Monsanto under the brand name Roundup. With the 1980s, glyphosate became one of the best-selling herbicides and was celebrated as a pesticide of the century also in scientific journals (Duke and Powles 2008).

Roundup sales skyrocketed even further when Monsanto developed genetically modified maize plants, so-called Roundup-ready crops, in the 1990s that were tolerant toward glyphosate. In field with Roundup-ready crops a glyphosate application would kill all plants despite the tolerant crop species. Global glyphosate consumption has increased 15-fold since the introduction of genetically modified plants, GM crops, in 1996 (Fig. 1.3). Moreover, also non-agricultural glyphosate use increased by 490% from 13,428 tons in 1994 to 79,224 tons in 2014; the share of non-agricultural use of the total glyphosate use is currently around 10%.

An unimaginably 826 million kg of glyphosate are applied every year on this planet (Benbrook 2016). Since 1974 in the USA alone, over 1.6 billion kg of glyphosate active ingredient has been applied, or 19% of estimated global use of glyphosate (8.6 billion kg, that is about 1 kg per inhabitant on our planet). In the first decades of its market presence, glyphosate was distributed through Monsanto's product Roundup only. However, with the expiration of patent rights, a large number of products from other companies around the world are now also available. In Germany, for example, more than

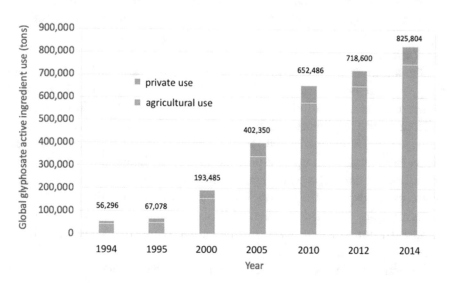

Fig. 1.3 Global use of glyphosate active ingredients in thousand tons. Data derived from Benbrook (2016)

100 products containing glyphosate in different concentrations and mixtures are approved (BVL 2019). By the way different concentrations—according to an agricultural chemist the information on the packaging of pesticides is quite inaccurate and 10% fluctuations of the concentrations have been measured.

The maximum amount of pesticides that can be applied in the fields is usually regulated for different crops and countries. Most glyphosate products are limited to a maximum of two applications within 1 year with a maximum quantity of about 4 kg/ha of glyphosate and year. This is at least the case in Austria and Germany. It is often stressed by agrochemical companies that in agricultural practice, for economic reasons alone, the dosage will be more precise than in the private sector. This might well be so, because in private use the motto "more helps more" prevails. Also, the dosage information provided by the manufacturer is often not easy to understand, contradictory, and sometimes impossible when milligrams of a product should be applied on a square meter base. Who has that accurate balances in the household? In addition, colorful product advertising conveys the feeling that the stuff is harmless anyway.

Glyphosate is approved for use in a wide range of crops and agronomical situations. Despite a ban on GM crops, glyphosate use has also substantially increased in Germany during the last couple of years (Dickeduisberg et al. 2012). A survey conducted among 896 German farmers covering around 250,000 ha of farmland revealed that about 87% of oilseed rape, 73% of pulses, and 66% of winter barley were treated with glyphosate. In total, glyphosate is applied on 39% of the Germany's arable land with predominating use for stubble management, pre-sowing application, and pre-harvest application.

Digitalization in agriculture and precision farming also includes new ways to apply pesticides. Switzerland is among the first countries to have approved drones for the application of pesticides (Heids Mist 2019). In contrast to helicopter spraying flights which are noisy, and the pesticides that are more prone to drift, unmanned aerial vehicles or drones are the new hope. First test flights were made high up in alpine pastures using herbicides against alpine docks (*Rumex alpinus*) a plant that can make quite extensive stands on places where cattle deposit their droppings and stimulate growth of this weed. However, using herbicides to fight this plant is only fighting the symptoms because the underlying reason is mainly wrong pasture management with too high nitrogen inputs (Zaller 2004).

The widespread use of glyphosate increases the chances of contamination of the environment, food, water, or the human body. Indeed, glyphosate residues can also be found in the urine of humans and animals—even if they have not been in direct contact with glyphosate (Liebrich 2012). Glyphosate was detected in the urine of humans from 18 European countries (BUND 2013). Also, members of the European parliament had an averaged 1.7 µg of glyphosate per liter in their urine (Chow 2016). This is seventeen times more than the European maximum limit of 0.1 µg per liter permitted for drinking water. Glyphosate is therefore apparently ingested via our drinks and food. Nevertheless, there is no regular official monitoring of glyphosate residues in food and feed in place. Often, the public is informed about pesticide residues in food by studies conducted by nongovernmental organizations (NGOs) or various consumer agencies. Glyphosate residues, for example, were also found in flour, bread, or cereal flakes (Ökotest 2013). Although the amounts of residues found are usually low, these tests show that glyphosate is not destroyed during the baking process.

In addition to cereal products, many drinks also contain glyphosate residues. Glyphosate was found in the 14 most popular beer brands in Germany (Umweltinstitut München 2016). The values discovered ranged from 0.5 to 30 µg per liter, so the legal limit value for drinking water exceeded 300-fold. Thus, the 500-year-old German purity law for beer, according to which beer may only consist of hops, malt, and water, could be challenged. It is suspected that the pesticide entered the beer via malting barley that has been sprayed shortly before the harvest. After that study was released, German brewers managed to severely limit the amount of glyphosate used in crops for brewing and improved the situation. Another study investigating glyphosate in beer and wine from the USA, Europe and Asia found similar results (Cook 2019). Of the tested 15 beers and five wines, 19 contained glyphosate. The highest level of glyphosate found was in wine from California with 51 µg per liter. Surprisingly, also three of four organic beers and wines contained glyphosate. Although glyphosate is not allowed or used in organic farming, several organic products were contaminated. While these levels of glyphosate are below risk tolerances of the US environmental protection agency EPA for beverages, it is possible that even low levels of glyphosate can be problematic. For example, in one study, scientists found that a concentration of 1 part per trillion (0.001 µg per liter) of glyphosate has the potential to stimulate the growth of breast cancer cells and disrupt the human endocrine system (Thongprakaisang et al. 2013).

Also in Swiss wines, residues of glyphosate and 32 other pesticides were found (Pinto 2018). These tests even found substances that are not permitted in Swiss viticulture any more. For example, the very persistent insecticide DDT, which has been banned in Switzerland since 1972, was detected in two wine samples 40 years after its ban. The use of herbicides such as glyphosate in viticulture is all in all a strange development. Until a decade ago, vegetation in the undergrowth of grapevines was mainly controlled mechanically, without the use of chemicals. Now, with the public discussions on glyphosate some winemakers begin to again tolerate plants beneath grapevines and even control wine yields with the undergrowth vegetation. Especially when producing high-quality wine, winemakers do not strive for high yield but rather control yields in order to reach top quality.

Switzerland is an example of a country with a pesticide-intensive agriculture with more products approved than in many other European countries. However, they are also very correct and responsible, hence warning signs in the landscape indicate that cherries in orchards might have been sprayed with pesticides. But maybe the signs only serve to deter cherry thieves. Over 2000 tons of pesticides are applied in Swiss agriculture every year, making Switzerland one of the world's leading countries in relation to its area under cultivation. To be fair it has to be noted that Switzerland also produces a great amount of vegetables, fruits, and wine which demand a higher pesticide input than cereals. In contrast to other countries, the Swiss authorities at least seem to be aware of the problem. Meanwhile there is a federal action plan for the reduction and sustainable use of pesticides according to which the quantity of pesticides should be reduced by about 25% in the next few years (BLW 2017).

Glyphosate residues were also found where they would not have been suspected, for example in raspberries picked in the forest (ORF 2016a). In a sample of forest raspberries, 290 mg of glyphosate per kg of raspberries were measured. Thus, after five raspberries (15 g) a child reaches the permitted daily dose of 0.5 mg per kg of body weight. An adult would reach this dose by consuming half a berry cup (62 g). But what are weed killers used for in the forest? Actually, herbicides are frequently applied to prevent young tree seedlings from being overgrown by grasses. When I was jobbing as an intern at the Austrian Federal Forests I had to remove these weeds with a small sickle. This was exhausting, but at least it did not contaminate the environment or interfered with human health. When weeds are treated with glyphosate and eaten by wild animals, the natural game meat would also be contaminated with glyphosate. Once again are we confronted with pesticide contaminations that we did not expect.

When farmers are asked whether they can imagine to farm without glyphosate, they often say that they perhaps could avoid glyphosate in arable farming, but not in forest management especially when so-called invasive alien tree species (e.g., in Central Europe robinia—*Robinia pseudoacacia*, tree of heaven—*Ailanthus altissima*) make problems. We will later see that these invasive species are often the reason to apply pesticides even in nature conservation areas. Fortunately, at least the Austrian Federal Forests have meanwhile pledged not to use glyphosate for weed control in their forests any more.

Speaking of glyphosate avoidance, a grotesque anecdote should end this chapter. Monsanto is now selling its blockbuster herbicide Roundup with a new active ingredient called AC. AC stands for acetic acid or vinegar. The products come in a nice spray bottle, of course at a pharmacy price. Interpreting this as a turn-away from glyphosate by the manufacturer itself is probably too naive, as it is perhaps simply a marketing trick for environmentally conscious consumers.

1.3.1.2 Herbicidal Active Ingredients Atrazine, Paraquat and Dicamba

The official pesticide list of the European Union lists 116 approved active ingredients for herbicides (European Commission 2019); in the USA the Environmental Protection Agency lists 74 herbicidal active ingredients (EPA 2019). However, unfortunately these databases are not very consumer-friendly and only reveal limited information. The German database is much more user-friendly and lists 763 herbicide products making a share of about 44% of all 1734 products in this database (BVL 2019).

Besides glyphosate, atrazine from the chlorotriazine family is among the most frequently used herbicide on US farms today (Philpott 2011). Although a commission of the US EPA has found associations between atrazine and thyroid and ovarian cancer as well as an influence on the human hormone system, still about 34 million kg of this herbicide is applied annually in US agriculture. Generally, atrazine is used to kill pre- and post-emergence broad-leaf and grassy weeds in crops such as sorghum, maize, sugarcane, lupins, pine, and eucalypt plantations, and in triazine-tolerant canola. Atrazine is also one of the most widely used herbicides in Australian agriculture (APVMA 2010). The use of atrazine was banned in the European Union in 2004, after frequently finding contaminations in groundwater, and the manufacturer (Syngenta) could not show that this could be prevented or that these contamination levels were safe (EC 2003). Although not member of the European Union, Switzerland has also banned atrazine. However, it is still produced there and exported around the world. Even years after the ban, atrazine is still regularly detected in groundwater samples throughout Europe and washed out of the soils where it was applied for example after floodings (BMNT 2019).

Another highly toxic herbicide to be mentioned here is paraquat. It is mainly used in tropical areas for weed control in orchards and vineyards as well as in coffee, tea, oil palm, and banana plantations. In arid regions, it is also used for so-called conservation tillage. There, herbicides are used instead of mechanical tillage in order to save fuel and working time as well as increase carbon storage of the soil. It is hard to believe but paraquat is also used in experimental medicine to induce Parkinson's disease in laboratory animals (Vaccari et al. 2017). Paraquat is one of the most important products of the Swiss-Chinese agrochemical giant Syngenta, although it is banned in 32 countries including Switzerland. Syngenta is advertising paraquat as an important contribution to sustainable agriculture, when only a few safety precautions are respected (MultiWatch 2016). The reference to safety precautions is quite cynical, as the product is often sold to farmers in bulk, without a label or instructions for its use. Even if labels are available on the pesticide containers, they are often not written in the local language, or the farm workers are illiterate and simply cannot read and do not even know that it is a poison. In India paraquat is sold under various names and also used for non-approved crops. Especially in developing countries, protective clothing is scarce or simply not used when it is too hot. Neither in industrialized countries is protective equipment normally used by pesticide applicators because applicators are afraid of the bad image this would give. A farmer told me that the main reason for switching to organic farming was that he was ashamed of his kids whenever he had to put on all these protective clothes for the pesticide spraying (Zaller 2018). Of course, manufacturers are eager to point out that their products are safe when used properly according to the manual including protective equipment.

Dicamba (3,6-dichloro-2-methoxybenzoic acid) is another active ingredient in broad-spectrum herbicides that is widely used to kill annual and perennial broadleaf weeds. It is on the market for about 50 years and used for household and commercial weed control. The primary commercial application of dicamba is in grain crops and turf areas but also to kill brushes and bracken in pastures, or legumes and cacti. Dicamba leads to an increased plant growth; eventually the plant outgrows its nutrient supplies and dies. Increasing use of dicamba has been reported with the release of dicamba-resistant GM plants by Monsanto. Especially older formulations of dicamba have been reported to easily drift after application and affect other crops not meant to be treated. Newer, less volatile formulations of dicamba are supposed to be less prone to vaporizing and inhibit unintended drift between fields.

1.3.2 Insecticides Kill Insects

Insecticides are ingested by insects as food, contact, or respiratory poisons. Some insecticides, for example, neonicotinoids, are also absorbed by the plant and distributed systemically throughout the plant. As soon as an insect nibbles on the plant or sucks plant sap, it takes up the poison and dies. The effect of insecticides is usually via a disturbance of the nervous system or a disruption of developmental processes in insects. Because of similar functions of the nervous system in insects and mammals, insecticides are considered the most critical group of active substances for humans in terms of their toxicity. Similar to herbicides, which do not differentiate between crops and weeds, insecticides do not differentiate between beneficial and harmful insects. This means that insecticides interfere with the interplay between beneficial insects such as parasitoids and pest insects that have been evolved over millions of years.

Global insecticide sales amount to more than 7 billion €; about 50% are against biting insects, 30% against sucking insects, and 10% against mites (Hallmann et al. 2009). Zoologically, mites are also arthropods but belong to the class of arachnids such as spiders, but this is a side issue. Among the biting insects, caterpillars of butterflies and moths account for more than two-thirds of the harmful effects, making them by far the most important pest group in the world, ahead of various beetles. Adult butterflies can no longer act as a pest because they have no biting mouthparts and can only consume liquid nectar with their long tongue.

The first synthetic insecticides were of the group of organochlorides, with dichlorodiphenyltrichloroethane (DDT), aldrin, dieldrin, endrin, heptachlor, chlordane, and endosulfan as most widely used substances. Some organochlorine compounds, such as sulfur mustards, nitrogen mustards, and Lewisite, have also been used as chemical weapons due to their toxicity to humans. DDT has been used since the beginning of the 1940s as a contact and feeding poison. Because of its good efficacy against insects and its simple manufacturing process, it was the most widely used insecticide in the world for decades. During the World War II, it was regularly used to control potato beetles or head lice. In forestry, it was applied with planes to control May beetles (around 400 species are called May beetles, *Phyllophaga* spp.), bark beetles (with around 6000 species, *Scolytus* spp.), and tussock moths (with about 2500 species; a famous pest species is the gypsy moth, *Lymantria dispar*).

The application rates of DDT in agriculture, especially in cotton cultivation, sometimes reached 35 kg/ha (BUA 1999). In comparison, the application rates of modern insecticides are sometimes less than one 100 g per application per hectare, so several 100 times lower than for DDT. These ultra-low doses also make it especially difficult for private people to apply the right dosage when recommendations require a few milligrams per square meter in their gardens. No household balance is that accurate, so overdosage is very common among private users. As early as the 1960s, especially with the publication of Rachel Carson's *Silent Spring* (Carson 1962), evidence was mounting on the accumulation of DDT in the fatty tissue of animals and humans. After all, it was also suspected of causing cancer. In the 1970s, it was thus banned in many countries.

In the 1990s, the global insecticide market was dominated by carbamates, organophosphates, and pyrethroids. Carbamates influence the nervous system of insects, but were also components of fungicides, and even sleeping pills. Organophosphates are also nerve poisons and originate from the same substance class as the supertoxic military combat gases sarin and VX. Pyrethroids are synthetically produced insecticides whose active ingredient is similar to the insecticidal ingredient pyrethrin contained in chrysanthemum plants. However, while pyrethrin disintegrates rapidly in the sunlight, pyrethroids are very long-lived and are associated with many serious health effects. However, although of natural origin, pyrethrin is also very toxic and harmful for the environment and humans.

Twenty years later, more than a quarter of the insecticide market is dominated by neonicotinoids (Simon-Delso et al. 2015).

1.3.2.1   Insecticide Active Ingredient Neonicotinoids

The name literally means new nicotine-like insecticides. However, despite a similar chemical structure and mode of action, the toxicity for mammals is about 700-fold lower than that of nicotine (Elbert et al. 1998). The name neonicotinoids are often shortened to neonics. Neonics are a group of active substances that act as a nerve poison against insect pests and are currently the most widely used insecticide class worldwide, with applications in agriculture, fruit growing, veterinary medicine, and fish farming. Neonicotinoids are either sprayed directly onto plants, biting and sucking insects, or they are applied onto the soil, or mixed in irrigation water. In addition, plant seeds are treated with these insecticides by seed dressing.

The neonicotinoid imidacloprid is the world's best-selling insecticide and the second best-selling pesticide after the herbicide glyphosate. The annual production of imidacloprid is estimated at around 20,000 tons of active ingredient. Because they are very easily soluble in water and their broad use, neonics are found everywhere in our environment—in soils, in water, and in the air. In just 6 years (2003–2009), sales of some neonicotinoids have increased 15-fold and created a market of approximately 2.3 billion € (Jeschke et al. 2011).

Seven neonicotinoid active ingredients are in use worldwide (Simon-Delso et al. 2015). These are imidacloprid (used on 140 different crops) and thiacloprid (50 crops), clothianidin (40 crops), thiamethoxam (115 crops), acetamiprid (60 crops), nitenpyram (12 crops), and dinotefuran (35 crops). Patent protection expired for most neonics and generic products are manufactured especially in India and China. Meanwhile, several hundred new neonics have been synthesized and are about to be approved.

Just like herbicide residues, residues of insecticide can be found in many food products. Neonicotinoid residues were for instance found in honey samples from across the world (Mitchell et al. 2017). Although the neonics occurred at levels considered safe for human consumption, this finding confirms the inundation of the environment with these pesticides. The concentrations found in honey samples (1.8 nanograms per gram honey) are below the maximum residue level authorized for human consumption; yet the coexistence of neonics and other pesticides may increase harm to bees. Bees come into contact with neonics when collecting nectar of crops, the seeds of which were treated with neonics. Because of their systemic mode of action, these substances are distributed across the treated plants. Residues of the neonicotinoid thiacloprid have even been found in honey samples from organic beekeeping. Theoretically, organic beekeepers are required to set up their hives in such a way that no significant impairment via pesticides is to be expected within a radius of 3 km. However, when a massive nectar crop such as oilseed rape is flowering, honeybees can easily fly up to 6 km. Of course, bees from organic beekeepers do not know that they are organic and take home this pesticide contaminated nectar and pollen. This immediately raises the question whether a viable coexistence between conventional, pesticide-dominated agriculture and organic agriculture is even possible.

It is very difficult to obtain country-specific information on the use of neonicotinoids, as authorities refuse to provide such information referring to the protection of manufacturers' trade secrets. A further complication is that countries that provide this information (such as the UK, Sweden, Japan, or California) report different measures. Some report financial sales, some consumption amounts, and some amounts imported or produced. However, trends can be seen, and they show a steady growth of amounts of neonics consumed. This will be sustained by an increase in agricultural intensification and the use of combination products that mix neonics with other pesticides. Of the estimated 20,000 tons produced worldwide, about 68% come from China (Simon-Delso et al. 2015). Neonicotinoids are used in 140 different crops in 120 countries worldwide, making it difficult to keep track of the exact figures. It is known from Great Britain that 91% of neonics are used in seed dressings (Goulson 2013); in Central Europe and other countries, it is probably similar.

Neonicotinoids are also used as a seed dressing in maize. Although maize is wind-pollinated and therefore not visited by honeybees, studies have found neonics can drift over kilometers of swirling dust during maize sowing under dry soil conditions (Pistorius et al. 2009). As a result, approximately 94% of honeybees in the US state of Indiana are exposed to the risk of neonicotinoid poisoning (Krupke et al. 2017). By the way, this study could not prove that neonicotinoid seed treatment has a positive effect on crop yields which is actually the original aim to apply pesticides. This was also observed for soybean production in the USA, after all the biggest soybean producer in the world. Studying almost 194 farms, yield benefits through neonicotinoid seed treatment were negligible at about 130 kg/ha at an average yield of about 3500 kg/ha (Mourtzinis et al. 2019). Thus, despite widespread use, this practice appears to have little benefit for most soybean producers.

Until 2013, five neonicotinoid insecticides were approved as active substances in the EU, namely clothianidin, imidacloprid, thiamethoxam, acetamiprid, and thiacloprid (EC 2019f). Due to the harmfulness of neonicotinoids for bees, the EU has restricted the use of three neonicotinoids (clothianidin, imidacloprid, and thiamethoxam) for the precautionary protection of bees in 2013. This measure was based on a risk assessment of the European Food Safety Authority (EFSA) in 2012. It prohibits the use of these three neonicotinoids in bee-attractive crops (including maize, oilseed rape, and sunflower) with the exception of uses in greenhouses, of treatment of some crops after flowering, and of winter cereals. At the same time, the pesticide manufacturers of the three substances were obliged to provide further data for each of their substances in order to confirm the safety of their products. Following another EFSA assessment, the remaining outdoor uses could no longer be considered safe due to the identified risks to bees and the three neonics were finally banned for outdoor uses in 2018.

For another neonicotinoid, acetamiprid, EFSA established a low risk to bees and a ban or further restrictions of this substance were considered not appropriate. A fifth neonicotinoid, thiacloprid, is a candidate for substitution, based on its endocrine-disrupting properties. Candidates for substitution are pesticides for which national authorities need to carry out an assessment to establish whether more favorable alternatives to using the plant protection product exist, including nonchemical methods.

A meta-study hypothesized that seed-applied neonicotinoids reduce arthropod natural enemy abundance, i.e., insect predators or spiders and parasitoids, but not as strongly as soil- and foliar-applied pyrethroids which neonics replaced in many cases (Douglas and Tooker 2016). However, after comparing nearly 1000 observations from North American and European field studies, this analysis revealed that seed-applied neonicotinoids reduced the abundance of arthropod natural enemies similarly to broadcast applications of pyrethroid insecticides. So, substituting pyrethroids with neonicotinoids seems to have no benefit for the abundance of natural enemies of pest insects.

A very little studied field is the effect of neonicotinoids on marine ecosystems. As mentioned, neonics are often used in commercial fish farms to protect lobster farms from natural shrimp, for example (Pisa et al. 2015). In Japan it was shown that since neonicotinoid application to agricultural fields began in the 1990s, zooplankton biomass has plummeted in a nearby lake (Yamamuro et al. 2019). The reduced zooplankton has led to shifts in food web structure and reduced yields of freshwater fish species—smelt (*Hypomesus nipponensis*) and eel (*Anguilla japonica*) to about 10% of the yields before neonics were applied. It is very likely that such disruptions also occur elsewhere.

Neonicotinoids are also used in veterinary medicine to control parasites such as fleas or ticks on dogs, cats, or other pets.

1.3.2.2 Insecticide Active Ingredient Fipronil

The phenyl-pyrazole fipronil is marketed since the 1990s as a reaction to frequent insect resistance to organophosphates, carbamates, and pyrethroids and because of its low toxicity to vertebrates. Similarly to neonics fipronil operates by disrupting neural transmission in the central nervous system of invertebrates; fipronil especially inhibits neuronal receptors and this continuous stimulation of neurons ultimately leads to the death of the insects (Simon-Delso et al. 2015). Fipronil is a very powerful insect venom and is used in doses of around 50 g/ha; in its toxicity to honeybees it is around 6000 times stronger than DDT. Even EFSA, which is not suspicious of being particularly strict, considers fipronil to be acutely dangerous for honeybees (EFSA 2015). The systemic properties of fipronil, combined with prophylactic applications, create strong selection pressure on pest populations, thus expediting evolution of resistance and causing control failure. Fipronil is also used in veterinary medicine against ticks and fleas in pets. Studies have shown that fipronil accumulates in the fatty tissue, the liver, and kidneys of pigs, cows, and chickens. It can be assumed that this is also the case in humans. Fipronil is also suspected of being carcinogenic and disrupting the hormone system. However, the problem has at best been discussed in specialist circles. Despite these nontarget effects, Austria has issued an emergency approval for fipronil as a reaction to the EU ban on neonicotinoids in order to combat wireworms in potatoes (Kainrath 2019).

In Europe, fipronil did not attract much public attention until the summer of 2017. Then fipronil residues were found in the eggs of hundreds of egg producers from Belgium and the Netherlands. The substance was illegally used there to clean the henhouses. Within a few weeks, millions of contaminated eggs were destroyed upon authority order, although authorities kept assuring that consumers have not been endangered. In the course of this scandal, remarkable trade flows of agricultural products became obvious. Fipronil-contaminated eggs from Belgium and the Netherlands were found in 45 countries, including the USA, Russia, and virtually all European countries. Tracing the egg flow is almost impossible because in Europe there is no obligation to label the origin of the eggs, although about 60% of eggs go into food processing. The food industry persistently rejects ideas of labeling the product origins fearing competitive disadvantages; perhaps they fear more that consumers would realize the sickness of the whole food industry.

1.3.2.3 Insecticide Active Ingredient Chlorpyrifos

Chlorpyrifos belongs to the group of organophosphates that were developed in the 1930s and 1940s for use as nerve gas agents—sarin was one of the most notorious one—and later adapted for use as insecticides at lower doses. In 1965, Dow Chemicals introduced chlorpyrifos as an insecticide for use in gardens and on fields. In agriculture, chlorpyrifos is used on crops, animals, and buildings, and in other settings, to kill a number of pests, including insects and worms. It acts on the nervous systems of insects by inhibiting the acetylcholinesterase enzyme. It is one of the most widely used organophosphate insecticides in US agriculture (EPA 2002).

After studies increasingly raised concerns about prenatal neurodevelopmental risks of chlorpyrifos, it was banned for home use in the USA in 2000 (Rauh et al. 2012). Recently an expert panel of public health experts provided evidence that all organophosphates threaten the health of children and pregnant women (Hertz-Picciotto et al. 2018). According to their report, exposure to organophosphates increases the risk of reduced intelligence (measured by IQs), memory and attention deficits, and autism for prenatal children. One wonders if government officials approving such pesticides listen to science or rather to lobbyists of the agrochemical industry.

More than 10,000 tons of organophosphate pesticides are sprayed in 24 European countries each year. There are dozens of human studies that exposures of pregnant women to very low levels of organophosphate pesticides put children and fetuses at risk for developmental problems that may last a lifetime (Nelsen 2018). The US regulators had already quietly banned 26 out of 40 organophosphate pesticides considered hazardous to human health; in Europe, 33 out of 39 organophosphates were banned. In Central America, organophosphates are still popular and ranked fourth among 24 chemical groups of imported pesticides.

An overview study showed that broad-spectrum insecticides such as carbamates, organophosphates, and pyrethroids can cause population declines of beneficial insects such as bees, spiders, or beetles which are important food insects for vertebrates such as bats or birds (Isenring 2010). Many of these species play an important role in the food web or as natural enemies of pest insects. In the UK for example, it was shown that of 95 incidents of severe bee poisonings between 1995 and 2001, organophosphates caused 42%, carbamates 29%, and pyrethroids 14% of cases (Fletcher and Barnett 2003). Insecticides which poisoned bee colonies included the carbamate bendiocarb and three pyrethroids: cypermethrin, deltamethrin, and permethrin. Synergistic effects between pyrethroids and fungicides (imidazole or triazole fungicides) can additionally increase the risk to honeybees (Pilling and Jepson 2006). Scientists speak of synergistic effects when the effect of two pesticides applied at the same time is greater than the sum of effects of each pesticide separately.

Terbufos and methamidophos are other popular organophosphate insecticides; however, they have been targeted to be phased out by the Rotterdam Convention (RC 2019). The objectives of the convention are to protect human health and the environment from hazardous chemicals. Terbufos is used in insecticides and nematicides and has been linked to lung cancer, leukemia, and non-Hodgkin's lymphoma. Methamidophos is used in potatoes and rice throughout Latin America, Spain, Japan, and Australia. Due to its toxicity, the use of pesticides containing methamidophos was phased out in Brazil, and voluntarily canceled in the USA (EPA 2009).

1.3.3 Fungicides Control Fungi

Fungal diseases are quite common in crop plants, especially in climatic regions with high relative humidity. The commercially important diseases often controlled with fungicides are leaf-spot diseases, late blight, downy mildew, rice diseases, fruit rots, storage rots, cereal seed-borne diseases, powdery mildews, cereal stem diseases, rusts, and smuts. Fungicides are also used to control many postharvest diseases that cause rapid and extensive breakdown of high-moisture commodities.

Fungicides are applied as dust, granules, gas, and, most commonly, liquid. Fungicide treatment of seeds, bulbs, roots of transplants, and other propagative organs is usually performed by the seed company. The goal here is to kill pathogens that are on the planting material or to protect the young plant from pathogens in the soil. Fungicides are also applied in the field during planting, for instance in-furrows, or after planting as a soil drench through drip irrigation. Often fungicides are sprayed at the foliage of crop trees or through trunk injection. The harvested produce is treated as a dip or spray in the storage house.

Fungicides are also used to reduce mycotoxin contamination in wheat affected by *Fusarium* head blight capable of causing severe illness or even death in humans and animals when consumed. But most fungicides developed so far have not been sufficiently effective to be useful for managing mycotoxins associated with other diseases (McGrath 2020).

In Europe, there are about 150 fungicidal active substances approved (EC 2019c). Major fungicides belong to the group of methyl benzimidazole carbamates, dicarboximide, phenylamide, phenylpyrrole, benzamide, and others. Besides, there are many biopesticides such as botanical oils (clove, garlic, peppermint, etc.), kaolin clay, and several bacteria and fungal species. Fungicides kill fungi by damaging their cell membranes, by inactivating critical enzymes or proteins, or by interfering with key processes such as energy production, respiration, or specific metabolic pathways (APS 2019). Some fungicides do not directly affect the pathogen itself but rather trigger plant defense mechanisms such as the production of thicker cell walls and antifungal proteins.

Like herbicides, also fungicides can have either a narrow or a broad-spectrum of activity. Narrow-spectrum fungicides are effective against only a few usually closely related pathogens, while broad-spectrum fungicides control a wide range of unrelated pathogens. Many of the early fungicides on the market were inorganic compounds based on sulfur or metal ions such as copper, tin, cadmium, and mercury. Copper and sulfur are still widely used in both conventional and organic farming. Other active ingredients in fungicides include neem oil, rosemary oil, jojoba oil, the bacterium *Bacillus subtilis*, and the beneficial fungus *Ulocladium oudemansii*.

Due to the specific situation, fungicide-intensive crops such as cereals, potatoes, rape, grapevine, apple, hops, and some vegetable crops differ from relatively fungicide-extensive crops such as maize or soybeans. Frequency of application of fungicides can range from 10 per season, as with potatoes, to more than 20 treatments per season in apple cultivation (Hallmann et al. 2009). Such frequent applications increase the risk of resistances where the pathogenic fungi become immune to the toxins of the applied fungicides. This is an increasing problem and will be described more closely later.

There are contact fungicides and systemic ones. Contact fungicides are not taken up into the plant tissue and protect only the plant part that is sprayed. Systemic fungicides are taken up and redistributed through the xylem vessels to all parts of a plant.

Like insecticides, fungicides are often used to treat seeds in order to protect them against fungal diseases. This treatment is effective as some plant diseases are transmitted exclusively via the seeds. In principle, this is an ancient method is known since 2500 years. At that time, olive pomace, ash, onion broth, or cypress juice was used to protect seeds from pathogens. At the end of the nineteenth century, the very effective but toxic chlorine phenolic mercury seed treatments were developed until they were banned in the 1980s.

Fungicide residues in food are generally analyzed less frequently. Food monitoring studies in Bavaria, Germany, have found pesticides in around 80% of 44 field salad (*Valerianella locusta*) samples examined; nine samples exceeded the legal limits (Ökotest 2017). Field salad is susceptible to fungal diseases and is therefore frequently treated with fungicides. Seven years later, 25 field salad samples were tested again, this time all below the permitted limits. There is at least some improvement, one might be affirmed. However, in fact in the meantime maximum residue levels (MRL) have been raised so that field salad may contain significantly more residues today than it did years ago. This increase in the MRLs is not an exception but rather the rule as will be elaborated in a later chapter. Particularly, leafy vegetables are susceptible to pesticide residues, as more chemicals accumulate on crops with many leaves than on fruit vegetables such as tomatoes or cucumbers with a relatively small surface area.

Another fungicide intensive crop are bananas, the most popular fruit in many countries. About 115 million tons of bananas are consumed each year, produced in more than 100 countries. Each American eats on average 4.5 kg per year, more than any other fruit. Also, in Europe, bananas are among the most popular fruits. Despite this massive demand for bananas, it cannot be grown in great amounts in the USA or Europe. The burden of supplying the banana hunger falls on countries in the tropics, including Costa Rica, India, Brazil, Ecuador, and some others. The warm and humid plantations are prone to the growth of an airborne fungus causing leaf-spot disease called Black Sigatoka (*Mycosphaerella fijiensis*) that can destroy an entire plantation in about a week (Chatterjee 2013). Plants with leaves damaged by the disease may have up to 50% lower yields, and control can take up to 40 sprays a year (Ploetz 2001). In the 1930s, chemical control was performed with the so-called *Bordeaux mixture*, a mixture of copper sulfate and slaked lime also used in viticulture at that time. Nowadays, mainly systemic fungicides (propiconazole, methoxyacrylate, azoxystrobin) are used. Additionally, broad-spectrum fungicides (dithiocarbamates, chlorothalonil) are applied, often with helicopters or aircrafts.

Researchers wanted to know whether pesticides are also ending up in animals that live near banana plantations (Grant et al. 2013). In particular they were interested in Spectacled caiman (*Caiman crocodilus*), fish-eating crocodilians that inhabit freshwater habitat in tropical regions of the Americas. These caimans are top predators in the ecosystem and considered a threatened species. Blood samples from 14 adult caimans were collected and analyzed for 70 different pesticides. The results were concerning as the samples contained nine pesticides, of which only two are currently in use. The remaining seven are historic organic pollutants, such as DDT, dieldrin, and endosulfan. Again, chemicals that have been banned for many years persisted in the environment. Researchers also found that caimans that were near the banana plantations were in a poorer health state than caimans in more pristine areas.

Fungicides based on copper are often used as alternatives to synthetic fungicides, also in organic farming. However, copper is toxic to aquatic organisms and the risk of copper bioaccumulation in fish and some other aquatic organisms may be high (EFSA et al. 2018b). Risks for birds and mammals, aquatic organisms, bees, and other nontarget arthropods, earthworms, and other soil macro-organisms are also reported.

Chlorothalonil is one of the world's most common broad-spectrum, nonsystemic fungicide widely used in arable farming on tomatoes, potatoes, and peanuts but also on golf courses. It is considered a "probable human carcinogen" and has been shown, among other fungicides, to make honeybees more vulnerable to the gut parasite *Nosema ceranae* (Pettis et al. 2013). It will soon be banned by the European Union after safety officials reported human health and environmental concerns (EFSA et al. 2018a). Chlorothalonil is also the most used fungicide in the USA and the UK; farmers in those countries call the ban "overly precautionary" (Carrington 2019).

Also, Switzerland banned chlorothalonil at the beginning of 2020 because of frequent groundwater contaminations. Now the manufacturer, the Swiss company Syngenta, announced that it would take legal action against the withdrawal (Schuller 2020). Syngenta criticizes that the authorities would contradict themselves about the danger of the metabolites of chlorothalonil found in drinking water. On the one hand, they have been classified as not dangerous for humans and the environment, and on the other hand, these were the main reasons for the ban. The manufacturer further argues that chlorothalonil has been used in Switzerland for 40 years and is needed to have a broad spectrum of different fungicides for avoiding resistances.

1.3.3.1 Fungicide Active Ingredients Captan and Folpet

Captan, from the group of phthalimides, is a frequently used fungicide in orchard crops, seed treatments, ornamentals, lawns, and turf and is also used as an in-can preservative, in adhesives and paint. Formulations include dust, emulsifiable concentrate, flowable concentrate, water-dispersible granules, wettable powder, and a variety of others. In the USA, captan is also applied as a postharvest dip to apples, cherries, and pears in order to prevent the spore germination of various fungal diseases. It also improves the outward appearance of many fruits, making them brighter and healthier looking.

The pesticide authority in Europe concludes that captan is of low toxicity by the oral and dermal routes, but it is toxic via inhalation (EFSA 2009). Besides being severely irritating to eyes, captan did not show genotoxic potential (toxic to DNA) and it is not teratogenic (disturbing the development of the embryo). However, captan was found to cause duodenal tumors in laboratory mice, a form of cancer in the first section of the small intestine. Interestingly, this expert report avoids the term cancerogenic throughout, although the manufacturer clearly states in the safety data sheet that it is suspected of causing cancer.

Captan was previously cited as a probable human carcinogen by the US Environmental Protection Agency (EPA 1999), but was reclassified in 2004. The EPA now states that captan is potentially carcinogenic at prolonged high doses many orders of magnitude above those likely to be consumed in the diet, or encountered by individuals in occupational or residential settings. A similar reclassification has been made for folpet, a structurally related fungicide, which shares a common mechanism of toxicity. Nevertheless, EPA is still concerned about post-application exposure to toddlers' hand-to-mouth activity on treated lawns. The manufacturer has agreed to voluntarily cancel this specific application of captan.

1.4   What Amounts of Pesticides Are Involved?

World pesticide expenditures at the producer level totaled about 50 billion € in 2012 (EPA 2017). Expenditures on herbicides consistently accounted for about 45% of total expenditures between 2008 and 2012, followed by expenditures on insecticides, fungicides, and other pesticides. In 2012, US expenditures accounted for 21% of world expenditures on herbicides (including plant growth regulators), 14% of insecticides, 10% of fungicides, and 23% of fumigants. However, agricultural crop area (97 million hectares) in the USA makes up only 6% of the world agricultural area (about 1.6 billion hectares).

China, the USA, Brazil, and the EU are not only four of the largest agricultural producers in the world, but are also some of the world's largest pesticide users—China using 1.769 billion, the USA 544 million, Brazil 377 million, and the EU using 375 million kg of pesticides in 2016 (Donley 2019).

The amount of pesticides applied each year is increasing despite industry claims to developing ever more effective pesticide formulations. Between 2002 and 2012 alone, the use of pesticides on German arable land rose from around 35,000 tons to over 45,000 tons, an increase of 30% (Klingenschmitt 2016). During the same time, the arable land has remained more or less unchanged. This development is becoming even more dramatic, as the new pesticides are more effective than the old ones.

The amounts of pesticides applied depend on a country' agricultural intensity and the prevalent climate. For instance, in Europe's climates, fungicides are more important than herbicides due to the strong infection pressure from fungal pathogens and account for more than a third of total pesticide sales, while insecticides are less important. Most intensive agriculture is mainly found in developed countries of North America, Europe, and Asia, where more than two-thirds of total pesticide amounts are applied. In small-scale subsistence agriculture in Africa and Latin America, on the other hand, pesticide use is still quite low (Fig. 1.1).

A widespread myth is that pesticides are only applied in cases of imminent danger and in amounts that are absolutely necessary. Even the German Agricultural Society (DLG) admits that the excessive use of pesticides not only damages the environment but also promotes the development of resistance of weeds and pests (Chmura 2017). It is estimated that every other pesticide is used at a wrong time (too early or too late) and is therefore not effective. This is quite remarkable, since self-criticism of agricultural societies is rare. The finding that farmers often apply too much pesticides is also due to incorrect advice from agroindustry representatives, who of course also have a conflict of interest here. During a public panel discussion, I myself witnessed how a representative of agrochemical industry openly admitted that his company's recommendations for fertilizer and pesticide use had been reduced by 50% in recent years. The reason for this was that the public pressure was too high, as more and more drinking water wells were polluted. This is well in line with findings from France showing that no yield losses would be expected if the input of fertilizers and pesticides was substantially reduced (Lechenet et al. 2017).

The amount of pesticide active ingredients sold in Germany amounted to 40,000 tons in 2006 and reached a record level of 48,600 tons in 2015 (Beste 2017). The global pesticide market is currently experiencing the strongest growth in Asia and South America, with pesticide expenditure rising particularly in China, India, Brazil, and Argentina (MultiWatch 2016). Looking at individual countries, Brazil is currently among the world's largest users of pesticides. Much is produced in Brazil for the international market. For instance, Brazil currently produces 23% of the so-called biofuel ethanol and 48% of the sugar available on the international markets.

Modern pesticides can be several thousand times more effective than the classic compounds (e.g., neonics vs. DDT), so that a much more intensive effect can be achieved with the same quantity (Nentwig 2005). Herbicides are now used in over 90% of all cereal, beet, and maize fields, in more than half of all vineyards and orchards, and less in rape and potato cultivation. Fungicides are mainly used in fruit, wine, and hop cultivation and in arable farming also in wheat and potatoes. Globally, a quarter of all insecticides are used in cotton crops. In Central Europe, insecticides are used in fruit, wine, and hop cultivation, but also in sugar beet, rapeseed, potatoes, and cereals. Pesticides are used less frequently only on meadows and pastures and in arable land in so-called cover crops or catch crops such as clover or alfalfa.

As mentioned before, the large monocultures in banana cultivation are very pesticide intensive. A typical banana plantation is sprayed with pesticides up to 50 times per season. The expenditure for chemical fungicides accounts for 30–50% of banana production costs (MultiWatch 2016). In modern banana cultivation, around 40 kg/ha of pesticides are used per year on a weekly interval usually from an airplane. The use of pesticides is also at the expense of the health of local workers under working conditions that are unethical in terms of medical or social standards (Lumetzberger 2016). Moreover, hardly anyone of the farmworkers knows about the risks involved in handling of pesticides and the majority does not use protective clothing or gloves. Workers' families are also affected if contaminated work clothes are taken off at home.

It is devastating to see that pesticide planes fly over small villages and schools with children playing outside. It is reported that some schools at least managed to get the planes to schedule their treatments after school. The pesticides sprayed nevertheless settle on the ground and are found in house dust, where they are an important source of pollution for children. These aspects at the production site have also to be considered in discussions on pesticide residue levels in imported food we eat. In organic banana production, biodegradable products to control pests and diseases are used; a fair-trade label additionally considers social aspects. Somewhat irritating is that these two aspects are not always combined.

Where do data on pesticide use come from? The most comprehensive and reasonably accurate pesticide use data are collected by the Food and Agriculture Organization (FAO) of the United Nations based in Rome, Italy (www.fao.org/faostat/). These statistics include data on the use of major pesticide groups (insecticides, herbicides, fungicides, plant growth regulators, and rodenticides) and of relevant chemical families. Data report tons of active ingredients of pesticides used in or sold to the agricultural sector for crops and seeds; information on quantities applied to single crops is not available. The primary data source is the standardized FAO questionnaire on pesticides use; official statistics may be complemented with government data sources such as yearbooks and ministerial data portals. If there are data gaps, they may be filled with secondary sources such as country studies from other international organizations. The collected data are routinely checked for internal consistency (e.g., outliers and significant variation in time series); any observed discrepancies are checked and validated with countries.

If one wants detailed data on specific countries, it is rather confusing, as countries publish figures of varying reliability. Generally, for Europe, only those pesticides that have been approved by the relevant authorities may be used in each European country and the amounts are published more or less in detail by the national authorities. The data only show amounts that came to farmers via official channels. Pesticides that are purchased in the Internet do not appear in official statistics. According to industry estimates, Internet purchases of pesticides can be substantial (Müller 2001). However, agrochemical companies also offer highly dangerous and often banned pesticides on their websites in countries of the global South; a quick look on the Internet can be quite eye-opening. Also not considered in these statistics are pesticide purchases of private users via the internet.

Pesticide purchases are usually controlled by authorities, however at a modest level. In Austria for example, 367 on-farm pesticide inspections were carried out in 2016 (BMNT 2018). With a total of 161,155 farms in this country, this represents an inspection rate of less than 1% of the farms. If other pesticide users such as landscape gardeners, golf courses, and municipalities are included, this proportion is even much lower. In any case, the probability of being controlled is extremely low. In comparison, organic farms are inspected annually at their own expense.

The German Federal Environment Agency is exemplary regarding the details of the published reports on pesticide sales as it also distinguishes between agricultural and private use. As of 2015, 48,611 metric tons of 277 pesticide active ingredients were used in Germany, around 35% of which were herbicides, 26% fungicides, 31% insecticides, and the remainder other active ingredient groups (UBA Berlin 2019). This amounts to an averaged 8.8 kg/ha pesticides per year or 2.8 kg/ha of active ingredients applied on arable land (calculation for 2015 without inert gases, with approx. 12.1 million hectares of arable land and permanent crops).

Claims that pesticide use is decreasing in the last few years need to be checked carefully. For instance, in the UK pesticide usage decreased from 34.4 million kg in 1990 to 16.9 million kg in 2016, equaling a reduction of about 51% (FERA 2020). However, at the same time the toxicity of pesticides has increased so dramatically that measuring their usage by weight becomes increasingly irrelevant and misleading because less pesticides are required to do the same job (PAN UK 2017). As an example, some neonicotinoid insecticides are 10,000 times more toxic than the most infamous insecticide in history, DDT (Goulson 2013). Second, the area of land being treated with pesticides has risen as well in the UK. For instance in 1990, the area treated with fungicides, insecticides, and herbicides was about 45 million hectares but increased in 2016 to about 73 million hectares, which is an increase by 63% (FERA 2020). Moreover, the number of times crops are treated with pesticides has gone up in the UK from averaged 2.5 treatments in 1990 to averaged 4.2 treatments in 2016 (PAN UK 2017). Thus, weight is a rather meaningless metric for measuring the use of pesticides.

Who is allowed to use and purchase pesticides? Farmers in Europe can purchase them only after presenting a certificate of training in the safe use of pesticides. This will be obtained after a two-day training—undoubtedly a major step forward. However, these trainings are partly offered by representatives of agrochemical companies. Farmers who do not want to do this training must hire professional contractors to perform the pesticide spraying for them or they can ask their trained colleagues to spray pesticides for them.

Private users can purchase and use pesticides without any proof of training. This is quite remarkable, as we are speaking of the same active ingredients that are contained in agricultural pesticides. Moreover, it is even less likely that private users follow the dosage recommendations and wear proper protective clothing while spraying. In my opinion, pesticides have no place in private gardens or homes and should therefore be banned for this use! I also cannot think of a reason why I should poison my home-grown fruits and vegetables or my home. At least in Europe, many pesticides are now only available in do-it-yourself stores after advice through the salesperson. Some do-it-yourself chains and garden centers have even taken all synthetic pesticides out of their assortment and are positioning themselves as a ecologically sensitive alternative to the competitors. In the Internet, however, most products are still available without proof of expertise. I also know from Austrian farmers in the border area to Hungary, the Czech Republic, or Slovakia who buy the "good old" pesticides now banned in Austria from back stocks in neighboring countries.

The consumption of pesticides of private persons is difficult to verify. In Austria not even the chemical industry cannot—or more precisely does not wish to—survey or estimate the difference between agricultural and non-agricultural use (Sattelberger 2001). In the interests of environmental and consumer protection and in order to estimate the exposure of humans and the environment, more transparency should be demanded from the responsible authorities. To this end, there is an urgent need to introduce a legal obligation to report the quantities of pesticide-active substances placed on the market and applied into our environment.

However, even with a proper training, failures in pesticide application can occur. Official checks without prior notice in Germany found a misconduct of pesticide applications by farmers in 50% of cases (Haas 2010). Ten years earlier, 89% of all pesticide applications in the German federal state of Hesse were not compliant and all regulations for keeping clear of water bodies were ignored by farmers. Pesticide contamination of water bodies and ultimately of drinking water is thus inevitable.

A very serious aspect is the faking of pesticides. With the global pesticide crop protection market valued at 60 billion US $, counterfeit pesticides are estimated to make up to 15% of that share, a lucrative business for international organized crime (UNEP 2018). Experts estimate that in Poland with intensive apple production up to 10% of all pesticides used could be counterfeit and in the Spanish region of Almería, a region with very intensive vegetable production, even around 25%. In addition, some agricultural companies in Almería have been suggested to be firmly in the hands of criminal organizations (Analytik-News 2006). Reportedly, most of the pesticide counterfeits come from China. The profit margins are probably as good as for luxury watches or designer handbags and are estimated at over a 1000% for counterfeit pesticides. Either the labels on the counterfeit pesticides are deceptively genuine or large containers are simply not labeled so checks at customs are difficult in these bulk samples. Random checks by the authorities are deliberately accepted and fines calculated (Sanderson 2006).

It is estimated that the smuggling of counterfeit pesticides causes the European economy an annual loss of 1.3 billion € endangering around 2600 jobs (Focus 2017). The port of Hamburg is considered one of the hubs for the criminal trade of pesticides (Zand-Vakili 2014). Often the chemicals from Asia reach their destination in Eastern Europe via Hamburg and are then transported on land across Europe. The majority are counterfeits of approved branded products. Copies differ from branded goods in their composition, but this is very expensive to investigate. In addition to the economic damage caused by these counterfeits, the costs for the environment and health are hard to calculate. Studies have shown that the criminal trade throws substances onto the market that are sometimes carcinogenic or genotoxic. After increased controls, the authorities in Hamburg confiscated a total of 196 tons of illegal pesticides in November 2016. This is certainly only the tip of the iceberg, as no authority is in a position to inspect the approximately 9 million containers handled annually in Hamburg. Once these pesticides have been smuggled into the EU, it is almost impossible to follow their distribution throughout the continent.

1.5   Patterns of Pesticide Use from Different Countries and Regions

Dependent on the agricultural structure, pesticide use varies considerably between countries (Fig. 1.4). From the 155 countries considered worldwide, Saint Lucia, a Small Island Developing State of the British Commonwealth, has the highest pesticide use with 19.6 kg/ha agricultural land. This reflects the very pesticide-intensive banana industry which is Saint Lucia's most important export commodity. The other top-ten countries regarding pesticide inputs are Hong Kong (16.6 kg/ha), Ecuador (13.9 kg/ha), Taiwan (13.3 kg/ha), China (13.1 kg/ha), Israel (12.6 kg/ha), Korea (12.4 kg/ha), the Seychelles (12.1 kg/ha), and Japan (11.8 kg/ha). The North American country with the highest pesticide input is the USA (2.5 kg/ha), in Central America it is Guatemala (10.0 kg/ha), and in South America it is Ecuador (13.9 kg/ha).

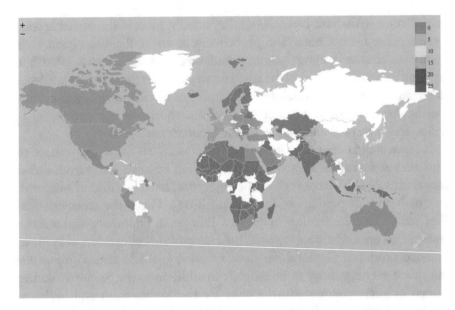

Fig. 1.4 Pesticide input in kilograms per hectare agricultural land in different countries of the world in 2017. Map based on data from FAOSTATS 2019. For white countries no data are available

Australia has an input of 2.0 kg/ha; the African country with the highest pesticide input is Mauritius (9.8 kg/ha) with its sugar cane industry. The European country with the highest pesticide input is Cyprus (8.2 kg/ha). It is important to note that this is only a very coarse comparison as pesticide inputs are averaged between pesticide intensive (e.g., fruit production) and low-input agricultural areas (e.g., grasslands) of a given country.

Because most of the examples in this book are from a European perspective, the following chapters are intended to provide some glimpses at pesticide use elsewhere on the globe. More in-depth research also regarding publications in local languages would give more details but was out of the scope of this book.

1.5.1 Africa

A very important exporter of agricultural products with the majority of citizens working in the agricultural area is Ethiopia. Principal crops include coffee, legumes, oilseeds, maize, cereals, potatoes, sugarcane, and vegetables. Many of these crops have long been produced by small-scale farm operations without much pesticide use but are nowadays cultivated with increasing pesticide intensity.

As a consequence, pesticide application in Ethiopia increased dramatically in the last few decades: since 1993, the year from when data are available in the FAO database, pesticide use has increased 17-fold from 242 tons of active ingredients to 4128 tons in 2017 (FAOSTAT 2019e). Of the total amount of pesticides used in Ethiopia in 2010, 75% were herbicides, 15% fungicides, 9% bactericides, and the remaining other pesticides. In the period 2014/15, about a quarter of the total agricultural area received pesticide applications, while 32% of cereal crops were treated (CSA 2016). Of the total pesticides imports 80% is used by commercial farmers and 20% by small-scale farmers, households, and for health and industrial purposes (Teklu 2016). The use of organochlorine pesticides like endosulfan is still allowed in Ethiopia, even though they are long banned in many other countries.

Environment and health issues hardly play a role in pesticides handling in Ethiopia (Mengistie et al. 2016). As a result, smallholder farmers are put at risk because they are refrained from training, support, or information provision on pesticides. A report shows that smallholder vegetable farmers in the Central Rift Valley often apply pesticides in violation of the recommendations by using unsafe storage facilities, ignoring safety instructions; empty pesticide containers are often just deposited in the landscape (Mengistie et al. 2017). Access to pesticides for farmers is facilitated because more shops sell pesticides. Not different to other more industrialized countries is a lacking monitoring of the actual pesticide use in the country. Several unregistered pesticides have been found on a tomato farm (Mengistie et al. 2017). Findings of a scientific project indicated that it is not only the large commercial farms but also the small-scale farms which need careful monitoring, as several pesticide residues exceeded the European drinking water standards in the studied areas (Teklu 2016).

A huge problem in Ethiopia is that farmers using pesticides are not instructed properly. Most users did not understand instructions on pesticide labels due to poor education, a high illiteracy rate, and the fact that the brand name and information on the packaging is written in a language unknown to applicators. An interview with 140 randomly selected farmers showed that 98% of them use pesticides, of which 45% purchase pesticides from open markets (Ocho et al. 2016). The herbicide 2,4-D was used by 57% of the farmers; half of them did not know the type of pesticides they used; only a third read the instructions; and less than 40% understood the signs on pesticide containers. Most of the studied households (98%) witnessed negative health effects of pesticides including nausea, vomiting, headache, and skin irritations. Farmers applying pesticides usually wear their normal clothes and do not wear shoes. Empty pesticide containers are usually dumped in the field, into irrigation canals or rivers, buried or burned at the farm, or used to store foods (Mengistie et al. 2015).

Smallholder farming households still make up the majority of the agricultural business in Ethiopia. Female smallholder farmers and their household members are usually not aware of the toxicity and negative effects of the pesticides they use (Seiwald 2019). Not using any personal protective equipment for postharvest use combined with the most common practice of applying pesticides with bare hands and feet on harvested maize laying on the ground inside the house is one illustrative example of pesticide "misuse".

1.5.2 Asia

1.5.2.1 China

China is the country with the largest pesticide production and use (Zhang et al. 2011). Reportedly by the end of 2016, China registered 665 active ingredients and a whopping 35,604 registered pesticide products. There are 9200 pesticides registered for rice production only, accounting for 29% of all pesticides registered, followed by cotton, wheat, citrus, apple, corn, cabbage, cucumber, health insecticides, and cruciferous vegetables (Zhang 2018). It is estimated that more than 2000 chemical companies produce pesticides in China (Zhang et al. 2011).

The use of pesticides in Chinese agriculture, despite substantial regional variation, accounts for one-third of the world total, proportionately more than either China's proportion of the world population (19%) or of world arable land (11%) might lead one to expect (Zhang et al. 2015). Reportedly, in 2015, China has launched a zero growth action in the use of pesticides and has promoted reduction and efficiency enhancement of pesticide use (Zhang 2018).

Among the top registered pesticides in China in 2016 were 1290 insecticides containing the neonicotinoid imidacloprid, 1064 insecticides containing chlorpyrifos, 956 herbicides containing glyphosate, and 776 herbicides containing atrazine. In 2016, China exported about 1.4 million tons of pesticides while importing only about 39 tons. Among the top exporting pesticides were glyphosate (477,067 tons), paraquat (172,614 tons), atrazine (40,014 tons), chlorpyrifos (22,215 tons), or imidacloprid (20,345 tons). The main exports of active ingredients were directed to the USA, Brazil, India, Australia, and Argentina. Besides, China also produces about 120,000 tons of less harmful biopesticides with about 90 active ingredients and about 3000 products (Zhang 2018).

Pesticide overuse and pollution is of course also a concern in China. A government report in 2016 stated that pesticide use of Chinese farmers reached three times the global average. The major pesticides for human poisonings were highly toxic organophosphorus pesticides, which accounted for 86% of the total cases (Zhang et al. 2011). A study reported that nearly 3000 children were poisoned by pesticides in eastern China's Zhejiang province between 2006–2015 (Fan 2017).

Chinese farmers spray about 14 kg/ha of pesticides annually, which is several-fold more than the amounts applied in the USA (2.2 kg/ha) and France (2.9 kg/ha) (Wang et al. 2017). Nevertheless, Chinese farmers also have limited knowledge on pesticide residues and are mainly worried about crop prices rather than food safety. In addition, consistent with large amounts of pesticide use, cases of increased pesticides residues in food and the environment have increased in China as well. It is estimated that only 10% of powdery and 20% of liquid pesticides remain on crops after application, while the rest of the pesticides end up in the environment (Liu and Guo 2019).

1.5.2.2 Japan

Japan is among the largest pesticide consumers and largest pesticide market in Asia (Zhang et al. 2011). Generally, environmental regulations in Japan are stringent comparable to the standards in Europe or the USA. Hence, pesticides such as DDT, HCH, organomercury fungicides/bactericides, and parathion have been banned in Japan decades ago. In Japan, pesticides are mainly used for rice production accounting for 41% of total pesticide use. Most of insecticides and fungicides/bactericides are applied to vegetable crops and most of herbicides (mainly glyphosate products) are applied to rice. Japan is also a great pesticide exporter to Southeast Asian countries.

After Japan implemented stricter pesticide residue limits, agricultural imports from China suffered considerably (Liu and Guo 2019). Research from Japan shows that neonicotinoid rice seeds coating are widely used, but more than 90% of these pesticides end up in soil or water (Sur and Stork 2003). Since the application of neonicotinoids to agricultural fields in Japan began in the 1990s, zooplankton biomass has plummeted in a lake surrounded by these fields (Yamamuro et al. 2019). This decline has led to shifts in food web structure and a collapse of two commercially harvested freshwater fish species: eel and smelt. Several alternative explanations such as invasive species, hypoxia, or changes in fish stocking were evaluated but could not explain the collapse in aquatic arthropods and fisheries.

1.5.2.3   South and South East Asia

Overall, Asia is the main producer of rice, the main staple food for half of the world's population (Van Nguyen and Ferrero 2006). Much of that production is from intensively cropped coastal regions in tropical South and South East Asia including the Philippines, Sundaland (including Java and Sumatra), Indo-Burma, South-Central China, and the Western Ghats. Many of these regions are also recognized biodiversity hot spots, with high abundances of endemic species (Myers et al. 2000). While traditional rice producing systems, such as the maavee system, have proved sustainable for decades without input of synthetical fertilizers or pesticides, modern rice production systems rely on more pesticide input (Settele et al. 2018).

In the last decade, outbreaks of key herbivore pests have increased significantly throughout Asia, mainly due to a lack of natural enemies. These natural enemies, such as damselflies and dragonflies, are important predators of rice pests such as plant- and leafhoppers which are significantly affected by increased pesticide applications (Heong et al. 2015a). The standard logic considers insects as the initial problem, and spraying insecticides as an adequate response in yield protection. However, studies have shown that insecticide treatments not only impact pest populations, but also dramatically reduce the number of biocontrol organisms (Gurr et al. 2016). Especially when insecticides are applied at the beginning of the rice cycle natural enemy numbers are reduced and damage by planthoppers during the later growth stages is likely (Spangenberg et al. 2015). Moreover, insecticide use is also harmful for pollinators, potentially leading to lower yields of insect pollinated crops within the rice landscape that provide additional income or nutritional diversity for farmers. Overcoming such misguided perceptions of agro-ecological interactions would be crucial when aiming to reduce pesticide use in many Asian countries (Settele et al. 2018).

1.5.2.4   Nepal

In Nepal, pesticide use increased almost 10-fold from 60 tons in 1990 to 574 tons in 2017 (FAOSTAT 2019g). However, Nepal for a long time lacked rigorous implementation of pesticide legislation and regulations to control pesticide sales and agricultural intensification (Atreya 2005). A study showed that women farmers make up the majority of the total farming labor force in Nepal's agriculture (Atreya 2007). Interviews of 325 male and 109 female farmers show that less than 8% had a training on the correct pesticide use.

Almost all interviewees were aware of negative impacts of pesticide use on human health and the environment; however, females had less awareness of pesticide risks. So, the study authors recommend gender-sensitive educational and awareness activities regarding the safe use of pesticides.

1.5.2.5   Laos, Indonesia, Malaysia, Sri Lanka, Vietnam

Farmer surveys in 10 Asian countries revealed that many farmers are still using insecticide compounds classified as extremely or highly hazardous to human health: methyl parathion, monocrotophos, and methamidophos (Heong and Escalada 1997). Herbicides were commonly used in rice farming in Indonesia, Malaysia, Sri Lanka, China, Vietnam, and the Philippines. Only farmers in China and Vietnam used fungicides regularly. The mean number of insecticide sprays in rice varied from 0.3 in Laos to 3.9 in Vietnam.

Not surprisingly, the situation in countries like Vietnam is similar or worse to that in more developed countries. For many Southeast Asian countries there is little knowledge on how frequently pesticides are applied and which pesticides are used. In order to get a better idea of the actual pesticides usage, researchers in Northern Vietnam collected pesticide packages deposited in the landscape close to rice fields, included observations of farmers spraying pesticides in the surroundings, and additionally interviewed local farmers (Sattler et al. 2018). They found 811 pesticide containers of 74 different active ingredients. Most often found were insecticide packages (39%), followed by fungicides (31%), herbicides (16%), and other active ingredients (14%). Active ingredients already banned in the European Union were applied on all sites. This approach appeared to be a very efficient and fast method to obtain at least some baseline information about pesticide application in a region. Of course, the prerequisite for this approach is a rather lax attitude toward environmental protection and trashing the landscape with empty pesticide containers.

In Laos, pesticide use is low, although there is an indication of recent increase in pesticides hazardous to human health, namely the commonly banned insecticides parathion and monocrotophos (Heong et al. 2002). Farm surveys of more than 5000 households in the Mekong, Vietnam, and paired farmer experiments showed that farm yields were not correlated with the number of insecticide sprays used in most cases (Heong et al. 2015b). Researchers suggest that rice farmers continue to apply insecticides despite the poor productivity gain because they overestimated losses caused by insects, and because of the aggressive marketing of pesticides.

Agriculture in Sri Lanka also increased in intensity. The country experienced a strong increase in chronic kidney diseases since it was first identified in the mid-1990s (Gunatilake et al. 2019). This disease primarily affects people in agricultural regions and similar patterns have also been reported from Mexico, Nicaragua, El Salvador, and the state of Andhra Pradesh in India. A global search for the cause of chronic kidney disease has not identified a single factor, but rather many factors that may contribute to the etiology of the disease. Besides the nutrition of people, exposure to the herbicides glyphosate and paraquat, were identified as very important factors for the establishment of this disease.

1.5.2.6   India

Pesticide use in India appears to decrease between 1990–2017 with an overall 30% decrease from 75,000 tons in 1990 to 52,750 tons in 2017 (FAOSTAT 2019f).

Reportedly, cancer rates in India are rather high and analyses revealed that pesticide use, alcohol consumption, and smoking were responsible for significantly higher prevalence of cancer cases in certain regions (Thakur et al. 2008). In a later chapter the difficulty in clearly linking cancer rates to pesticide use will be addressed in more details.

The formulation of organic farming policies has just begun in India after its *Green Revolution* during which the country adopted modern farming methods of high-yield seeds, synthetical fertilizers, and pesticides. Vandana Shiva is undoubtly one of the worlds most famous environmental activist, ecofeminist, and food sovereignty advocate from India with several publications on these topics (Shiva 2013). In 1993, she even received the Alternative Nobel Prize for her activities in these matters. In her books and talks, she brings devastating examples of pesticide poisoning through the introduction of genetically modified crops in India while farmers who pursue organic agriculture earn up to 10 times more than conventional farmers (Shiva et al. 2013). In India, near the city of Bhopal, also the greatest disaster in agrochemical industry happened in 1984 with thousands of victims and where children even today are being born with disabilities. The sensitivity regarding pesticides is high in India and some years ago the Indian state of Sikkim decided to phase out pesticides on every farm in the state (Gowen 2018). Officials say that the switch to all-organic has had health benefits for the people, who are getting more nutritious food, while restoring soil health and biodiversity. The state's move to all-organic also stimulated its tourist industry, with a growing market for ecotours and on-farm vacations.

1.5.3   Australia

With one of the strongest agrarian economies, it will come as little surprise to learn that Australia is a also prolific producer and consumer of pesticides. Pesticide use in Australia increased 3.5-fold from 17,866 tons in 1990 to 63,416 tons in 2017 (FAOSTAT 2019b). Pesticide classes such as insecticides, herbicides, and fungicides increased by at least 250% during this time. Australia has a long history of using pesticides in agriculture and forestry, and as a result, pesticide contamination is widespread across both rural and urban areas. Australia's pesticide import is much larger than the export (Zhang 2018). Herbicides accounted for 47% of the total import in 2006. Of the pesticides imported, the products from China are increasing, including the herbicides glyphosate, paraquat, and glufosinate-ammonium (Zhang 2018). Also, widely used in Australian agriculture are organophosphates, despite being banned in both the EU and parts of the USA. About 80 pesticides banned internationally are still used in Australia and even farmers warned that the use of these pesticides will handicap overseas marketing of agricultural products (Burton 2019). There is also an issue with overshooting maximum residue levels of pesticides in the Australian environment.

Like in most countries, there is no centralized database in Australia that allows one to assess impacts of pesticides across the landscape. However, an Australian NGO provides map, showing accidents, fauna impacts, human health impacts, spray drift, and water pollution (FoE 2019). Since 2015, citizen science has helped cataloguing decades of pesticide incidents across Australia in the hope of shedding new light on what has happened and is happening on a national basis. Also, several Freedom of Information requests have confirmed that low-level pesticide contamination of water supplies is quite frequent.

According to these data the herbicide triclopyr is the most commonly detected pesticide. It is widely used to kill blackberries (*Rubus* species). Other commonly detected pesticides are the herbicides hexazinone, sprayed in pine plantations, and MCPA, and atrazine used to kill broadleaved weeds. NGOs found almost 200 pesticides in over 3500 locations in Australian waterways over a couple of years and recommend bans and label changes for pesticides commonly detected in waterways, particularly the herbicides atrazine and simazine, which represent 20% of pesticides entering waterways across Australia.

1.5.4 The Americas

1.5.4.1 United States of America

Pesticide use in the USA remained quite stable between 1990 and 2012 at an averaged 406,645 tons annually (FAOSTAT 2019h). In this time span, amounts of herbicides increased by 23% from 206,384 tons in 1990 to 255,826 tons in 2012 (the most recent FAO dataset); amounts of fungicides and bactericides were 22,680 tons in 1990 and increased to 24,040 tons in 2012, which is an increase by 6%. Insecticides decreased at the same time by 24% from 86,182 tons in 1990 to 65,771 tons in 2012. However, something is suspicious in this FAOSTATS database for the USA as exactly the same total amounts of pesticides are reported between 2012 and 2017.

In the USA, pesticide regulation is largely overseen by the US Environmental Protection Agency (US EPA). Unlike the safety threshold afforded by the EU, the pesticide industry in the USA only has to demonstrate that its products "will not generally cause unreasonable adverse effects on the environment," which is partially defined as "any unreasonable risk to man or the environment, taking into account the economic, social, and environmental costs and benefits of the use of any pesticide…" (Donley 2019). Hence, harm to the environment and to humans from occupational exposures remains mainly a cost–benefit analysis.

In the USA, some 40,000 people have brought legal action against Bayer-Monsanto. They claim that the use of glyphosate has caused them to develop various types of cancer.

There are several cases where these differences between the European and US-American pesticide regulations become obvious. A recent case is the insecticide chlorpyrifos. The Trump administration's endorsement of the pesticide comes years after the EPA under president Obama moved to restrict use of the chemical (Levin 2019b). While European authorities recommend to ban chlorpyrifos for mounting toxicity evidence of brain damage in children, the US EPA rejects a proposed ban thereby denying the conclusions of the agency's own experts. In the meanwhile, California announced its own state-level ban on chlorpyrifos after researchers found that pregnant women who lived near farms that sprayed it had increased risks of having a child with autism, memory problems, and lower intelligence.

Regarding the use of neonicotinoids (acetamiprid, clothianidin, dinotefuran, imidacloprid, and thiamethoxam) that will be banned in Europe, the US EPA appears to be less ambitious. It recommended additional measures to prevent the neonicotinoids from escaping the field and called for the use of more protective equipment to shield workers but did not further restrict their use (Heller 2020). Although ecologists and the civil society urged for a ban, EPA has rejected these calls because of the important role in agriculture and reduced risks if neonicotinoids are used according to label instructions.

1.5.4.2 Brazil

Brazil developed within a few decades to one of the most important producers of agricultural products in the world. Hence, also the pesticide use in Brazil increased 7.6-fold from totally 49,695 tons in 1990 to 377,176 tons in 2017 (FAOSTAT 2019c). Insecticide usage increased 3-fold in this time, herbicides 10-fold due to widespread cropping of GMOs, and fungicides 7-fold.

Brazil is the most important coffee producer in the world with about 2.3 million hectares of coffee plantations (dosReis et al. 2015). Although integrated pest management is getting more and more popular, still many conventional coffee farms have a heavy pesticide use. Pesticides in coffee plantations are applied directly to the soil or onto the leaves. It is estimated that only 30–40% of the applied pesticides actually reach the target pest organisms. In general, pesticides are less persistent in tropical than temperate climates, with temperature and soil humidity being their major influencing factors. Neurotoxic insecticides (organophosphates, pyrethroids, carbamates, cyclodiene chlorinated, organotins) are among the most used (95%) in coffee crops. It is estimated that two-thirds of Brazil's population are exposed, at different levels, to the harmful effects of pesticides due to consumption of contaminated food or farm work (dosReis et al. 2015).

Brazil has approved hundreds of new pesticides since the presidency of Jair Bolsonaro starting in 2019, and more than 1000 since 2016 during president Michel Temer (Phillips 2019). Many of those pesticides are banned in Europe and contain active ingredients classified as highly hazardous. Brazilian toxicologists raise concern and claim that the approval processes have been accelerated and complain about links between members of the government and big agribusiness.

Pesticide use in Brazil is of great concern for countries that import Brazilian products, especially when Free Trade Agreements such as the one between the European Union and the South American Common Market (Mercosur) are implemented (Weiss 2019). The geographer Larissa Mies Bombardi compiled an atlas of agricultural poisons in Brazil and suggested that EU consumers will get back dangerous pesticides exported by EU agrochemical companies through a free trade agreement between Europe and South American Mercosur states (Mies Bombardi 2019).

According to the above-mentioned atlas, Brazil uses about one million tons of pesticides annually while the official FAO statistics only mentions about 377,000 tons (Mies Bombardi 2019). Over 500 pesticides are authorized in Brazil, 150 of which are banned in the EU. Glyphosate is by far the best-selling pesticide in Brazil. Increase in pesticide use is also the result of increased areas under cultivation advancing further and further from the central savannah into the Amazon forest. The area under soybeans, for example, has almost doubled from 18 million hectares in 2002 to 33 million hectares in 2015. It was documented that in the south of Brazil, where the large agricultural areas are, between 12–16 kg/ha of pesticides are sprayed. The majority of pesticides (about 70%) are used to grow genetically modified soybeans, maize, and sugar cane.

When it comes to pesticide residual limits, the authorities in Brazil seem to be generous. In the case of soybeans, the glyphosate maximum residue level in Brazil is 10 mg/kg, 200 times higher than that permitted in the EU (0.05 mg/kg). In drinking water, Brazil allows 5000 times more glyphosate residue than Europe. Moreover, in contrast to Europe, there is no precautionary principle in Brazil. Also, once a pesticide has been registered in Brazil, the license never expires and is not subject to periodic reevaluations, as is the case in the EU.

1.5.4.3   Argentina

Argentina increased its pesticide use 7.5-fold from 1990 until 2017, from 26,126 tons in 1990 to 196,009 tons in 2017 (FAOSTAT 2019a). More than 50 pesticide active ingredients banned in the European Union for safety reasons are still in use in Argentina. Thus, pesticide-related health and pollution aspects are a great concern for the Argentinean public, especially when living in agricultural areas (Arancibia et al. 2019).

A dietary risk assessment for 308 pesticide residues for Argentina revealed that 27 active ingredients exceeded the acceptable daily intake for 6–23-month-old children, 22 active ingredients for 2–5-year-old children, 10 substances for pregnant women (Maggioni et al. 2017). Milk, apples, potatoes, and tomatoes were the foods that contributed most to the intake of these pesticides.

However, not only areas with intensive agriculture are affected by pesticides. The Central Andes are considered an area of high environmental relevance in South America and its glaciers are the main freshwater sources for many countries. A study found residues of DDT, PCBs (polychlorinated biphenyls), hexachlorocyclohexanes (HCHs), chlordane compounds (CHLs), and hexachlorobenzene (HCB) in fish and sediments of the Central Andes (Ríos et al. 2019). Some of these substances are banned since years but still persist in the environment and are taken up by humans eating this contaminated fish.

The Argentinean Pampas is an area with intensive cropping of herbicide-tolerant GMOs and therefore an intensive use of glyphosate. Consequently, glyphosate and its main metabolite AMPA can be found in 83–100% in soil samples of the region (Primost et al. 2017). Maximum concentrations found were among the highest reported in the world, in soil samples with 8.1 mg glyphosate per kg soil and 38.9 mg AMPA per kg soil. Because of these findings, researchers consider glyphosate and AMPA as quasi persistent pollutants.

1.5.4.4 Chile

Pesticide use for Chile is hard to assess because the official FAOSTATS database gives exactly the same amounts of pesticide use for the years 2004–2017, namely 9830.7 tons (FAOSTAT 2019d). Exactly the same amounts for different years are just not possible considering different weather conditions and given the increase in agricultural intensification in Chile.

Chile hazelnut farmers were recently in the news in Germany because of pesticide-contaminated hazelnuts used for a famous nougat cream (Boddenberg 2018). Hazelnuts are one of the most important ingredients in this product and most of the hazelnuts for this product are cultivated in Chile. Hazelnuts in Chile are grown on several thousand of hectares, under heavy use of the herbicides glyphosate and paraquat. Paraquat is banned in Europe because it has been associated with kidney failure, shortness of breath, lung pain, visual and liver damage, severe skin injuries, damage to the nervous system, and Parkinson's disease (Stykel et al. 2018). Nevertheless, paraquat is still allowed in Chile and the USA, and consumers of this hazelnut cream in Europe therefore support its application.

Another critical aspect is the Chilean salmon production which uses antimicrobials and pesticides to prevent bacterial infections and sea lice (Quinones et al. 2019). Several different pesticides (emamectin benzoate, diflubenzuron, teflubenzuron, and cypermethrin) have been found in sediments near salmon cages in southern Chile. These results were similar to data reported for the Northern Hemisphere (Scotland, Norway, Canada) and raise serious concerns about nontarget effects in aquatic ecosystems.

1.5.4.5 Central America

Agriculture in Central American countries goes with the general trend of intensification. For 20 years, experts have observed a conspicuous accumulation of chronic kidney failure in the sugar cane growing regions of Central America; more than 20,000 people have already died (INKOTA 2019). Recent scientific findings found a connection with the use of the weed killers paraquat and glyphosate (Gunatilake et al. 2019). In Nicaragua and El Salvador, chronic kidney failure is among the most frequent causes of death among men in the last 20 years, depending on the region. Normally, this disease mainly affects people over the age of 60 with high blood pressure and high blood sugar levels. However, most victims in Central America are young men with low blood pressure and normal sugar levels who work on sugar cane plantations. Among the main factors the study authors mention the exposure to glyphosate and paraquat. As mentioned above paraquat has been banned in the European Union and many other countries for many years due to the risks to human and animal health. In Central America, however, it is still used on a massive scale. It is just difficult to understand how it is allowed to produce and sell pesticides that are clearly considered hazardous.

1.5.5 Europe

Europe is probably the continent where the public is most sensitized toward pesticide, perhaps also because several of the biggest agrochemical companies are still based in Europe. As many examples given throughout this book are from Europe anyway, I will keep this chapter short.

Regarding the amount of pesticides used between 1990 and 2017, Europe is the only world region where pesticide application remained at a constant level or even slightly decreased (Fig. 1.1). The main pesticide-related topic in Europe in recent years centered around glyphosate, especially also since the German agrochemical company Bayer bought the US company Monsanto. Glyphosate is not only used when growing food but also when growing corn and sugar beets for biogas, which generates electricity and produces biomethane (Graupner 2019). The use of glyphosate is based on an agricultural cultivation strategy to kill the green fields in spring in order to prepare the fields for the following crop and get rid of everything that could harm the future harvest.

In the summer of 2019, the Austrian parliament voted to ban glyphosate and Austria was set to become the first EU country to completely ban this weed killer (DW 2019a). Meanwhile, European authorities rejected this law for formal reasons; a new request is under consideration. Of course, Bayer immediately responded stating that the decision by the Austrian National Council contradicts extensive scientific results on glyphosate although investigations showed that European regulators were copy-pasting manufacturer's assessments on the safety of glyphosate.

Meanwhile also Luxembourg proposed to ban glyphosate by the end of 2020 (MAVRD 2020), and also Germany declared that it will ban glyphosate from the end of 2023 after a phased effort to reduce its application (DW 2019b). This reduction plan includes a systematic strategy that would initially prohibit its use in private gardens and on the edge of farmers' fields. Also, hundreds of mayors across Europe banned glyphosate from their municipalities—in defiance of their national government. Whether the phasing-out of glyphosate in EU countries will be seriously pursued by the responsible parties remains to be seen; experience makes one doubt.

1.5.6 Antarctica

One would think that a continent that has no permanent human settlements (apart from research bases), no agriculture, and therefore no pesticide spraying is free from these manmade chemicals. Unfortunately, this is not the case.

Already back in the 1980s, PCBs and chlorinated hydrocarbon pesticides such as DDTs and HCHs (BHCs) were measured in air, water, ice, and snow samples collected around the Japanese research stations in Antarctica and adjacent oceans (Tanabe et al. 1983). While the atmospheric concentrations of chlorinated hydrocarbons decreased in the transport process from northern lands to Antarctica, the compositions of PCBs, DDT compounds, and HCH isomers were relatively uniform throughout this process.

We know since Rachel Carson's *Silent Spring* that persistent organic pollutants can travel long distances and biomagnify through the food web accumulating in higher trophic level predators. Researcher found that the insecticide DDT, banned decades ago in much of the world, still shows up in penguins in Antarctica, probably due to the chemical's accumulation in melting glaciers (Geisz et al. 2008). Moreover, DDT is still approved by the World Health

Organization for indoor use to fight malaria. Researchers found that ratios of DDT in Adélie penguins have declined significantly since 1964 indicating current exposure to old rather than new sources. However, DDT has not declined in Adélie penguins from the Western Antarctic Peninsula for more than 30 years suspecting that current sources of DDT for the Antarctic marine food web. Estimates suggest an annual 1–4 kg of DDT that is currently being released into coastal waters along the Western Antarctic Ice Sheet due to glacier melting.

An accumulation of DDT has also been found in eggs of resident Adélie and Emperor penguins and migrating snow petrels (*Pagodroma nivea*) and South polar skuas and brown skuas (both predatory seabirds; *Catharacta* species) nesting in East and West Antarctica, suggesting long-range transport to the polar regions from certain malaria-endemic countries where DDT is still in use (Corsolini et al. 2011).

1.6   Danger from Toxic Accidents and Landfills

Pesticides are not only applied openly in our landscape, but are of course produced in very large industrial plants. This production generates a lot of toxic waste, which is either burned in incinerators or deposited in landfills. The past has shown that tragic accidents can occur during production. Unfortunately, in this context, the human brain turns out to be not very reliable and catastrophic events are quickly forgotten, especially if they happened far away. Hence, it follows a short reminder of a few serious accidents; smaller accidents usually do not make it into the public.

The most serious chemical accident in history occurred 1984 in the Indian city of Bhopal. Several tons of methyl isocyanate were released into the atmosphere at a pesticide plant belonging to the US chemical company Union Carbide. As many as 25,000 people died as a result of the direct consequences, about 500,000 people were injured, and some are still suffering from the consequences today (Mandavilli 2018). The inaccurate number of people affected is due to the fact that there was a slum area around the pesticide factory where an unknown number of people lived. The remediation of the area, which had been poisoned with mercury and carcinogenic chemicals, was a major step forward. More than 25 years after the accident, an Indian court found eight executives of the operating company guilty of negligent homicide and sentenced them to 2 year's probation and a fine of 1800 €. It is unbelievable that the biggest industrial disaster in the world was therefore treated like an ordinary traffic accident. As a compensation payment, families with members injured or killed by this accident received a one-off payment amounting to a few hundred € (Amnesty International 2009).

Chemical accidents happen not only in developing countries, but also in the heart of Europe. Among the biggest and most severe one was the so-called Seveso disaster that happened in 1976 in a small chemical manufacturing plant about 20 km north of the city of Milano in the Lombardy region of Italy. It resulted in the highest known exposure to dioxin—tetrachlorodibenzo-p-dioxin (Eskenazi et al. 2004). Pure dioxin is a colorless solid with no distinguishable odor at room temperature. It is usually formed as a side product in organic synthesis and burning of organic materials. Dioxin became known as a contaminant in Agent Orange, a herbicide used by US army in the Vietnam War. The substance is very toxic, carcinogenic to rodents, and has led to the death of several thousand poultry and rabbits near the Seveso disaster. To prevent dioxin from entering the food chain over 80,000 animals had been slaughtered by 1978. Several monitoring studies have been conducted since the accident and still confirm an excess risk of lymphatic and breast cancer in the most exposed zones (Pesatori et al. 2009).

In Europe, the Seveso accident prompted the adoption of legislation on the prevention and control of such accidents, the so-called Seveso-Directives I–III (EC 2019e). The directive applies to more than 12,000 industrial facilities in the European Union where dangerous substances are used or stored in large quantities. The directive is widely considered as a benchmark for industrial accident policy and has been a role model for legislation in many countries worldwide.

In 1986, when Chernobyl was the scene of the biggest nuclear accident in a nuclear power plant, one of the biggest chemical accidents occurred in Switzerland. First, an accident caused several 100 liters of the herbicide atrazine to enter the Rhine via wastewater. Then 1351 tons of pesticides burnt in a warehouse of the chemical company Sandoz near Basel (Schmider 2016). The result was the death of fish in the Rhine and a cloud of poison that spread across the entire region at that time. Near the source of the fire, the chemical phosgene was stored, a nerve gas used as a chemical warfare agent in World War I. Thirty years later, this chemical catastrophe even worries the authorities and the population (Sda 2016). Although the contaminated soil was excavated and cleaned to a depth of 11 m, soil measurements at the site of the fire still show traces of the pesticide oxadixyl, a widely used fungicide that has since been banned.

Toward the end of 2014, an environmental scandal also dominated the Austrian media. Hexachlorobenzene (HCB) was found in milk and meat products from a picturesque valley in the Austrian province of Carinthia.

HCB is a fungicide that was used in cereals, as a disinfectant in grain storage, and also as a wood preservative. Since 1981, HCB has been banned in Germany, since 1992 in Austria, and since 2004 almost worldwide within the framework of the Stockholm Convention. Hexachlorobenzene is among the so-called dirty dozen of persistent, highly toxic organic pollutants (UNEP 2017). These chlorine compounds are strongly suspected of being carcinogenic and mutagenic. Their danger results primarily from possible accumulation in the human body, extreme longevity, high toxicity, and the possibility of long-distance transport. Individual substances are also known as endocrine disruptors, i.e., chemicals that influence the human hormone system.

How can a pesticide that has been banned in Austria since 1992 be found in milk products 22 years after its ban? Well, a cement factory in which HCB-contaminated waste from an agrochemical company was burned was regarded as the presumed polluter. Cement plants are often used for waste disposal, because it is assumed that the high temperatures required during cement production will defang the toxic products. Since HCB was used as a pesticide throughout Austria until the 1980s, soil pollution can probably still be found throughout the country (ORF 2014). In addition to the population at the site of the pollution, the victims are, as so often in these disputes, the farmers. In such cases, the pesticide manufacturers usually fly the coop, refer to inappropriate dosages and applications of their products and deny any responsibility.

Pesticide loads have also been found in the Central Asian Republic of Tajikistan, a former state of the Soviet Union (Möseneder 2008). Approximately 8000 tons of pesticides, half of them DDT, were buried in an area of 12 ha. While in the Soviet times the area was fenced in and guarded, now cows and donkeys graze on the pastures. Local farmers even dig up the poisonous pesticide cocktail unhindered to keep their fields free of vermin. The rehabilitation of the landfill would cost tens of million €, money that simply does not exist. The consequences of drinking water pollution are not foreseeable.

Also in many developing countries, too, thousands of tons of highly toxic old pesticides such as DDT, endrin, and lindane are still around (Asendorpf 2003). Sometimes local authorities are aware of the problem and do something. In Botswana, for example, the Ministry of Agriculture supports collections of the pesticides lying around on farmland. The collected pesticides are then shipped in containers to England for disposal. The pesticides came to Botswana in the mid-1980s after a plague of locusts as a generous donation from the UN Food and Agriculture Organization FAO. The FAO estimates the stock of old pesticides that have passed their sell-by date or have meanwhile been banned at around 500,000 tons worldwide. Not only do these pesticides pose a serious health risk to humans, they also contaminate water and soil and can render entire areas unusable for agricultural use. About a third of these old pesticides belong to the *Dirty Dozen* of persistent chlorine compounds mentioned earlier and now banned worldwide. The financing of such disposal activities is still a matter of dispute, at least the agrochemical companies are holding back discreetly. These old pesticides are also a consequence of the *Green Revolution*, in the course of which pesticides were distributed to developing countries over decades through loans by developmental aid or the World Bank.

According to UN estimates, more than 20% of the world's pesticide reserves are highly toxic, persistent, organic chemicals that degrade extremely slowly in the environment (UNHRC 2017). These unused pesticides can spoil and accumulate in the environment. Some developing countries reportedly also have purchased too large quantities of pesticides, while others have purchased pesticides unsuitable for the region from other developing countries or received as donation from industrial countries. Much of this was due to pressure or wrong advice from the agrochemical industry or corrupt governments in these countries. There is also a problem with existing pesticide stocks when pesticides are banned. The FAO recommends transitional periods during which the pesticides to be banned can be used up before they are finally banned. This is a highly problematic regulation, since pesticides are usually banned because of acute health risks and the products do not simply become less toxic during this transitional period. Once again, economic interests prevail over health or environmental issues.

The possible dangers that arise when pesticide factories or depots are flooded are rarely addressed. A flooding for instance after a hurricane may be gone, but the danger it left behind can still be hazardous. Experts warn that flood waters can cause pesticide containers to leak or spill, contaminating surrounding water bodies (Hallman 2018). Homeowners, pesticide applicators, and businesses should be aware of this and take steps to protect themselves and their neighborhoods from inundated pesticide storage sites.

However, the problem of old pesticides deposited in landfills is not limited to developing countries. Swiss agrochemical companies, which are among the world's top-selling agrochemical businesses, have left behind at least 18 hazardous waste landfills in the area of the city of Basel, where several chemical companies are located (MultiWatch 2016). From the end of World War II until the mid-1960s, the large corporations disposed of sometimes highly toxic waste from their factories in abandoned gravel pits in the groundwater-rich Rhine plain. After that, some of the chemical waste was landfilled in neighboring Germany and France. Of these 18 landfills, only two have been remediated to date, while the remaining sites still pollute the groundwater and probably also the drinking water. In any case, an old pesticide landfill is located in the immediate vicinity of a drinking water area near Basel, from which 230,000 people obtain their drinking water. Analyses of drinking water samples found mutagenic substances. The responsible authorities and industry have known about these contaminants for decades, but have done little about them. The latest plans are now to set up an asylum center for 500–900 refugees in the contaminated buildings of a former toxic waste landfill (Cassidy 2016). But let us close this chapter before it becomes even more disgusting.

1.7 Pesticide Registration: Very Complex But Still Not Adequate

Let us now focus on another well-cultivated myth, namely that pesticides are rigorously tested before they are widely used. Admittedly, pesticides are perhaps more rigorously tested than many other industrial chemicals because potential side effects of the majority of synthetic chemical substances are not investigated at all. We will see, however, that many of these tests are based on scientifically shaky ground and very intransparent.

In 2015, the Chemical Abstract Service, the worldwide registry for synthetic chemicals, celebrated the 50th anniversary with the entry of its 100 millionth chemical substance (CAS 2015). The substance is patented for the treatment of leukemia. This is remarkable, because for most chemicals there is no practical application known. With no signs that chemical innovation is slowing, the current pace of substances added to this registry over the next 50 years would suggest registration of more than 650 million new chemical substances. If the past growth trajectory is any indication, that number may well be significantly higher.

It is estimated that humans living in industrialized countries come into contact with tens of thousands of different chemical products in the course of their lives (Marquardt and Schäfer 2004). Hard to believe, but as far as the effects on health and the environment are concerned, almost nothing is known about the majority of those 100 million chemicals. In the USA, only a few hundred of the more than 80,000 chemicals used have been tested for their side effects (PCP 2010). Only about 1500 substances are included in a list of hazardous substances.

Another issue are the around 2000 untested chemicals that can be found in conventional packaged foods in our supermarkets (Hoeffner 2019). These are additives such as sodium nitrate, antioxidants such as butylated hydroxyanisole (BHA), or a variety of other chemicals commonly found in food packaging such as polypropylene or bisphenol A, but this is out of the scope of this book.

We have seen before that there are differences in the approval procedure of pesticides in different countries. It is fair to say that the European Union currently has the most comprehensive and protective pesticide regulations of any major agricultural producer in the world (Donley 2019). The European Commission oversees pesticide approval, restriction, and cancelation in the EU. The applicant of a registration, usually the pesticide manufacturer, must ensure that the pesticide to be placed on the market do not have any harmful effect on human or animal health or any unacceptable effects on the environment. The burden of proof lies on the pesticide industry that should demonstrate that the product does not result in harm to humans or the surrounding environment (EP 2005, 2009). Theoretically, the EU prohibits the approval and continued use of pesticides that are recognized as mutagens, carcinogens, reproductive toxicants, or endocrine disruptors unless exposure to humans is considered negligible. The problem here is that the manufacturer commissions the relevant studies and that the studies are not public. It sounds sarcastically, but this would be similar to a situation where you go to your mechanic for car inspection and tell him/her that you already inspected your car at home and found everything alright.

In Europe the application for approval of a pesticide-active substance is submitted to the European Commission (EC). The EC assigns a member state, called a Rapporteur Member State, to verify if the application is admissible and prepare a draft assessment report. Then the European Food Safety Authority (EFSA) issues its conclusion based on this report. A standing Committee for Food Chain and Animal Health votes on approval or non-approval. This is usually adopted by the European Commission and then published in the EU official journal. The whole procedure takes between 2.5 and 3.5 years from the date of admissibility of the application to the publication of a regulation approving a new active substance. This time varies greatly depending on how complex and complete the dossier is. For the authorization the EU is divided into three zones representing the main climatic regions: North, Central, and South. EU countries assess applications on behalf of other countries in their zone and sometimes on behalf of all zones.

In the USA, pesticide regulation is largely overseen by the US Environmental Protection Agency (EPA). Unlike the safety threshold afforded by the EU, in the USA before EPA may register a pesticide, the applicant "must show, among other things, that using the pesticide according to specifications will not generally cause unreasonable adverse effects on the environment." "Unreasonable adverse effects on the environment means any unreasonable risk to man or the environment, taking into account the economic, social, and environmental costs and benefits of the use of any pesticide" (US EPA 2019). Thus, in the USA, harm to plants, animals, the broader environment, and humans from occupational exposures remains mainly a cost–benefit analysis (Donley 2019).

China, despite being the world's greatest user and producer of pesticides, has long suffered from scattered data, complex laws, and lack of transparency regarding implementation and compliance (Snyder and Ni 2017). However, recently China has passed regulations updating certain aspects of pesticide use, including establishing licensing requirements for sellers of pesticides, record-keeping requirements for users, and committees in charge of evaluating pesticide safety. China has also progressed in recent years with banning or phasing out highly hazardous pesticides.

Brazil's pesticide regulations are overseen by three governmental agencies. However, multiple factors have severely limited the effectiveness of human and environmental health safeguards, including the protection of the agrochemical industry, and massive budget and personnel shortfalls of the registration agencies (Pelaez et al. 2013).

Environmental risk assessments (ERA) of pesticides follow in Europe certain standards and must be carried out by accredited laboratories. If an active substance meets the criteria, it is included in a "positive list" following a joint decision by all EU member states and is therefore suitable for use as a plant protection product. The approval of a pesticide is only granted for special applications, i.e., against certain pathogens on corresponding crops. This all sounds very meticulous and trustworthy. However, especially in the case of glyphosate many issues of scientific misconduct surrounding in the registration process became public (Burtscher-Schaden 2017). The problems range from private laboratories that have doctored risk studies on behalf of the manufacturers, the nondisclosure of study results, and the manufacturers' influence on authorities and scientists to the agreement between manufacturers and authorities to copy-paste relevant assessments written by the manufacturers.

For each pesticide, a no-observed effect level (NOEL) is assessed in terms of milligram of pesticides per kg of body weight per day. MacBean's classic compendium defines NOEL as "the highest dose in an animal toxicology study at which no biologically significant increase in frequency or severity of an effect is observed" (MacBean 2013). A 100-fold uncertainty factor is applied to extrapolate the NOEL for human beings: it includes a tenfold interspecies uncertainty factor to take account of the fact that humans may be more sensitive than other animals, and a tenfold intraspecies uncertainty factor to take account of the fact that some human beings are more sensitive than others; an additional tenfold uncertainty factor may be added, for example for children (Snyder and Ni 2017). This then allows calculation of an acute reference dose (RfD), which is "an estimate, with uncertainty spanning perhaps an order of magnitude, of a daily oral exposure to the human population (including sensitive subgroups) that is likely to be without an appreciable risk of deleterious effects during a lifetime" (EPA 2011). This is the amount of oral pesticide intake which does not produce any observable effects. It does not take account of other types of contact with the pesticide, or of unobservable effects, or the possible effect of an accumulation of different pesticides.

Pesticide classification is quite complicated because there are various international, regional, and other governmental organizations using different classification systems for the toxicity of pesticides. The World Health Organization (WHO) and the FAO estimate the acute toxicity of pesticides along a scale from Ia—extremely hazardous, Ib—highly hazardous, II—moderately hazardous, III—slightly hazardous, U—product unlikely to present acute hazard in normal use, O—not classified, and publishes regular reports on the status of knowledge (FAO and WHO 2019). In contrast, the US EPA classifies pesticides into four classes, with Class I being the most and Class IV being the least toxic. The EU classifies pesticides according to a system of nine risk symbols, going from "corrosive" to "harmful," combined with a system of almost 50 associated risk phases (EP 2015).

Additionally, pesticides are also classified by several international nongovernmental organizations (NGOs). The most influential NGO in the field of pesticides is the Pesticide Action Network (PAN), established in 1982 and now comprising more than 600 organizations in more than 90 countries (PAN 2019). Their database brings together a diverse array of information on pesticides from many different sources, providing human toxicity (chronic and acute), ecotoxicity, and regulatory information for more than 6400 pesticide active ingredients and their transformation products, as well as adjuvants and solvents used in pesticide products.

A very unethical practice in relation to pesticide authorizations is the fact that pesticides classified as unsafe, cancerogenic, or toxic in one country may continue to be produced and exported in other countries where it is not banned. A study compared this situation between the USA, Brazil, the EU, and China. Results show that 72 pesticides approved for agriculture in the

USA are banned in the EU (making up 146 million kg); 17 pesticides approved in the USA are banned in Brazil (18 million kg) and 11 pesticides approved in the USA are banned in China (18 million kg) (Donley 2019). The majority of pesticides banned in at least two nations have not appreciably decreased in the USA over the last 25 years and almost all have stayed constant or even increased over the last 10 years. Also, Australian farmers still are allowed to use 80 pesticides that are already banned in other countries (Leu 2014).

But let us more closely consider two heavily debated cases, the herbicide glyphosate and the insecticide chlorpyrifos.

1.7.1 The Glyphosate Case

Glyphosate is marketed as safe for animals and humans because it is designed to disrupt a biochemical pathway that only exists in plants and some microorganisms. Simply put, it inhibits an enzyme that promotes plant growth; once exposed, plants cannot produce amino acids properly and die. Since that enzyme (EPSP synthase) is not present in humans, agrochemical companies tend to argue that glyphosate poses few risks to human health.

A dispute has arisen over glyphosate in recent years especially after experts from the International Agency for Cancer Research (IACR) a division of the World Health Organization of the United Nations classified glyphosate as probably carcinogenic to humans (classification 2A). However, the EU authority EFSA and the German Federal Institute for Risk Assessment (BfR), the agency particularly responsible for glyphosate registration in Europe, and several other national agencies did not confirm this IACR classification. When it came to decide over a 10-year extension of glyphosate, the European Parliament was unable to come to an agreement; thus, a provisional approval was granted until the end of 2017. Vociferous international protests organized by several international NGOs formed against a new approval, and in only a few months, over 1.3 million signatures were collected from across Europe to persuade the EU Commission to engage in a process of reflection. The main impetus for the debate was the different interpretation of a handful of lab studies on mice and epidemiological studies on humans that were differently interpreted.

Also, the US EPA considers glyphosate as "not likely to be carcinogenic to humans." EPA asserts that there is no convincing evidence that "glyphosate induces mutations in vivo via the oral route." IARC concludes there is "strong evidence" that exposure to glyphosate is genotoxic through at least two mechanisms known to be associated with human carcinogens (DNA damage, oxidative stress). How can it be that national agencies and IARC reach such different conclusions?

An outstanding expert in this matter, Charles Benbrook, addressed this question and formulated three primary reasons, why EPA/EFSA and IARC reached diametrically opposed conclusions on glyphosate genotoxicity (Benbrook 2019). Firstly, in the core tables compiled by EPA and IARC, the EPA relied mostly on registrant-commissioned, unpublished regulatory studies. Of these studies, 99% of which were negative, while IARC relied mostly on peer-reviewed studies of which 70% were positive. Secondly, EPA's evaluation was largely based on data from studies on technical glyphosate, whereas IARC's review placed heavy weight on the results of formulated glyphosate-based herbicides, the commercial products that are applied in the field. Thirdly, EPA's evaluation focused on typical general population dietary exposures assuming legal, food crop uses, and did not address generally higher occupational exposures and risks. In contrast, IARC's assessment encompassed data from typical dietary, occupational, and elevated exposure scenarios. By the way, the journal editors of this paper were aware of the toxic issue of this review and asked for 10 anonymous reviews for this paper (Hollert and Backhaus 2019), while usually 2–4 reviewers are considered sufficient to judge the quality of a paper.

The debate on a potential cancerogeneity of glyphosate makes clear that it is important to distinguish between the exposure to the active ingredient glyphosate, which occurs via residues in food and feed, and occupational exposure to the formulated glyphosate-based herbicide. Both aspects are relevant. It was also striking to see that the outcome of the very same experimental studies was differently interpreted by the environmental risk assessments. The dilemma is that all these studies and the underlying data that are used during the authorization process are not disclosed and cannot be assessed for independent scientific scrutiny. For me the secrecy around these studies is difficult to understand and clearly leaves a strange aftertaste as if something should be hidden. The inadequate information of the many pesticides we spread into our environment also prompted us to do our own experiments. Meanwhile, there are some indications in the EU that pesticide study results might become better available to independent scientists in the future. Considering the two approaches, the hazard-based (IARC) and risk-based analysis (EFSA), it appears wise to favor the precautionary principle as the safest way to proceed.

Besides the potential cancerogeneity of glyphosate, numerous other severe effects on the environment and humans have been documented. We will specifically elaborate on those in Chap. 2. Questions related to hazards and corresponding risks identified in relation to glyphosate and its formulated herbicide preparations divide scientific circles and official health and environmental authorities and organizations and touch upon fundamental aspects of risk assessment and product regulation (Székács and Darvas 2018).

The question how dangerous glyphosate is has also a legal dimension, especially when people who applied glyphosate got sick. Since the IARC's decision, tens of thousands of people across the USA have sued Monsanto, alleging that Roundup caused them to develop non-Hodgkin's lymphoma, a type of blood cancer (Zimmer 2018). Then, the so-called Monsanto papers suggested a cozy relationship between EPA officials and Monsanto and that EPA had possibly prevented a review of the chemical from taking place. At the same time, IARC was criticized for selectively "editing out" data from its assessment, removing evidence that showed no links between glyphosate and cancer.

When each concerned party accuses the other of distorting the facts, and even independent organizations and regulatory bodies have struggled to stay above the fray, it can be hard to know what to believe. But maybe that is also part of the game for agrochemical lobbyists—after all it is important for them that some doubts remain about alleged safety issues with a given product (Oreskes and Conway 2010). We know this mechanism from the debate about the danger of tobacco smoking or man-made climate change.

While some studies have found that glyphosate itself is probably not that toxic at most environmental concentrations, other research has focused on the commercial herbicides. Meanwhile, it should be evident that glyphosate and the commercial product are not the same. However, the complete list of ingredients of the commercial product is considered a company secret and not published and therefore a hindrance for scientific verification. In the past, individual components in those recipes have been found to be quite toxic— for instance, a chemical known as POE tallowamine which is toxic for aquatic organisms and was banned in the EU in 2016.

Another issue which can yield different results are the many forms of glyphosate that are used. This was found by repeating of a 1978 experiment on the toxicity of glyphosate chemicals on water-fleas *Daphnia magna*. Back then glyphosate was tested nontoxic to *Daphnia* and this has become a scientific fact. However, these historic results could not be subsequently reproduced by other researchers repeating the experiments. It seems that the toxicity of glyphosate had apparently become 300 times stronger (Cuhra 2015). The study author suggests that this could lie in the fact that there exist several chemicals named glyphosate, and in herbicide formulations, various glyphosate salts (diammonium salt, isopropylamine salt, potassium salt) are used. If different forms of glyphosate with different water solubility are tested one gets different toxicities.

One analysis goes beyond the comparison of the assessments made by the EFSA and IARC and shows that not classifying glyphosate as a carcinogen by European authorities appears to be not consistent with the applicable guidance and guideline documents (Clausing et al. 2018). Researchers used the authorities' own criteria as a benchmark to analyze their weight of evidence approach and concluded that glyphosate is "probably carcinogenic." Another critical article revealed that most of the science used in the risk assessment process to support the safety of glyphosate was conducted more than 30 years ago (Vandenberg et al. 2017). For instance, in the US EPA's 1993 registration review of glyphosate, 73% of the almost 300 references were published prior to 1985; importantly, only 11 were peer reviewed! It is unacceptable that safety assessments of the most widely used herbicide on the planet rely largely on fewer than 300 unpublished non-peer-reviewed studies while excluding the vast modern literature on glyphosate effects. More than 3000 studies have been published alone between the years 2000–2019.

A comprehensive review of existing literature focused on the group of people most highly exposed to glyphosate and showed that the link between glyphosate and non-Hodgkin's lymphoma is stronger than previously reported (Zhang et al. 2019). For this study, the research team examined epidemiological studies in humans but also laboratory animals published between 2001 and 2018 and found that exposure to glyphosate increased the risk to get this cancer by 41%. Therefore, these findings are aligned with IARC 2015 classification that glyphosate is a "probable human carcinogen."

Glyphosate is perhaps not the most inherently dangerous of pesticides on the market, but its broad use affects our environment and lives more than other agrochemicals. Although we will later see that glyphosate can be found almost everywhere in our environment, it stands out as the one widely used pesticide that has not been included in years of annual government surveys of pesticide residues in food, water bodies, or soils (Gillam 2017).

The same industry manager who helped hide the carcinogenic potential of PCBs in the 1970s (Beiles 2000) has now been shown to have also influenced the US Environmental Protection Agency (EPA) regarding the carcinogenic potential of glyphosate in the 1980s (SustainablePulse 2017). This 30-year cover-up was confirmed by court documents released by the US District Court in San Francisco. The documents show that industry influenced the EPA to change the classification of glyphosate as a class C carcinogen (suggestive potential of carcinogenic potential) to a class E category (evidence of non-carcinogenicity for humans) in 1991. This change in glyphosate's classification occurred during the same period that Monsanto was developing its first Roundup-ready (glyphosate-resistant) GM crops.

In January 2020, the US EPA published an interim registration review decision that they did not identify any human health risks from exposure to glyphosate (EPA 2020). They identified potential ecological risk to mammals and birds, but these risks are expected to be limited to the application area or areas near the application area. The EPA also identified potential risk to terrestrial and aquatic plants from off-site spray drift, consistent with glyphosate's use as an herbicide. They continue that glyphosate is a versatile herbicide that provides a broad spectrum of weed control across numerous agricultural and non-agricultural sites. Glyphosate is generally inexpensive in agricultural settings. All in all, the EPA concludes that the benefits outweigh the potential ecological risks when glyphosate is used according to label directions.

In Europe we also have the REACH Regulation (Registration, Evaluation, Authorization and Restriction of Chemicals) that is intended to ensure that there are no chemical substances on the European market whose hazard potential is not sufficiently described (ECHA 2019). This will at least improve chemical safety. REACH is a regulation of the European Union, adopted to improve the protection of human health and the environment from the risks that can be posed by chemicals, while enhancing the competitiveness of the EU chemical industry. It also promotes alternative methods for the hazard assessment of substances in order to reduce the number of tests on animals. In principle, REACH applies to all chemical substances; not only those used in industrial processes but also in our day-to-day lives, for example in cleaning products, in paints, as well as in articles such as clothes, furniture, and electrical appliances. REACH places the burden of proof on companies. To comply with the regulation, companies must identify and manage the risks linked to the substances they manufacture and market in the EU. They have to demonstrate how the substance can be safely used, and they must communicate the risk management measures to the users. If the risks cannot be managed, authorities can restrict the use of substances in different ways. In the long run, the most hazardous substances should be substituted with less dangerous ones.

In principle, REACH seems like the way to go for authorities; however, there are also some critical points (DNR 2018). REACH was adopted in 2006; however, in the first 10 years only about 191 chemicals have been identified to have serious effects on human health and the environment. A total of 21,121 substances have been registered under REACH. The European Commission identified the poor quality of the data submitted by the chemical industry in the registration dossiers as the biggest problem in the implementation of REACH to date. This would impede the identification of hazardous substances and thus their regulation. Important safety information on the chemicals required for evaluation is often not available. In total, two-thirds of the registrations submitted under REACH between 2008 and 2018 were not complete. ECHA and the member states often have to spend resources themselves to obtain additional information allowing them to assess the substance. This represents a partial return to the previous system where the full burden of proof lay on the authorities and would also delay the evaluation process. Furthermore, the precautionary principle would not be applied so far. Another problem is that only about a quarter of the registration dossiers are regularly updated, and it would be necessary to more strictly implement the principle of "no data, no market."

The fact is, however, that the vast majority of the chemicals we use worldwide have not been tested for their effects on humans and the environment. Considering that the majority of cancers can be linked to chemicals in the environment, this situation is an intolerable ignorance on the part of the competent authorities (PCP 2010).

The development of a pesticide is very time-consuming and quite expensive for companies. Dozens of thick file folders full of data have to be submitted to the regulatory authorities, and the industry needs to test about 100,000 to 160,000 chemical substances in order to find a suitable active ingredient (Strassheim 2019). Research for this takes up to 4 years, the registration process another 5–6 years leading to costs of about 250 million € for each pesticide.

The desired efficacy of the pesticide is only one of many basic prerequisites that have to be fulfilled for approval. There are separate books that give advice on the complicated registration process, so we will deal here with environmental behavior and health issues only. For further discussions on side effects of pesticides are few definitions of the terms hazard, risk and exposures are necessary. According to the International Code of Conduct on Pesticide Management (FAO WHO 2014), a hazard is the inherent property of a pesticide having the potential to cause undesirable consequences, e.g., properties that can cause adverse effects or damage to health, the environment, or property.

The risk of a pesticide is the probability and severity of an adverse health or environmental effect occurring as a function of a hazard and the likelihood and the extent of exposure to a pesticide. Exposure is the concentration or amount of a pesticide that reaches a target organism. Therefore, in a risk assessment, the hazard (e.g., toxicity) of a pesticide and the level of exposure are evaluated. Data on hazard will determine the acceptable exposure level of humans or nontarget organisms in the environment; the exposure assessment will show whether this acceptable level will be exceeded or not. The same principle is applicable both to the assessment of risks to human health and to the environment.

1.7.2 The Chlorpyrifos Case

We have seen for the case of glyphosate that independent academic studies and industry-sponsored studies may lead to fundamentally different conclusions (Tweedale 2017). This has also been observed for the insecticide chlorpyrifos. Chlorpyrifos has been shown to be brain-harming for children and disrupting the normal functioning of thyroid hormones. However, its approval was instead based on just one single study, which was commissioned by industry and came to a different result.

In the USA, there are heavy discussions between the administration and the EPA whether chlorpyrifos should be banned or not. In the meantime, it seems as if it will be banned at least in California by the end of 2020 (Levin 2019a). This move comes as the manufacturers have reached an agreement with Californian authorities to withdraw their products there. Chlorpyrifos is used to control pests on a variety of crops, including alfalfa, almonds, citrus, cotton, grapes, and walnuts.

The European EFSA has identified concerns about possible genotoxic effects as well as neurological effects during child development, supported by epidemiological data indicating effects in children. This means that no safe exposure level can be set for the substance (EFSA 2019b). Thus, due to health risks, chlorpyrifos will be banned throughout Europe in 2020. This is good news for consumers and farmers.

Prior to its ban in Europe, shortcomings in the authorization of chlorpyrifos have been identified in an industry-funded toxicity study that concluded that no selective effects on neurodevelopment occur even at high exposures. In contrast, the evidence from independent studies points to adverse effects of current exposures on cognitive development in children. In reviewing the industry-funded neurotoxicity test data on chlorpyrifos, researchers noted treatment-related changes in a brain dimension measure for chlorpyrifos at all dose levels tested, although those have not been reported in the original test summary (Mie et al. 2018). Hence, the authors critically suggest that conclusions in test reports submitted by the producer may be misleading. The difference between raw data and conclusions in the test reports indicates a potential bias that would require regulatory attention and possible resolution.

Meanwhile, chlorpyrifos is still in use in many southern countries affecting many low-income farm families in agricultural communities (Rauh 2018).

1.7.3 Other Shortcomings of Environmental Risk Assessments

We have already mentioned that the pesticide products used in the field, the so-called formulations including adjuvants or co-formulants, can show a much higher toxicity than the active substance alone (Defarge et al. 2016; Klátyik et al. 2017; Mesnage and Antoniou 2018). At present, the health risk assessment of pesticides in the EU and the USA focuses mainly on assessing the effects of active ingredients. Hence, adjuvants are not subject to an acceptable daily intake, and they are not included in the health risk assessment of dietary exposures to pesticide residues.

The difference between active ingredients and formulations containing additionally adjuvants and co-formulants is not always clearly stated in scientific papers resulting in some confusion about pesticide effects (Mesnage and Antoniou 2018). Environmental risk assessments (ERAs) currently do not address the presence of adjuvants and co-formulants in food, in our bodies, as well as in the environment. Some co-formulants are very toxic. For glyphosate-based herbicides, it was shown that the first generation of polyethoxylated amine (POEA) surfactants (POE-tallowamine) contained for instance in Roundup are markedly more toxic than glyphosate itself (Mesnage et al. 2019). In the mid-1990s, first-generation POEAs were progressively replaced by other POEA surfactants (ethoxylated etheramines), which exhibited lower nontarget toxic effects. However, as the composition of the commercial pesticide product is legally classified as confidential commercial information it is hard to fully assess their environmental impact.

In agricultural practice many pesticides are used at the same time for instance in tank mixtures. Neither the products used in practice nor cross effects of different products are routinely considered in ERAs. Generally, pesticide formulations are only further investigated in the environmental risk assessment if the toxicity of the substance cannot be determined on the basis of the active substances alone. From a legal point of view, in most cases, it is sufficient to study the active substance of the pesticide in the most sensitive species. As a result, many side effects have not been sufficiently tested. The isolated examination of the active substances does not tell anything about the effect of the finished product. Some co-formulants have been shown to be hormonally active, some influence human reproduction, and others have negative effects on the nervous system. It is therefore not surprising that several studies have shown that pesticide formulations are much more toxic than pure active substances (Mullin et al. 2016). Some pesticide products, including active ingredients and adjuvants, were up to a 1000 times more toxic in experiments with human cell lines than their active substances alone (Mesnage et al. 2014).

For birds, effects are mainly tested on a quail or duck species and thus any interaction with other bird species is underestimated. For birds, mammals, and fish, long-term tests are only planned if the acute short-term tests show a higher acute toxicity; thus, chronic effects that develop over a longer period are systematically underestimated. For aquatic invertebrates, the large water flea (*Daphnia magna*) is used for testing, but this species is not the most sensitive species for all pesticide classes, for example, neonicotinoids (Gibbons et al. 2015). For other invertebrate organisms, long-term tests (including reproduction, behavior, and juveniles) are performed only for the honeybee or an earthworm species. However, worldwide we know about 16,000 other bee species and about 3600 earthworm species. Thus, findings for a few surrogate species are transferred to all other species. Effects on other insects will only be studied if the tests with arthropod indicator species show a risk; this means that effects on a majority of beetles, other insects, and spiders will not be tested. Effects on amphibians such as toads, frogs, or salamanders are not tested regularly. The same refers to reptiles. Hormonal (endocrine) effects which are suspected for many pesticides are regularly not tested.

I do not want to go into more details, as the message should have been clear by now: The registered active pesticide substances are only tested on very few selected organisms. Nevertheless, manufacturers and lobbyists keep claiming that pesticides are the best tested substances on earth!

Some doubts were raised regarding the responsible authorities as well. A study showed that more than half of the 209 scientists working for European Food Safety Authority (EFSA) have direct or indirect links with industries they should control (CEO 2016). EFSA and EU member state authorities also appear to apply double standards in assessing scientific studies. Studies that do not show negative health effects for pesticides are more likely to be accepted, while studies that show negative health effects are more likely to be criticized. For the herbicide glyphosate, the suspicion that the entire approval procedure for pesticides is tailored to the interests of industry has been suggested. Studies provided by the manufacturers remain almost always undisclosed as confidential trade secrets and can therefore not be verified by independent scientists.

Without stressing the case of glyphosate even more, it is worth mentioning that the European Chemicals Agency ECHA had its positive assessment of glyphosate commented in advance by a lobbying association of the agrochemical industry (MDR 2017). Pesticide manufacturers have also tried to influence government decisions in a targeted manner—in particular through purchased studies. That this is not an isolated case has been shown by the Monsanto Papers released by a court in California. According to these, Monsanto supported members of the glyphosate cancer assessment team of the US Environmental Protection Agency EPA. In addition, internal emails showed that Monsanto also hired scientists to publish ready-made articles as independent studies. These studies should serve to refute the classification of glyphosate as potentially carcinogenic by the World Health Organization's cancer research agency IARC.

In October 2017, the European Parliament was also informed of these Monsanto Papers by American lawyers. When Members of the European Parliament wanted to talk to representatives of the US corporation, they were rejected by Monsanto. As a result, the EU parliament, in an unprecedented action, withdrew Monsanto lobbyists' admission for the parliament (Gruber 2017). It is suspected that these lobbyists are trying to influence amendments to laws and also risk assessments for pesticides. The German Federal Institute for Risk Assessment (BfR) has now also been alleged of scientific plagiarism and to have copied the manufacturer's evaluation for dozens of pages for the risk assessment of glyphosate.

In November 2017, after several votes, a 5-year extension of glyphosate in Europe was decided instead of the 10-year extension demanded by the EU Commission. A qualified majority of 18 member states voted for an extension; nine countries (France, Italy, Belgium, Austria, Greece, Cyprus, Malta, Croatia, and Luxembourg) voted against. Portugal abstained. In Germany, the approval of the Minister of Agriculture has caused a brief storm of indignation as the minister voted contrary to the Federal Government's rules of procedure against the Ministry of Environment in the same governmental coalition. Personnel consequences were not drawn. Industry representatives and agricultural officials were impressed by the minister's "courageous" step. The fact that the glyphosate producers had warned the EU Commission of possible claims for damages if the approval was not renewed was taken was also a matter of democratic concern.

Finally, the biodiversity crisis might also give some indications that ERA is not working properly. Biodiversity declines are evident especially in agricultural landscapes across Europe and pesticides are discussed as among the primary responsible factors. The question is how it can be that the most heavily regulated group of chemicals, whose environmental impacts are evaluated in very complex and expensive ecotoxicological studies, can nevertheless have such a negative impact? One claim is that the environmental risk assessments (ERA) during pesticide approval are simply inadequate to protect biodiversity despite the enormous effort (Brühl and Zaller 2019). The main reason for this failure is because ERA does not assess what is actually happening in agricultural practice. First, ERA ignores the fact that almost always several pesticides are used in the fields at the same time. Second, ecological interactions between organisms that are disturbed by pesticides are not considered. Third, biodiversity in the treated fields is not considered. Although political concepts on sustainable farming strive to protect biodiversity, the current ERA system is just not designed to protect biodiversity. As a consequence, agriculture is blamed for the decline in biodiversity, although only pesticides approved to be safe are in use. In Germany and other countries, many farmers protest against ever stricter environmental laws, and to some extent this is even understandable. Because for decades, farmers were advised to use the very same pesticides that are now made responsible for the biodiversity crisis. Rather than making ERA even more complicated, an easier solution would be a drastic reduction in pesticide use by more agro-ecological approaches, a return to the very core of integrated pest management, and an increase in semi-natural habitats such as hedgerows and field margins in the landscape to promote natural enemies of pests and ultimately a further expansion of organic farming or other farming methods with reduced pesticide input.

1.8   Pesticide Residue Limits

When pesticide residues are found in food, it is often reassured that they are on average below the legal limits. The general philosophical problem of creating threshold values has long been discussed (Beck 1986). An example of the dilemma when threshold values are averaged refers to the nutrition of the world's population with the hundreds of millions undernourished people on the one hand and the hundreds of millions of obese people on the other hand. One could come to the cynically conclusion that on average all people on this earth are well nourished. Similarly, it is cynical to say that the average pesticide load for a region is low.

We know that there are regions or occupational groups of people with above-average levels of pesticide exposure. Regions with intensive fruit and wine production and farmers or farm workers are examples for this. On the other hand, however, we do not know whether low levels of pesticide contaminations are really harmless to sensitive or sick people or children. Those who argue with the low average pesticide contamination also do not take into account socially unequal risk situations.

The setting of threshold levels already accepts to poison nature and people at least a little—up to the threshold. As for pesticides, threshold levels are always set for single active substances not for the actually used product containing active substances and adjuvants. However, in reality the environment and humans are confronted with hundreds, perhaps even thousands of more or less harmful substances at the same time revealing the limitation of the current threshold approach.

1.8.1 Pesticide Residues in Food Products

Given their widespread use, it might not surprise that pesticide residues can be found anywhere. As EFSA's annual report on pesticide residues shows, over 25% of food consumed in Europe contains residues of two or more pesticides (EFSA 2019a). The report, which is based mainly on more than 11,000 fruit and vegetable samples collected in 2017 across all EU member states, reconfirms the results from previous years: just over half (54%) of the food tested was free of detectable pesticide residues, whereas over one in four (27.5%) contained two or more pesticide residues within the legally permitted maximum residue levels (MRLs) (Fig. 1.5). Between 1999–2017 there is a trend towards fewer pesticide-free food samples and more samples with multiple pesticides residues.

The maximum number of residues in a single sample (peppers) was 30 and a total of 353 pesticides were detected in food across the EU. Two in three (62%) EU fruits and nuts from conventional farming contain pesticides. Higher rates of pesticide mixtures may be found in our summer salads: 70% of the currants and blackberries and over 60% of cherries, strawberries, lettuce, rocket, and bananas were found to have two or more pesticide residues.

In response to this EFSA report, PAN Europe highlights that the absence of a safety assessment for pesticide mixtures in food fails to address EU law requirements, but it also puts consumer health at risk in a clear violation of human rights (PAN Europe 2019).

Pesticide residues are not only found in raw products but also in processed food, for instance in chocolate bars (Global2000 2016b) and chocolate Easter bunnies (Global2000 2016a)—up to four different pesticides in one bar and up to 12 different pesticides in one chocolate bunny. After such reports, the competent authorities are quickly ready with appeasements. They are assuring

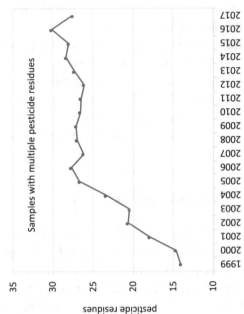

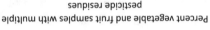

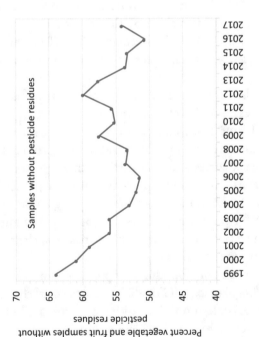

Fig. 1.5 Percentage of vegetables and fruits in European shops without detectable pesticide residues (left graph) and with multiple pesticide residues (right graph). Data for graphs derived from PAN Europe (2019)

that the quantities found in the chocolates were below the legal limits and no danger to health should be expected. However, some of the pesticides frequently found (e.g., endosulfan, chlorpyrifos, cypermethrin, deltamethrin, and permethrin) are endocrine disruptors which act in quantities far below the legal limits.

We will discuss later how the limit values were created. Then it will become clear that these limit values have a legal meaning, but say little about health safety. The proven pesticide residues in chocolate indicate a massive use of pesticides in production. The consequences for the health of cocoa farmers and the local environment during production are deliberately ignored.

Often in debates, the citizen's reservations about pesticides are explained in a somewhat derogatory way with an increasingly urban way of life that is too far away from agricultural reality. These people then find it difficult, it is suggested, to make a proper assessment between the benefits and risks of chemical plant protection (Hallmann et al. 2009). Agreed, many people have little recollection of how food is produced. However, it is also understandable that suspicion arises when dubious activities of industry influence come to light.

The standard response to pesticide residue scandals in food is that pesticides would not have been allowed if they were indeed as harmful as the activists say. The industry representatives then quickly emphasize that practically everything can be measured with today's analytical methods and that all residue reports are below the legal limits. One becomes somewhat suspicious though when it is mentioned that the legal residue levels for pesticides in food have in many cases tremendously increased in recent years. This will be the topic of a later chapter. Residue level limits were not raised because the substances have turned out to be less harmful, but rather because common production methods, such as the earlier mentioned ripening spraying or the cropping of genetically modified herbicide-tolerant crops, make it impossible to remain below the lower limit values.

Increased pesticide residue levels may also stem from improper use. Checks without prior notice by the Federal Environmental Agency in Germany revealed that 50% of the checked farmers used their pesticides incorrectly (Haas 2010). Reports on pesticide monitoring performed in Germany show that in 2017 a total of 5260 farms in agriculture, forestry, and horticulture were inspected where a total of 2594 samples of soil, plants, seeds, or treatment liquids were examined (BVL 2017a). If we consider a total of 275,400 agricultural holdings in Germany (as of 2016), the control rate was around 1.9%. In the same year, 2277 pesticide trading companies were inspected which is about 19% of these companies. These control rates show that there would be plenty of room for improvement. To be fair it has to mentioned that several supermarket chains require pesticide residue levels below the official ones and also require additional controls of their producers.

In less well-organized countries, the situation is most likely more worrying. In Romania I observed that glyphosate was applied in the vineyard for understock treatment with a garden hose. It is impossible to apply correct dosages with this method. The distribution of the pesticides was performed by Sinti and Romanies as day laborer. When I asked why these people worked without protective clothing, I was told that they did not ask for it (Zaller 2018).

Serious deficiencies in the sale of pesticides were the topic of a monitoring report by the World Health Organization (ECCHR 2016). Germans agrochemical giant Bayer distributes hazardous pesticides in India but failed to adequately inform the users of both the dangers of pesticides and the necessary protective measures. The report states that through this failure, the health and lives of tens of thousands of people is endangered. Often proper labels on the pesticides sold in India were missing. Pesticide labels must contain warnings about health and environmental risks. However, Bayer sends the pesticide to India in so-called big bags which do include a warning about the risks, e.g., to unborn children, but the warning is absent from the products as sold in India.

Since humans and the environment worldwide are affected by poisonings caused by highly dangerous pesticides, it was examined which highly dangerous pesticide-active substances are exported from Germany and whether these include pesticides that are banned in the EU. According to official data, in 2017 a total of 233 different pesticide active ingredients, totaling 59,616 tons of active ingredients, were exported from Germany to numerous countries around the world (PAN Germany 2019). When comparing a list of 62 pesticides, more than a quarter of all pesticides exported from Germany in 2017 were highly dangerous (PAN Germany 2019). Nine of these pesticides had no EU approval, including the herbicides cyanamide, acetochlor, and tepraloxydim. These three active substances are classified as carcinogenic (category 2) and toxic for reproduction (category 2). The insecticide cyfluthrin, which was classified by the World Health Organization as highly dangerous with regard to its acute toxicity and thus in the second highest hazard class for acute toxicity (WHO Ib), was also exported. How can it be justified that people and their environment are endangered by these pesticides?

The particular risks of pesticide use under conditions of poverty include:

- Lack of or inadequate protective equipment and clothing for mixing and applying the pesticides.

- Improper storage of pesticides in the living room, bedroom, or kitchen besides food and easily accessible for children or pets.
- No take-back systems for used pesticide containers which leads to littering of pesticide containers and reuse of empty containers to store drinking water or food.
- Low level of training and language barriers: users are often not sufficiently informed about the pesticides and their toxicity and information on toxicity and application is not available in all national languages; illiterate users cannot read the information on the labels; pictograms can be misinterpreted.
- Illegal repackaging of pesticides.
- People weakened by malnutrition or chronic diseases with limited access to health care are also especially vulnerable to pesticide poisoning.

1.8.2 How Are Pesticide Residue Limits Set?

In Europe, EU legislation harmonizes and simplifies pesticide maximum residue levels (MRLs) and sets a common EU assessment scheme for all agricultural products for food or animal feed (EC 2019b). Applicants seeking approval of a pesticide must submit scientific information about the minimum amounts of pesticide necessary to protect a crop from pests and diseases and the residue level remaining on the crop after such treatment. The European Food Safety Authority (EFSA) then verifies that this residue is safe for all consumer groups, including vulnerable groups such as babies, children, or chronically ill people. When there is a risk established for any consumer group, the MRL application will be rejected and the pesticide may not be used on that crop. Food safety thus has priority over plant protection. At least that is the noble intention.

How and when the pesticide may be used is defined by the relevant national authorities and can be found on the label of the pesticide. These authorizations on a national basis are included to account for specific local and environmental conditions that might demand different uses of pesticides across Europe (EC 2008). For example, in the southern member states where it is warmer, there are more insects and thus more insecticides are needed. In other parts of the EU, it is more humid—conditions that suit fungal infestation, and thus more fungicides are needed. So, we get already a feeling that the agricultural practice is a very important factor for setting MRLs, not only the safety for consumers.

Farmers, traders, and importers are responsible for food safety, which includes compliance with MRLs and member state authorities are responsible for their control and enforcement. If pesticide residues are found at a level of concern for consumers, the Europe-wide Rapid Alert System for Food and Feed (RASFF) circulates the information and measures are taken to protect the consumer. For crops grown outside the European Union, MRLs are set on request of the exporting country. MRLs apply to 315 fresh products and processed products. Legislation covers 1100 pesticides currently or formerly used in agriculture in or outside the EU.

A general default MRL of 0.01 mg per kg food product applies where a pesticide is not specifically mentioned; the MRL for drinking water is 0.0001 mg (0.1 µg) per liter of water. These are very small quantities, which on the one hand indicate the strict regulations, but on the other hand also indicate the hazardous nature of the substances.

In discussions about the significance of pesticide residues in food, it is often stated that the MRLs would be particularly safe. Commonly experiments with animals are carried out to assess the toxicity of a pesticide or other harmful substances to humans. This is ethically questionable, but a discussion about it would go beyond the scope of this book. My esteemed colleague Peter Weish, an award-winning Austrian ecologist and environmental activist, takes a radical view in this context by stating that "every industrial production of pesticides is in itself a crime towards life" and adapts a quote of the philosopher Adorno saying that "you can't do something right in the wrong."

Risk assessments of pesticides using animal experiments uses healthy laboratory mice or rats that are fed with pesticide contaminated food or directly sprayed with a single active substance under controlled conditions. Observations on the animals (so called endpoints) are then used to calculate toxicological limits for humans such as the acceptable daily intake (ADI). One can imagine that the transfer of the results from laboratory animals to humans is tricky. Historically, the threshold concept was introduced in the 1950s when the ADI concept was proposed to calculate the risk of chemical contaminants in human food (NRC 1993). This concept was then adopted by the FAO and WHO. Formally, the ADI was defined as the no-observed-effect level (NOEL) in toxicity studies divided by a safety factor. Weight loss, reduction in weight gain, alteration in organ weight, and inhibition of cholinesterase activity are indicative of specific adverse effects that may be considered when establishing the NOEL. The uncertainty factor makes allowances for the type of effect, the severity or reversibility of the effect, and variability among and within species.

Since it is not known how sensitive a human being is compared to a laboratory animal, a factor of 10 was introduced. So, the determined NOEL values are divided by 10 to account for this aspect. Furthermore, it is generally assumed that there are differences in sensitivity within the human population. Healthy people react differently than sick people and old people differently than small children. Therefore, another factor of 10 was derived from this uncertainty. In other words, a safety factor of 100 is used, assuming that humans are 10 times more sensitive than the most sensitive test animal and that the most sensitive humans are 10 times more sensitive than the average human. The interspecies safety factor is based on studies in adult laboratory animals and, therefore, is applicable to adult human beings, but studies in adult laboratory animals are not necessarily good for predicting the response of human infants and children. The other tenfold intraspecies safety factor is meant to cover variations within human populations, including genetic predisposition, poor nutrition, disease status, and age. A factor of 10 for intraspecies variation in susceptibility may be sufficient for any one element of interpersonal difference but may not be sufficient for multiple elements. Thus, as presently determined, a safety factor of 100 may not be sufficient to account for the potential increased sensitivity of infants and children.

An additional factor of any height can be applied if the authorities know too little about the toxicity of a substance due to insufficient data for instance when no tests are available. Often a safety factor of 1000 may be applied when insufficient toxicity data are available. For considering potential carcinogenesis, a safety factor of 10 was proposed, when studies in humans involving prolonged ingestion have been conducted with no indication of carcinogenicity (NRC 1993). When chronic toxicity studies have been conducted in one or more species with no indication of carcinogenicity and data on humans are either unavailable or scanty, a factor of 100 was proposed.

Overall, this safety factor concept appears scientifically sound, doesn't it? But we need to remember that this was originally created because no one knew better back in the 1950s (Neumeister 2016). Indeed, its limitations are widely recognized among critical scientists. By the way, the term safety factor is mainly used by authorities and pesticide sympathizers. The critical scientific literature correctly calls them uncertainty factors (Dorne 2010). Also, the US EPA has recommended using the term uncertainty factor rather than safety factor in recognition of the fact that the ADI does not guarantee absolute safety (NRC 1986). However, would we talk about uncertainty instead of safety in connection with pesticide limits, it would make us feel a little uneasy and, in any case, interfere with the narrative of the agrochemical industry. Environmental experts interviewed by Marie-Monique Robin about the creation of these safety factors admitted rather bluntly that it was initially decided by the BOGSAT method: a bunch of guys sitting around a table. Another interviewed expert confirmed with refreshing clarity that the safety factor is a figure that fell from out of nowhere and was scribbled on a napkin (Robin 2014).

Assuming that this uncertainty factor concept is well meant, there are still many real-life aspects that are not considered. I already mentioned the focus on active substances while ignoring adjuvants and the broad assessing of the applied formulations and ready-to-use mixtures, the ignorance of multiple pesticide applications and interactions between tank mixtures, the ignorance of long-term effects, lifelong chronic exposure, or impacts on future generations (so-called multi-generation effects) and interactions with other stressors such as global change factors.

In one of the first studies assessing multi-generation effects of pesticides, researchers saw descendants of rats exposed to glyphosate developing prostate, kidney and ovarian diseases, obesity, and birth abnormalities (Kubsad et al. 2019). These diseases and other health problems were seen in the second- and third-generation offspring of rats. In the experiment, pregnant rats were exposed to glyphosate between their 8th and 14th days of gestation. The second generation had significant increases in testis, ovary, and mammary gland diseases, as well as obesity. In the third generation, males had a 30% increase in prostate disease and females a 40% increase in kidney disease. More than one-third of the second-generation mothers had unsuccessful pregnancies. Additionally, two out of five male and female rats in the third generation were obese. This study strikingly demonstrates that the ability of pesticides (here glyphosate) to impact future generations. However currently, this aspect is completely ignored in the approval of pesticides.

We have learned that in agriculture many pesticides are used simultaneously. The mutual influence of different pesticides, so-called cocktail effects, is systematically excluded and not investigated. There are numerous studies which show that pesticides in combination with other pesticides are much more toxic than if they were used individually (Laetz et al. 2009). These are called synergistic effects. Meanwhile, in human medicine there is increasing sensitivity about possible interactions between different medicines. When we take claims seriously that pesticides for agriculture are just as medicine for humans we should also consider interactions between different pesticides applied onto a field. The investigation of possible cross-activities would be extremely complicated and costly, as combinations of thousands of pesticides are involved. But ignoring these issues just because we do not know how to deal with it is grossly negligent.

How is it assessed whether a product is dangerous for the environment? The difficulty here is that organisms and ecosystems can react in an extremely wide variety of ways. For humans, we know that we react individually to chemical substances, depending on gender, time of day, age, psychological status, diseases, and the presence of other stressors. For all other organisms that are tested, for example, it is assumed that a tested earthworm reacts in the same way as all its fellow 4000 species, one surrogate bird species for the test is meant to represent the behavior of all other bird species. The pragmatic solution for these deficiencies in the assessment of pesticide effects is extrapolation. This means that the results of an experiment are used to infer other species and other environmental situations.

Most of these extrapolations are scientifically difficult to understand or clearly inadmissible and often wrong (Seok et al. 2013). For example, results obtained with adult honeybees are transferred to all other life stages and all bee species, sometimes even to all insect pollinators. Such extrapolations clearly raise doubts on the rigorousness of pesticide risk assessments. A colleague who worked for the approval of pesticides in an agrochemical company mentioned that in many cases the quality of the studies carried out for pesticide registration would not meet the quality standards required in peer-reviewed scientific journals. Unfortunately, this cannot be verified because the studies are not accessible.

Another important shortcoming in the testing of pesticide effects is that, for budgetary and practical reasons, only certain endpoints are tested. These endpoints are freely chosen by the investigator. Typical endpoints are the mortality of the test animal, stages of cancer, or the weight of certain organs. But endpoints can also be much subtler in reality, such as neurotoxic effects. These include, for example, learning defects, malfunctions of the immune system, reproductive organs, or multigenerational effects. Most of these aspects are not considered in the investigation of possible effects of pesticides. So much again for the supposedly best-studied substances!

The interaction of different toxins which is reality in an agricultural field is not considered either (Goodson et al. 2015). In actual situations, organisms are constantly exposed to several pesticides simultaneously. The problem that pesticides are always only used in formulations, but in most cases only the active substances are evaluated in tests, was explained earlier (Kortenkamp 2014).

It should also be noted that the assessment of the effect of pesticides is based on linear dose–response relationships. This sounds complicated, but is in principle the application of the Paracelsus phrase, according to which "all things are poison, and nothing is without poison, only the dose makes a thing not a poison." In other words, the higher the dose, the stronger the effect;

therefore it is assumed to deduce effects for other doses from the tested dose-response relationships. The doctor and philosopher Paracelsus, who worked in Switzerland, Germany, and Austria in the 16th century, is often quoted and still plays an important role in the approval of pesticides. Of course, we now know much more about the effects of toxins on humans and the environment. The assumption of a linear relationship between dose and response has rarely been questioned, but for many toxins, especially those with endocrine, i.e., potentially hormone-like effects, these linear relationships are not valid. Consequently, all pesticides should also be tested for their endocrine activity, because if this is the case then the substance is already effective in the smallest concentrations (Vandenberg et al. 2012).

The transfer of findings from simple laboratory tests to real-world situations is in principle also made for medical/pharmacological tests. However, in the case of medical drugs there is also a network of doctors who can provide feedback on any side effects observed on their patients. Such a feedback process with additionally including environmental effects on is lacking for pesticides. We see, there is a great deal of uncertainty associated with risk assessments, nevertheless manufacturers and also many ecotoxicologists still claim that pesticides are among the most intensively tested substances, comparable only to drugs (Strubelt 1996).

Despite (or because?) all these uncertainties and unscientific aspects, the EU member states advocated for the precautionary principle. According to this any substance should be withdrawn from the market at the slightest suspicion regarding effects on our health. The fact that many pesticides with clear detrimental health effects are still on the market shows that this principle has often been ignored. Often, I hear that my insistence on the precautionary principle is exaggerated and only causes panic in the public. Well, don't we all practice the precautionary principle in our daily lives when we use safety belts in our cars or wearing helmets when cycling, for example? Of course there are fellows around that refuse to use the seat belt or helmets because they did not need them so far as they were not involved in an accident. By the way, the precautionary principle is nothing new and already took root in the 1970s in Germany and was adopted in the Maastricht Treaty of the European Union 1992, and globally implemented in the Rio Declaration of the UN Conference on Environment and Development (Agenda 21) in 1992.

In addition to the pure dose-effect relationships, the duration of exposure, the time of an exposure in relation to an organism's developmental stage, any additional exposure, stress, or illness, the sex of the organism that comes into contact with the toxins, or concomitant damage or illness of the test organism play a role how the toxicity of a substance will be manifested in an organism.

Moreover, and most importantly and already stressed above, there is no dose–effect relationship for pesticides (and other chemicals) with hormonal effects, the so-called endocrine disruptors or EDCs—they are effective even in the lowest concentrations (Vandenberg et al. 2012). Data are mounting showing that endocrine disruptors have effects on laboratory animals, wildlife, and humans at doses that are considered safe by traditional toxicology testing. Further, it is important that many endocrine disruptors produce nonmonotonic dose–response curves, which challenges the assumption that safe doses can be extrapolated from traditional high-dose testing. Toxicologists are heavily debating these low-dose effects and recommend to include these concepts in environmental risk assessments of pesticides (Vandenberg et al. 2013).

1.8.3 Elasticity of Residue Limits

Of course, pesticide limits cannot be set arbitrarily high. The relevant EU regulation explicitly states that the levels should be set as low as is compatible with good agricultural practice in order to protect particularly vulnerable groups such as children and the unborn. It is the manufacturers who apply for maximum residue limits. To do this, they must submit documentation of residue tests that have taken place under the worst-case scenario.

Some say that there is no safe dose of pesticides in our food or drinking water because the toxic substances can accumulate in certain depots of our body and it can take decades for health problems to emerge (Daunderer 2005). The MRL of pesticides is the highest level legally permitted in food or feed after pesticides have been used in accordance with good agricultural practice. Adhering to the precautionary principle one would think that MRLs should be particularly low, at least for food products that are frequently consumed.

It is interesting to see, though, that MRL levels for food and feed products are not stable over the years. I am mainly referring to the situation in Europe, but it appears to occur also in the USA and Brazil and perhaps in other countries as well (Bøhn et al. 2014). We have heard that the default pesticide MRL in Europe is 0.01 mg/kg of food product. However, for some frequently used pesticides there are specific MRLs in place. The example of glyphosate shows that the permissible MRLs for foodstuffs vary widely (most recent status 2013): 20 mg per kg for soybeans, sunflowers, barley, oats; 50 mg per kg for wild mushrooms; 10 mg per kg for rye, wheat, linseed, lupins, rape, peas, lentils; 2 mg per kg for beans; 1 mg per kg for corn; 0.1 mg per kg for most vegetable products; 0.05 mg per kg for meat (except kidney), milk, and eggs.

This means that the pesticide load tolerated by law for a veggie burger made of soybeans is 400 times higher than for a meat burger of the same weight. Also, interesting is the development of MRLs over the years (Table 1.3). For instance, soybeans showed a 50-fold glyphosate MRL increase from 1999 to 2013 in Brazil and a 200-fold increase in the USA and Europe. This is assumed to be the consequence of ripening sprays before harvest and the cultivation of Roundup-ready soybeans that make it impossible to stay below the formerly stricter residue levels (Bøhn et al. 2014). Scientifically nobody can justify this, but I am sure there are economic and political explanations for this.

Table 1.3 Development of maximum residue levels (MRL) for some pesticide active ingredients in selected food and feed products. MRL is the highest level of a pesticide residue that is legally tolerated in or on food or feed when pesticides are applied correctly (Good Agricultural Practice). The default MRL is 0.01 mg/kg food or feed product

Country	Crop	Maximum residue level MRL mg/kg				Change (lowest-highest)
Glyphosate (herbicide)						
		1999	2004	2008	2013	
Brazil	Soybean	0.2	10	10	10	50-fold increase
USA	Soybean	0.1	20	20	20	200-fold increase
Europe	Soybean	0.1	20	20	20	200-fold increase
	Oranges			0.5	0.5	Unchanged
	Table grapes			0.1	0.5	5-fold increase
	Wine grapes			0.1	0.5	5-fold increase
	Bananas			3.0	4.0	1.three-fold increase
	Oat			20.0	20.0	Unchanged
	Rice			0.1	0.1	Unchanged
Chlorpyrifos (insecticide)						
		2008	2016	2018		
Europe	Oranges	0.3	0.3	1.5		5-fold increase
	Apples	0.5	0.01	0.01		50-fold decrease
	Table grapes	0.5	0.01	0.01		50-fold decrease
	Wine grapes	0.5	0.5	0.01		50-fold decrease
	Cranberries	0.05	0.05	1.0		20-fold increase
	Bananas	3.0	3.0	4.0		1.3-fold increase
	Barley	0.2	0.2	0.6		3-fold increases
Captan (fungicide)						
		2008	2011	2013	2019	
Europe	Oranges	0.02	0.02	0.02	0.03	1.5-fold increase
	Apples	3.0	3.0	3.0	10.0	3.3-fold increase
	Table grapes	0.02	0.02	0.02	0.03	1.5-fold increase
	Wine grapes	0.02	0.02	0.02	0.02	Unchanged
	Raspberries	3.0	3.0	10.0	20.0	6.6-fold increase
	Blueberries	0.02	0.02	15.0	30.0	1500-fold increase
	Hops	0.05	0.05	0.05	150.0	3000-fold increase

Data from Brazil and the USA according to Bøhn et al. (2014); European data extracted from EU pesticide database (EC 2019d)

The changes of MRLs for the herbicide glyphosate, the insecticide chlorpy-rifos, and the fungicide captan during the last few years are really striking. The reasons behind these changes are difficult to understand. While an increase in MRLs for soybeans could be argued with the widespread use of ripening spraying or GM soybeans with heavy glyphosate use, it is unclear why glypho-sate MRLs in table grapes and wine grapes also have been raised fivefold or that for bananas 1.3-fold. Positively, MRLs for the insecticide chlorpyrifos that has been shown to affect the brain development of children have been decreased 50-fold between 2008 and 2013 for apples and grapes, but at the same time increased 20-fold for cranberries. The most significant MRL changes can be observed for captan, which is according to the manufacturers safety datasheet "probably cancerogenic": MRL for apples has been raised 3.3-fold from 2008 to 2019, for raspberries 6.6-fold, for blueberries 1500-fold, and for hops an unbelievable 3000-fold.

To be fair it also has to be noted that in some cases, approvals for pesticides were revoked because the MRL could not comply with good farming practice (BVL 2017b).

In an earlier chapter, we already saw that it is not easy to get rid of all the pesticide residues on fruits and vegetables. Therefore, particular fruit cleaners are offered on the US market. However, it seems rather strange to use an addi-tional product for cleaning instead of avoiding the application of pesticides right from the beginning. Anyway, the US National Pesticide Information Center informs how to wash pesticides from fruit and veggies (NPIC 2019). On the website the story of a fictive person named Kaye is told. Kaye was at the grocery store buying fruit for a salad when she noticed a bottle of fruit-washing soap on display next to the produce. Her fruit salad was going to be part of a potluck lunch, so Kaye wanted to clean the fruit thoroughly from germs or pesticide residues. She then wondered if she needed to purchase such an expensive fruit cleaner, or if water-rinsing was enough. Kaye thought per-haps her antimicrobial dish soap would work just as well as any fruit wash. We learn from the website that washing fruits with water reduces dirt, germs, and pesticide residues remaining on fruit and vegetable surfaces. Holding the fruit or vegetable under flowing water removes more than dunking the produce. Peeling or scrubbing produce like potatoes with a stiff clean brush or rubbing soft items like peaches while holding them under running water works best to remove residues. However, pesticide residues can stick better to waxy or soft-skinned fruits. If the produce was treated with wax, pesticide residues may be trapped underneath the wax. Obviously, some fruit and vegetable washing products can be effective at removing dirt or residues, but they have not been proven to be any more effective than water alone. Dish soap or bleach can get trapped or absorbed by the pores and become difficult to rinse off the fruit once they have been applied and using cleaning products may actually add residues to the produce.

Washing fruit and vegetables with running water is good for at least reducing the amount of pesticides. When testing 196 samples of lettuce, strawberries, and tomatoes, it was found that rubbing the fruit rather than only running water over the fruit worked best to eliminate pesticides (Krol 2000). Pesticide residue was reduced for nine out of the 12 pesticides tested when produce was washed with running water. However, the remaining three pesticides, the fungicide vinclozolin, and the insecticides bifenthrin and chlorpyrifos were not reduced. We remember, that chlorpyrifos was banned in Europe because it is shown to affect brain development in children. It seems that the water solubility of pesticides does not play a significant role in the observed decrease. The majority of pesticide residue appears to reside on the surface of produce where it is removed by the mechanical action of rinsing.

In another study, vegetables were soaked in vinegar for 20 min and also in a salt and water solution to remove chlorpyrifos, DDT, cypermethrin, and chlorothalonil pesticides (Zhang et al. 2007). Both methods worked well. The vinegar effectively removed pesticides, but left a residue that affected taste. A 10% saltwater solution also worked really well, but the most effective method still seems to be a baking soda solution. Hence, the only real way of ensuring your produce is pesticide free is to peel it. However, peeling would not protect from systemic pesticides contaminating the whole fruit and additionally reducing the intake of fiber and nutrients. Probably, the best option is to eat organically produced food.

What is the position of pesticide manufacturers on MRLs? On a manufacturer's website, pesticides for plants are compared with medicine for human health (BASF 2016). Accordingly, they protect plants before and/or after harvesting from diseases, pests, and weeds and dissolve as soon as they have served this purpose. The argumentation consistently refers to MRLs as trade standards, not as something related with health issues. It is interesting to see that the manufacturer admits on the website that the residue limits are not safety limits. The justification states that "although residues above the MRL are not economically acceptable, they do not automatically pose a risk to the consumer." The manufacturer also complains that there are already retail chains that demand even lower than the legally accepted MRLs, and this would undermine confidence in the approval authorities and the legal requirements. This concern for the authorities is really nice, especially as the manufacturers are usually complaining about too strict rules by authorities.

Let us briefly reconsider the comparison between pesticides and pharmaceuticals. The overarching principle for pharmaceutical governance is to ensure that medicines are effective, safe, and marketed on the basis of need. Responsible for the global governance of pharmaceuticals is the World Health Organization (WHO). For pesticides, there is no such global governance; rather individual national standards are set around the world. Pharmaceuticals are also monitored involving the collection, detection, assessment, monitoring, and prevention of adverse effects of medicines. Monitoring continues throughout the lifetime of each product, building a well-developed safety database for that product. This governance and especially continued surveillance for indicators of negative effects (called vigilance) of pharmaceuticals and pesticides differ widely once market authorization has been granted (Milner and Boyd 2017). For pesticides, the process of substance discovery and early testing is similar than for pharmaceuticals involving tests for effectiveness, toxicology, fate, and behavior in the environment using a combination of laboratory and field trials. If approved by the regulatory authority, the pesticide can be used in accordance with the label for the duration of the license (typically 10–15 years in the European Union and 15 years in the USA).

But there is no such a long-term monitoring for pesticides. Who provides feedback to agrochemical companies when a species disappears in nature due a pesticide application? Food is monitored for MRLs to help protect human health from the effects of consuming pesticides, but we have seen that international MRLs vary widely. Importantly, these MRLs only exist for food but not for contamination of the environment. Moreover, there are no MRLs for the air we breath. Pesticides are used without proper knowledge of safe environmental limits; the total environmental dose of pesticides is mainly governed by the demand of applicators and markets rather than by a limit on what the environment can endure.

For pesticides there is no accessible information about where, when, and why pesticides have been used, making it impossible to quantify potential environmental effects. An essential step for pesticide regulation in the future would be to develop an equivalent procedure to pharmaceuticals with clear responsibilities for monitoring the use and effects of pesticides on manufacturers and growers providing data open to the public.

When talking about pesticides and residue limits, genetically modified organisms (GMOs) must also be mentioned. In Chap. 2, I will briefly elaborate on the potential of GMOs in order to reduce the pesticide load for our environment. So far, the use of GMOs is mandatorily combined with the use of herbicides, because otherwise the supposed advantages cannot be exploited. I have already mentioned that changes in MRLs appear to be made in adaptation to agricultural practices rather than environmental and human health (Bøhn and Cuhra 2014). The examples so far have mainly considered plant cultivation; however, animal breeding is also affected by pesticides. The quality of food and feed is essential for animal health and quality does not only mean a balanced content of nutrients, minerals, vitamins, or important fatty acids, but also the absence of toxins such as those from pesticides.

Surprisingly, even in the scientific literature there is little data on herbicide residues in GM plants, although they have been on the market for more than 20 years. One study examined soybeans grown in different agricultural systems, organic, GMO soybeans that were resistant to glyphosate, and conventional soybeans as a third variant (Bøhn et al. 2014). The analyses showed that all GMO cultivation samples contained residues of 9 mg/kg of soybeans of glyphosate and AMPA. AMPA (aminomethylphosphonic acid) is the main degradation product of glyphosate and indicates the degree of decomposition of the active substance glyphosate. In contrast, conventionally produced and organically produced soybeans showed no glyphosate residues. This is explained by the fact that glyphosate is applied to GMO fields more often and at higher dosages, as many weeds are now resistant to glyphosate-based herbicides (Shaner et al. 2012). Globally, GMO soy is the number one GMO crop. Soybeans represented 61% of the world's oilseed production. Global soybean production in 2011 was around 252 million tons, with the USA (33%), Brazil (29%), Argentina (19%), China (5%), and India (4%) as the largest producers. In 2018, soybean was cropped on 36.1 million hectares; in the USA, 95% of GMO soy is grown (ASA 2019).

Especially in countries where GMOs are grown or where these products are imported, the authorities have increased the legally accepted amount of glyphosate in food and feed (MRL) in recent years (Table 1.3). These MRL adjustments are of great economical help as otherwise the GM soybeans would no longer be marketable. It seems as if glyphosate is considered less harmful in GMO soybeans than in normal soybeans. These changes of residue limits also underpin the importance of labeling food containing GMOs. However, in the USA, this demand by consumer associations has been prevented for decades by the industrial lobby; only individual federal states have introduced GMO labeling. Europe appears to be more consumer-friendly and has a labeling regulation for GMO-free products; only a technically unavoidable and accidental contamination of 0.9% is tolerated.

If you now say that you are not concerned with soya, since you are not a vegetarian and avoid tofu or other soya products anyway, then you may be interested to know that around 70% of the proteins used in meat production in Europe are covered by imported soya. Only about 2% of soya is used for human nutrition (OESF 2016). It is these global networks in food production that makes the pesticide subject even more complex and confusing. Ultimately, however, we are almost always individually confronted with the pesticide problem, whether we like it or not.

These scientifically unsubstantiated changes in MRLs and the numerous shortcomings of pesticide risk assessments make it harder to trust authorities in this matter. In any case, we have seen that pesticide residues in food are on the increase and limits are still being set according to outdated standards without taking into account new scientific research (Myers et al. 2016).

If MRL changes happened out of blind trust in pesticides, it would be sad. But if these adjustments had come about as a result of pressure or lobbyism by agrochemical companies, it would be deeply worrying for our democratic societies.

In fact, there is substantial evidence that a great deal of lobbyism is interfering an objective assessment by national authorities. Some mechanisms were aired by the former mentioned so-called Monsanto papers.

But much influencing is probably not so obvious. For instance in Germany, studies state that glyphosate is important for agriculture and a ban would cause billions of € damage per year across Europe. Later these studies turned out to be cosponsored by the agrochemical industry without declaring this conflict of interest (SZ 2019). Of course, studies financed by industry more often lead to industry-friendly results. These publications did not deal with the question of whether glyphosate is harmful or not but focused on agronomical aspects. The authors forecast yield losses of up to 10% if glyphosate is banned (Schmitz and Garvert 2012). These pro-glyphosate statements circulated in the political debate for years and are still used by glyphosate advocates as objective proof of the indispensability of the substance. In the meantime, the journal editors retracted the article because of nondisclosure of a financial source supporting this study and a potential conflict of interest by author. Editors state that correct disclosure might have influenced the peer-review process.

In wrapping up this residue level chapter it seems clear that one cannot be certain that residues below MRLs are ineffective, and it is problematic when reports on pesticide residues in food or drinking water keep reassuring their harmlessness. The fact is that MRLs are primarily legal limits, rather than toxicological thresholds (Jezussek 2012).

1.9   War Rhetoric Sets the Mood

In the words of farmers but in also in various media reports, we find ourselves in a constant war and fight against weeds, pests, or so-called alien species that invade us from foreign countries. A textbook on pest management openly speaks of a perpetual war against pests that mankind must fight to survive (Fletcher 1974). It also states that pests, especially insects, are our main competitors on Earth.

This rhetoric is probably no coincidence; after all, many modern pesticides actually originate from former arsenals of chemical weapons. The same agrochemical companies that today produce pesticides have also played a major role in the production of poisonous war gases (Mimkes and Pehrke 2016). Among the producers of these toxins was also Mobay, a cooperation between Monsanto and Bayer the same two companies that merged in 2018 to form the world's largest agrochemical and seed company.

Military imagery is still used in this industry today (Gehrmann 2016). The advertising of an insecticide refers to war profits. Swarms of mosquitoes appear in aircraft formation as in World War II. In addition, there are quotations from famous warlords such as Winston Churchill.

By the way, the name Roundup of the world's most widely used glyphosate-herbicide, translates to raid—hopefully, only a raid against weeds and not against suspicious or unpopular people.

Also, in public media, where product advertisement should not be dominating panic is made against annoying organisms such as the for people harmless marmorated stink bugs (*Halyomorpha halys*) that are called a plague (ORF 2018). The bugs were brought to Europe from Asia and now can be found on balconies and terraces, often in cities. When it gets colder in autumn, they look for a warm place in the houses in order to reproduce there. In the Internet, pest control companies additionally heat up the mood by stating that these bugs are extremely disturbing in the immediate vicinity, but also admitting that such pests do not pose a real threat. Nevertheless, it is further warned, people should take the infestation not too easy but better fight it right away until it is too late.

War rhetoric is also used when it comes to control so-called neobiota. Neobiota are new organisms that occur in previously unknown regions. They are called neophytes if they are plants and neozoa if they are animals. Prime examples of neophytes in Europe are the rose-flowering Himalayan balsam (*Impatiens glandulifera*) originally from Asia, which makes very extensive stands along rivers in Central Europe, or the yellow-flowering Canadian goldenrod (*Solidago canadensis*), which grows on drier places near railways and road margins. In the USA, important alien plant species are for example Kudzu (*Pueraria lobata*), common tumbleweed (*Kali tragus*), Privet (*Ligustrum* spp.), or the multiflora rose (*Rosa multiflora*). Well-known neozoa are the brownish-orange Spanish slug (*Arion vulgaris*) feared by every garden lover or, more recently, the box tree moth (*Cydalima perspectalis*) a species native to eastern Asia and invasive in Europe since about 2006. Many of these species are referred to as invasive and consequently should be fought at least in the narrative of people who like to be under control of their surroundings. The xenophobic climate in our societies dramatizes the situation even more. Accordingly, heavy chemical guns will be fired from which not even nature reserves, national parks, or, as a colleague told me, the Galapagos Islands will be spared.

In addition to some desire for war, we are also steadily reminded to our sense of order and tidiness. Standard farmer journals are usually full of advertisements for pesticides. For example, there is advertising with a female cleaning lady. She is equipped with green rubber gloves, a scrubbing brush, wearing a blue spotted apron, and attached hair and announces with her thumbs pointing upward: "Your field has never been so clean!" (Syngenta 2017). A complete solution and safety against all weeds is promoted for the product. This supposed longing for cleanliness and safety and the social pressure to do so should not be underestimated, as it also represents an important reason for pesticide use in private gardens. But also, in discussions with farmers, it is regularly emphasized that pesticides are often applied so that everything looks nice in the field. At the end one does not want to risk the image of the sloppy farmer of the village.

Farmers magazines also advertise a "new, flexible herbicide with a broad action spectrum for potatoes." The product is "user-friendly" and has "good crop compatibility." It is up to the reader to interpret this; anyway it is strange that an herbicide is applied to a crop. Finally, the same magazine of course also advertises fungicides. Interesting in this context is the open warning for resistances: "High-performance fungicides today are also endangered by resistance through intensive use." However, fortunately, the advertised fungicide has a "protective and curative effect and is an excellent resistance breaker for efficient anti-resistance management securing long-term success for the grower." The resistance problem is huge in pesticide-intensive agriculture will be handled in a later chapter.

The historian Michelle Mart wrote an outstanding book with broad coverage on America's enduring embrace of dangerous chemicals and points out that this military speech were already established in the 1950s (Mart 2015).

This rhetoric permeated mainstream newspaper and magazine articles emphasizing that this war was not only fought by farmers and other experts but also by private people. So, the intentions were to make consumers believe that the products are good for home use and no special expertise to use these pesticides was necessary. The goal was also to deemphasize the hazards of pesticides by referring to "crop protection substances" and avoiding attributions such as "hazardous" or "toxic."

1.10 Agroecosystems and Biodiversity Also Benefit Human Societies

Before we examine the concrete effects of pesticides on the environment and humans, it is important to review what is actually at stake. The agricultural land, sometimes also called cultural landscape when grown over centuries as in Europe or Asian countries, is forming an ecosystem, the so-called agroecosystem. Ecosystems function through the interaction of biotic and abiotic components. Animals, plants, and microorganisms interact in a variety of ways, for example by exchanging nutrients, energy, or information. Of course, the primary goal for agriculture is to produce food. However, it is often forgotten that agroecosystems additionally provide many services for the society so-called ecosystem services. An agroecosystem is used to produce food for humans, animal feed, livestock, fiber, wood, or other raw materials.

The concept of ecosystem services is well established in science and almost used inflationary. Roughly speaking, this refers to the effects of ecosystems on humans. Besides food production, other important agroecosystem services include the pollination of fruit blossoms by insects or other pollinators, natural pest control by antagonistic organisms, the provision of clean drinking water through natural filtration of precipitation and groundwater, the provision of fresh air, or an aesthetic cultural landscape that can be used for recreational purposes.

These services can be evaluated for example by calculating what it would cost to artificially pollinate the field crops. This is not so far-fetched because in some Chinese provinces people pollinate the fruit trees because not sufficient insect pollinators are around to do this job. In the documentary *More than Honey* by Markus Imhoof, for example, it can be seen how human pollinators equipped with small brushes are climbing around in fruit trees in order to take over the role of the missing insect pollinators. Calculations on the worth of ecosystem service are sometimes criticized because they give nature a monetary value. On the other hand, however, it is easier to communicate about the benefits of nature for society especially to people who mainly think in economic terms; hopefully, it also helps to explain politicians to more appreciate the value of nature.

The fundamental difference between natural ecosystems such as forests, grasslands, steppes, and prairies and agroecosystems is that the latter are more strongly managed by humans. Hence, agroecosystems do not have closed material and nutrient cycles, because the removal of the harvested material also removes nutrients contained in harvested crops from the system. To keep the cycle going, these nutrients must be added regularly in the form of fertilizers; otherwise the productivity of the agroecosystem would be exhausted over time. With the same reasoning the use of pesticides is justified because agricultural systems are often monocultures where the natural interplay between beneficial organisms and pests cannot develop.

In an agroecosystem, there is usually a complex interplay between the organisms living there. Also, natural pest control is taking place constantly without notice. Aphids are eaten by ladybirds; gluttonous caterpillars are parasitized and rendered harmless by ichneumon wasps, just to mention a few interactions. Also, spiders, which are generally not very popular, contribute significantly to maintaining the ecological balance of nature. With more than 45,000 species and an occurrence of up to a 1000 individuals per square meter, spiders are among the most species-rich and widespread predatory animal species. It is estimated that spiders worldwide kill up to 800 million tons of insects annually, more biomass than humans consume in the form of meat and fish (Nyffeler and Birkhofer 2017).

Hence, the objective should be to embed agricultural fields in a richly structured landscape with many undisturbed retreat areas for these beneficial organisms and a diverse vegetation.

Chronically ignored in this respect, not only in the public debate but also by agroecologists is the immense importance of the soil and biota living therein. Soil is basically a nonrenewable resource and cannot be restored, as can be seen from the many once wooded karst landscapes in the Mediterranean region. The soil is often only seen as a medium for the growth of crops. But of course, the soil can do more: it stores water, contributes to climate protection by storing carbon, is important for the supply of plant nutrients, and is a habitat for soil organisms. The interplay between these aspects is commonly referred to as soil health. As a dynamic system soil reacts to change and the great challenge for agriculture is to maintain and if necessary improve soil functions and soil health.

A very important component of soil health is an active soil life. The diversity of soil organisms is overwhelming: it is estimated that a handful of arable soil (about 200 g) contains more organisms than there are humans on earth (Jeffery et al. 2010). The majority make microorganisms such as bacteria and fungi, but also bigger creatures such as earthworms, insects, springtails, and lots of mites and even mammals (moles, voles, gophers) live in the soil. It is estimated that about 25% of all species living on earth are found in the soil; most of them are not even scientifically classified. Soil life is also important for the sustainable health of crops and that brings us back to the core topic of this book. In our own work, we found a link between earthworms in the soil and aphids that suck on the plant aboveground mainly because earthworms provide nutrients to plants (Grabmaier et al. 2014) and that earthworms can even influence slug herbivory by helping plants to produce protective chemicals against slugs (Zaller et al. 2013). Moreover, earthworms can reduce mycotoxins, natural toxins produced by fungi on cereals (Wolfarth et al. 2016) and decompose soil-borne plant fungal pathogens such as white mold (*Sclerotinia sclerotiorum*) which can affect more than 400 crop species worldwide (Euteneuer et al. 2019). Thus, any detrimental effects of pesticides to soil life might have ramifications on crop health as well.

So we see that soil is much more than a factory for plant production, but an extremely biodiverse entity, the place of endless reactions that control a host of services of use to humanity and to the natural environment (Brussaard et al. 2007). Cautious estimation on the financial worth of soil life services to us humans is an unimaginable 1300 billion € per year (van der Putten et al. 2004).

Although it is not as appreciated, our soils are just as important to the human society than air and water. At least a former US President, Franklin D. Roosevelt, seemed to have acknowledged this when stating that "A nation that destroys its soils destroys itself."

The importance of earthworms has already been mentioned; they will continue to play a role later, as we have tested the effects of pesticides on earthworms in our own work. Between five and ten million earthworms live in the healthy soil of one hectare of grassland in Central Europe, which, with an average weight of one gram per worm, corresponds to around 5000–10,000 kg of earthworm biomass per hectare. There are about 4000 different species worldwide, the more abundant earthworm are, the better the soil fertility is (van Groenigen et al. 2014). When the soil is too intensively and improperly cultivated, earthworms can disappear almost completely. Even in climates with freezing winters, earthworms can be active all year long, even under a snow cover as revealed in a research project in a meadow where underground activity of earthworms and growth of plant roots were observed using endo-scopes (Arnone and Zaller 2014). Earthworms can dig up to 150 tunnels per square meter or 900 m of channels per cubic meter (Kretzschmar and Aries 1990), thereby improving water uptake, water retention, infiltration, and soil drainage. Surface runoff and erosion after heavy rainfalls can thus be reduced. The earthworms incorporate several tons of dead plant material per hectare into the soil each year. Organic and mineral parts are well mixed in the excre-ments of worms, and the nutrients are readily available and enriched for plants. The term excrement is perhaps a bit disrespectful, since earthworm droppings do not stink at all; on the contrary, they have a pleasant, earthy smell. An average earthworm population in a meadow can produce up to 100 tons of these so-called castings per hectare and year.

Nowadays, earthworms enjoy a very good reputation among farmers and hobby gardeners. But this was not always the case. Until the middle of the nineteenth century, earthworms were generally regarded as pest. Charles Darwin, the famous British natural scientist, made a major contribution to improving the image of the earthworm. He has not only done groundbreak-ing work on evolutionary theory and the origin of species, but in his last years he has also devoted himself to the earthworm (Darwin 1881). In his very last book entitled *The Formation of Vegetable Mold through the Action of Worms with Observations on Their Habits*, he published several decades of detailed observations and measurements on earthworms and the natural sciences. The work was considered a "bestseller" at the time, with 3500 copies sold imme-diately and 8500 in less than 3 years which, at the time, even rivaled the sale of his most well-known book *On the Origin of Species* (Feller et al. 2003). The book covers the importance of earthworm activity on a variety of topics: soil formation and weathering processes, soil horizon differentiation and the for-mation of the topsoil, the role of earthworm burrowings and castings in soil fertility and plant growth, the burial of organic materials and soil enrichment with mineral elements, the cycle of erosion, and the protection of archeologi-cal remains through earthworm burial. Finally, Darwin also performed a series of original experiments to determine if earthworms possessed, or not, a cer-tain "intelligence." This part of the book was, among others, one of the main reasons for its success. Despite Darwin's clear demonstrations of the impor-tance of biological activities of earthworms in the maintenance of soil fertility, his book on worms has been mostly neglected by agronomists and soil scien-tists, primarily due to the predominant soil fertility and management para-digms of the nineteenth and twentieth centuries. Finally, Darwin comes to an eminent conclusion in his book: "One can probably doubt whether there are many other animals which have played such an important role in the history of the earth as these low-organized creatures."

In addition to earthworms, an almost infinite number of other organisms live in the soil: nematodes, springtails, mites, spiders, beetles, and above all countless microorganisms. These soil organisms are just as threatened with extinction as the aboveground animals and plants, but this aspect has received little attention in the public debate on the biodiversity crisis.

References

agrarheute (2012) Frankreich: Parkinson als Berufskrankheit anerkannt. In: agrarheute. http://www.agrarheute.com/news/frankreich-parkinson-berufskrankheit-anerkannt

Almeida RM, Han BA, Reisinger AJ, Kagemann C, Rosi EJ (2018) High mortality in aquatic predators of mosquito larvae caused by exposure to insect repellent. Biol Lett 14:20180526

Amnesty International (2009) Dodging responsibility. Corporations, Governments and the Bhopal Disaster. https://www.amnesty.ch/de/themen/wirtschaft-und-menschenrechte/fallbeispiele/bhopal/30-jahre-bhopal/0905_digest_bhopal_english.pdf. Accessed 04 Oct 2019

Analytik-News (2006) Gefälschte Pestizide in der Landwirtschaft bedrohen Gesundheit. https://www.analytik-news.de/Presse/2006/498.html. Accessed 04 Oct 2019

APS (2019) What are fungicides? In: American Phytopathological Society (APS) (ed) https://www.apsnet.org/edcenter/disimpactmngmnt/topc/Pages/Fungicides.aspx. Accessed 07 Dec 2019

APVMA (2010) Atrazine toxicity: analysis of potential modes of action. Australian Government. https://apvma.gov.au/sites/default/files/publication/14366-atrazine-toxicity.pdf. Accessed 01 Dec 2019

Arancibia F, Motta RC, Clausing P (2019) The neglected burden of agricultural intensification: a contribution to the debate on land-use change. J Land Use Sci 15:235–251

Arnone JA, Zaller JG (2014) Earthworm effects on native grassland root system dynamics under natural and increased rainfall. Front Plant Sci 5:152. https://doi.org/10.3389/fpls.2014.00152

ASA (2019) 2019 Soystats. A reference guide to soybean facts and figures. http://soystats.com/wp-content/uploads/2019-SoyStats-Web.pdf. Accessed 05 Oct 2019

Asendorpf D (2003) Gift für die Armen. Die Zeit. http://www.zeit.de/2003/35/U-Altpestizide/komplettansicht?print. Accessed 05 Oct 2019

Atreya K (2005) Health costs of pesticide use in a vegetable growing area, central mid-hills, Nepal. Himal J Sci 3:81–84

Atreya K (2007) Pesticide use knowledge and practices: a gender differences in Nepal. Environ Res 104:305–311

Balser M, Mischke T, Ritzer U, Thurn V (2019) Landwirtschaft. Fragwürdige Verflechtungen. Süddeutsche Zeitung, 28 April 2019

BASF (2016) Lebensmittelsicherheit und Rückstände in Lebensmitteln. Unsere Position. https://agriculture.basf.com/de/Pflanzenschutz/Neuigkeiten-und-Veranstaltungen/Positionen/Lebensmittelsicherheit.html. Accessed 05 Oct 2019

Beck U (1986) Risikogesellschaft. Auf dem Weg in eine andere Moderne. Suhrkamp, Frankfurt am Main

Beiles N (2000) The people v Monsanto. The Guardian. https://www.theguardian.com/theguardian/2000/jun/05/features11.g2. Accessed 31 Jan 2020

Benbrook CM (2016) Trends in glyphosate herbicide use in the United States and globally. Environ Sci Eur 28:3

Benbrook CM (2019) How did the US EPA and IARC reach diametrically opposed conclusions on the genotoxicity of glyphosate-based herbicides? Environ Sci Eur 31:2

Beste A (2017) Vergiftet. Pestizide in Boden und Wasser - das Beispiel Glyphosat. In: AgrarBündnis eV (ed) Der kritische Agrarbericht 2017. Schwerpunkt Wasser. ABL Bauernblatt Verlags-GmbH, Konstanz, pp 204–208

BLW (2017) Aktionsplan zur Risikoreduktion und nachhaltigen Anwendung von Pflanzenschutzmitteln. https://www.blw.admin.ch/blw/de/home/nachhaltige-produktion/pflanzenschutz/aktionsplan.html. Accessed 18 Jan 2020

BMNT (2018) Grüner Bericht 2018. Bericht über die Situation der österreichischen Land- und Forstwirtschaft. In. Bundesministerium für Nachhaltigkeit und Tourismus, Vienna, Austria, p 268

BMNT (2019) Pestizide im Grundwasser. Was sind Pestizide und wie bzw in welchen Konzentrationen gelangen diese in das Grundwasser? https://www.bmnt.gv.at/wasser/wasserqualitaet/grundwasser/pestizidegrundwasser.html, 17 April 2019, Accessed 03 Oct 2019

Boddenberg S (2018) Haselnüsse für Nutella vergiften Chile. DW Made for Minds, https://www.dw.com/de/haselnüsse-für-nutella-vergiften-chile/a-46488884. Accessed 28 Nov 2018

Bøhn T, Cuhra M (2014) How "extreme levels" of roundup in food became the industry norm. Independent Science News, https://www.independentscience-news.org/news/how-extreme-levels-of-roundup-in-food-became-the-industry-norm/. Accessed 02 February 2017

Bøhn T, Cuhra M, Traavik T, Sanden M, Fagan J, Primicerio R (2014) Compositional differences in soybeans on the market: Glyphosate accumulates in Roundup Ready GM soybeans. Food Chem 153:207–215

Brown DL (2016) Dead end path: How industrial agriculture has stolen our future. Moab Book Works, Moab, UT

Bruckner M et al (2019) Quantifying the global cropland footprint of the European Union's non-food bioeconomy. Environ Res Lett 14:045011

Brühl CA, Zaller JG (2019) Biodiversity decline as a consequence of an inadequate environmental risk assessment of pesticides. Front Environ Sci 7:75

Brussaard L, de Ruiter PC, Brown GG (2007) Soil biodiversity for agricultural sustainability. Agric Ecosyst Environ 121:233–244

BUA (1999) DDT und Derivate – Modellstoffe zur Beschreibung endokriner Wirkungen mit Relevanz für die Reproduktion. S. Hirzel Verlag, Stuttgart

BUND (2013) Determination of glyphosate residues in human urine samples from 18 European countries. BUND. http://db.zs-intern.de/uploads/1371118293-130612_gentechnik_bund_glyphosat_urin_analyse.pdf. Accessed 30 Sept 2019

Bundestag D (2009) Kleine Anfrage: Einsatz von Pestiziden auf Strecken der Deutschen Bahn. In: 16/13918 D (ed). Deutscher Bundestag. http://dipbt.bundestag.de/dip21/btd/16/139/1613918.pdf. Accessed 28 Sept 2019

Burton L (2019) Australian farmers warned the misuse of pesticides will see overseas markets turn away. ABC News, https://www.abc.net.au/news/rural/2019-07-26/chemicals-on-crops-jeopardise-export-markets/11270722. Accessed 24 Nov 2019

Burtscher-Schaden H (2017) Die Akte Glyphosat. Wie Konzerne die Schwächen des Systems nutzen und damit unsere Gesundheit gefährden. Kremayr & Scheriau, Wien

Butte W (1999) Occurrence of biocides in the indoor environment. In: Salthammer T (ed) Organic indoor air pollutants: Occurrence-measurement-evaluation. Verlag Wiley-VCH, Weinheim

BVL (2017a) Jahresbericht Pflanzenschutz– Kontrollprogramm 2017. BVL-Report 131 Berichte zu Pflanzenschutzmitteln

BVL (2017b) Widerruf der Zulassung der Pflanzenschutzmittel Pirimor Granulat, PIRIMAX und Calypso hinsichtlich bestimmter Anwendungen an Kohlrabi. Bundesamt für Verbraucherschutz und Lebensmittelsicherheit. http://www.bvl.bund.de/DE/04_Pflanzenschutzmittel/06_Fachmeldungen/2017/2017_05_30_Fa_Widerruf_Pirimicarb_Thiacloprid_Kohlrabi.html?nn=1400938. Accessed 05 Oct 2019

BVL (2018) Beistoffe in zugelassenen Pflanzenschutzmitteln. Bundesamt für Verbraucherschutz und Lebensmittelsicherheit. https://www.bvl.bund.de/SharedDocs/Downloads/04_Pflanzenschutzmittel/zul_info_liste_beistoffe.html. Accessed 29 Sept 2019

BVL (2019) Verzeichnis zugelassener Pflanzenschutzmittel. Bundesamt für Verbraucherschutz und Lebensmittelsicherheit. https://apps2.bvl.bund.de/psm/jsp/index.jsp. Accessed 30 Sept 2019

Carrington D (2019) EU bans UK's most-used pesticide over health and environment fears. The Guardian, 29 Mar 2019. https://www.theguardian.com/environment/2019/mar/29/eu-bans-widely-used-pesticide-over-safety-concerns?CMP=Share_iOSApp_Other. Accessed 01 Feb 2020

Carson R (1962) Silent spring. Houghton Mifflin, New York, NY

CAS (2015) CAS assigns the 100 millionth CAS registry number® to a substance designed to treat acute myeloid leukemia. American Chemical Society. https://

www.prnewswire.com/news-releases/cas-assigns-the-100-millionth-cas-registry-number-to-a-substance-designed-to-treat-acute-myeloid-leukemia-300106332. html. Accessed 05 Oct 2019

Cassidy A (2016) Das Asylheim auf dem Sondermüll. Tagesanzeiger, 28 Jan 2016. https://www.tagesanzeiger.ch/schweiz/standard/das-asylheim-auf-dem-sonderm-uell/story/26897618. Accessed 05 Oct 2019

CEO (2016) Scientific scrutiny on EFSA's work, at last? Corporate Europe Observatory. https://corporateeurope.org/efsa/2016/12/scientific-scrutiny-efsa-work-last. Accessed 05 Oct 2019

Chatterjee R (2013) Global love of bananas may be hurting Costa Rica's crocodiles. NPR All Things Considered. https://www.npr.org/sections/the-salt/2013/09/24/225793450/global-love-of-bananas-may-be-hurting-costa-ricas-crocodiles?t=1570179909309. Accessed 04 Oct 2019

Chmura T (2017) "Weiter wie bisher" wird nicht funktionieren. B5 Aus Landwirtschaft und Umwelt. Bayerischer Rundfunk

Chow L (2016) Results of glyphosate urine test: 'it's not good news'. Glyphosate in water and food. Global Research. https://www.globalresearch.ca/results-of-glyphosate-urine-test-its-not-good-news-glyphosate-in-water-and-food/5525529. Accessed 30 Sept 2019

Clausing P, Robinson C, Burtscher-Schaden H (2018) Pesticides and public health: an analysis of the regulatory approach to assessing the carcinogenicity of glypho-sate in the European Union. J Epidem Comm Health 72:668–672

Cook K (2019) Glyphosate in beer and wine. CALPIRG Education Fund. https://calpirg.org/feature/cap/glyphosate-pesticide-beer-and-wine. Accessed 30 September 2019

Corsolini S, Borghesi N, Ademollo N, Focardi S (2011) Chlorinated biphenyls and pesticides in migrating and resident seabirds from East and West Antarctica. Environ Int 37:1329–1335

Cribb J (2014) Poisoned planet. How constant exposure to man-made chemicals is putting your life at risk. Allen & Unwin, Sydney

CSA (2016) Agricultural sample survey 2015/2016. In: Volume VII. Report on crop and livestock product utilization. Private Peasant Holdings MS (ed). Central Statistical Agency, Addis Ababa, Ethiopia

Cuhra M (2015) Glyphosate nontoxicity: the genesis of a scientific fact. J Biol Phys Chem 15:89–96

Darwin C (1881) The formation of vegetable mould through the action of worms. John Murray, London

Daunderer M (2005) Gifte im Alltag. Wo sie vorkommen. Wie sie wirken. Wie man sich dagegen schützt., 2. Auflage. C.H. Beck, München

Defarge N et al (2016) Co-formulants in glyphosate-based herbicides disrupt aroma-tase activity in human cells below toxic levels. Int J Environ Res Pub Health 13:264. https://doi.org/10.3390/ijerph13030264

Dickeduisberg M, Steinmann H-H, Theuvsen L (2012) Erhebungen zum Einsatz von Glyphosat im deutschen Ackerbau. Nr 434 (2012): Tagungsband 25 Deutsche Arbeitsbesprechung über Fragen der Unkrautbiologie und -bekämpfung/Neue Entwicklungen in der chemischen Unkrautkontrolle/Recent Developments in Chemical Weed Control

DMR (2011) 'Bed bug pesticide poisoning' killed Californian woman and six other tourists in Thailand. Daily Mail. https://www.dailymail.co.uk/news/article-1385518/Bed-bug-pesticide-poisoning-caused-death-California-woman-tourists-Thailand.html. Accessed 30 Nov 2019

DNR (2018) REACH-review. Ergebnisse und Bewertung. In: EU-Koordination DN (ed) https://www.dnr.de/fileadmin/Publikationen/Steckbriefe_Factsheets/18_07_26_EUK_Steckbrief_REACH_Review_Aktualisiert.pdf.   Accessed 05 Oct 2019

Donley N (2019) The USA lags behind other agricultural nations in banning harmful pesticides. Environ Health 18:44

Dorne J (2010) Metabolism, variability and risk assessment. Toxicology 268:156–164

dosReis MR, Fernandes FL, Lopes EA, Gorri JER, Alves FM (2015) Pesticide residues in coffee agroecosystems. In: Preedy VR (ed) Coffee in health and disease prevention. Academic, Amsterdam, NL, pp 235–244

Douglas MR, Tooker JF (2016) Meta-analysis reveals that seed-applied neonicotinoids and pyrethroids have similar negative effects on abundance of arthropod natural enemies. PeerJ 4:e2776

Drobny HG, Schulte M, Strek HJ (2012) 25 Jahre Sulfonylharnstoff-Herbizide – ein paar Gramm veränderten die Welt der chemischen Unkrautbekämpfung. Julius-Kühn-Archiv, 25th German Conference on Weed Biology and Weed Control, March 13–15, 2012, Braunschweig, Germany 434:21–33

Duke SO, Powles SB (2008) Glyphosate: a once-in-a-century herbicide. Pest Manag Sci 64:319–325

DW (2019a) Austrian parliament votes to ban glyphosate weedkiller. https://www.dw.com/en/austrian-parliament-votes-to-ban-glyphosate-weedkiller/a-49450418. Accessed 05 Oct 2019

DW (2019b) Germany set to ban glyphosate from end of 2023. Deutsche Welle, 04 Sept 2019. https://www.dw.com/en/germany-set-to-ban-glyphosate-from-end-of-2023/a-50282891. Accessed 05 Oct 2019

EC (2003) Review report for the active substance atrazine. https://ec.europa.eu/food/plant/pesticides/eu-pesticides-database/public/?event=activesubstance.detail&language=EN&selectedID=972. Accessed 01 Dec 2019

EC (2008) Factsheet: new rules on pesticide residues in food. In: European Commission DgfHaC (ed). https://ec.europa.eu/food/sites/food/files/plant/docs/pesticides_mrl_legis_factsheet_en.pdf. Accessed 05 Oct 2019

EC (2019a) COMMISSION DELEGATED REGULATION (EU) .../... of 13.3.2019 supplementing Directive (EU) 2018/2001 as regards the determination of high indirect land-use change-risk feedstock for which a significant expansion of the production area into land with high carbon stock is observed and the certification of low indirect land-use change-risk biofuels, bioliquids and biomass fuels.   https://www.transportenvironment.org/sites/te/files/PART-2019-142068 V1.pdf. Accessed 24 Nov 2019

EC (2019b) EU legislation on MRLs. In: Commission E (ed) https://ec.europa.eu/ food/plant/pesticides/max_residue_levels/eu_rules_en. Accessed 05 Oct 2019

EC   (2019c)   EU   pesticides   database.   https://ec.europa.eu/food/plant/ pesticides/eu-pesticides-database/public/?event=homepage&language =EN. Accessed 23 Nov 2019

EC (2019d) EU pesticides database. MRL evolution. https://ec.europa.eu/food/ plant/pesticides/eu-pesticides-database/public/?event=pesticide.residue. displayMRL&language=EN. Accessed 21 Dec 2019

EC (2019e) Major accidents hazards. The Seveso Directive – Technological Disaster Risk Reduction. https://ec.europa.eu/environment/seveso/. Accessed 20 Dec 2019

EC (2019f) Some facts about neonicotinoids. In: Commission E (ed) https://ec. europa.eu/food/plant/pesticides/approval_active_substances/approval_renewal/ neonicotinoids_en. Accessed 26 Nov 2019

ECCC (2016) Canadian environmental sustainability indicators: household use of chemical pesticides and fertilizers. Environment and Climate Change Canada. https://www.ec.gc.ca/indicateurs-indicators/default.asp?lang= en&n=258BC62B-1. Accessed 29 Sept 2019

ECCHR (2016) BAYER: double standards in the sale of pesticides. https://www. ecchr.eu/en/case/bayer-double-standards-in-the-sale-of-pesticides.   Accessed   28 September 2019

ECHA (2019) Understanding REACH – registration, evaluation, authorisation and restriction of chemicals. In: Agency EC (ed) https://echa.europa.eu/regulations/ reach/understanding-reach. Accessed 05 Oct 2019

EFSA (2009) Conclusion regarding the peer review of the pesticide risk assessment of the active substance captan. EFSA J 7:296r

EFSA (2015) Conclusion on the peer review of the pesticide risk assessment of the active substance glyphosate. EFSA J 13(3):4302

EFSA (2019a) The 2017 European Union report on pesticide residues in food. EFSA J 17:e05743

EFSA (2019b) Chlorpyrifos: assessment identifies human health effects. https:// www.efsa.europa.eu/en/press/news/chlorpyrifos-assessment-identifies-human- health-effects?utm_source=EFSA+Newsletters&utm_campaign=0b782cf0bc- EMAIL_CAMPAIGN_2019_07_04_09_03_COPY_01&utm_ medium=email&utm_term=0_7ea646dd1d-0b782cf0bc-63968509.   Accessed 18 Oct 2019

EFSA et al (2018a) Peer review of the pesticide risk assessment of the active substance chlorothalonil. EFSA J 16:e05126

EFSA et al (2018b) Peer review of the pesticide risk assessment of the active substance copper compounds copper(I), copper(II) variants namely copper hydroxide, copper oxychloride, tribasic copper sulfate, copper(I) oxide, Bordeaux mixture. EFSA J 16:e05152

Elbert A, Nauen R, Leicht W (1998) Imidacloprid, a novel chloronicotinyl insecticide: biological activity and agricultural importance. In: Ishaaya I, Degheele D (eds) Insecticides with novel modes of action: mechanism and application. Springer, Berlin, pp 50–74

EP (2005) Regulation (EC) No 396/2005 of the European Parliament and of the Council of 23 February 2005 on maximum residue levels of pesticides in or on food and feed of plant and animal origin and amending Council Directive 91/414/EEC. Off J Eur Union L70:1–16

EP (2009) Regulation (EC) No 1107/2009 of the European Parliament and of the Council of 21 October 2009 concerning the placing of plant protection products on the market and repealing Council Directives 79/117/EEC and 91/414/EEC. Off J Eur Union https://eur-lex.europa.eu/legal-content/EN/TXT/?uri=CELEX:32009R1107. Accessed 21 Dec 2019

EP (2015) Regulation(EC) No 1272/2008 of the European Parliament and of the Council of 16 December 2008 on classification, labelling and packaging of substances and mixtures, amending and repealing Directives 67/548/EEC and 1999/45/EC, and amending Regulation (EC) No 1907/2006. https://eur-lex.europa.eu/legal-content/EN/TXT/PDF/?uri=CELEX:02008R1272-20150601&from=EN. Accessed 21 Dec 2019

EPA (1999) R.E.D. facts Captan. https://archive.epa.gov/pesticides/reregistration/web/pdf/0120fact.pdf. Accessed 15 Dec 2019

EPA (2002) Interim reregistration eligibility decision for chlorpyrifos. In: Agency USEP (ed) Prevention, pesticides EPA 738-R-01-007 and toxic substances February 2002, (7508C). https://nepis.epa.gov/Exe/ZyPDF.cgi/200008BM.PDF?Dockey=200008BM.PDF. Accessed 06 Dec 2019

EPA (2009) Disulfoton and methamidophos; Product cancellation order. https://www3.epa.gov/pesticides/chem_search/reg_actions/reg_review/frn_PC-032501_23-Sep-09.pdf. Accessed 17 September 2019

EPA (2011) Integrated Risk Information System (IRIS) Glossary. https://iaspub.epa.gov/sor_internet/registry/termreg/searchandretrieve/glossariesandkeywordlists/search.do?details=&vocabName=IRIS%20Glossary. Accessed 21 Dec 2019

EPA (2017) Pesticides industry sales and usage. 2008-2012 market estimates. In: Agency USEP (ed) https://www.epa.gov/pesticides/pesticides-industry-sales-and-usage-2008-2012-market-estimates, https://www.epa.gov/sites/production/files/2017-01/documents/pesticides-industry-sales-usage-2016_0.pdf. Accessed 29 Sept 2019

EPA (2019) Pesticide chemical search. Conventional, antimicrobial and biopesticides active ingredients. https://iaspub.epa.gov/apex/pesticides/f?p=chemicalsearch:1. Accessed 21 Sept 2019

EPA (2020) Glyphosate. Interim registration review decision case number 0178. https://www.epa.gov/sites/production/files/2020-01/documents/glyphosate-interim-reg-review-decision-case-num-0178.pdf. Accessed 31 Jan 2020

Eskenazi B et al (2004) Relationship of serum TCDD concentrations and age at exposure of female residents of Seveso, Italy. Environ Health Perspect 112:22–27

European Commission (2019) EU pesticides database. https://ec.europa.eu/food/plant/pesticides/eu-pesticides-database/public/?event=activesubstanceselection&language=EN. Accessed 21 Sept 2019

Eurostat (2018) Agriculture, forestry and fishery statistical book. European Union. https://ec.europa.eu/eurostat/statistics-explained/index.php?title=Agriculture,_forestry_and_fishery_statistics. Accessed 25 September 2019

Eurostat (2019) Pesticide sales. http://appsso.eurostat.ec.europa.eu/nui/show.do?dataset=aei_fm_salpest09&lang=en. Accessed 30 Nov 2019

Euteneuer P, Wagentristl H, Steinkellner S, Scheibreithner C, Zaller JG (2019) Earthworms affect decomposition of soil-borne plant pathogen Sclerotinia sclerotiorum in a cover crop field experiment. Appl Soil Ecol 138:88–93

Fan L (2017) China founds pesticide office to combat pollution, overuse. Sixth Tone. http://www.sixthtone.com/news/1000987/china-founds-pesticide-office-to-combat-pollution%2C-overuse. Accessed 14 Dec 2019

FAO (2000) The energy and agriculture nexus. Environment and natural resources working paper no. 4. Department NRMaE. http://www.fao.org/docrep/003/X8054E/x8054e05.htm. Accessed 17 October 2019

FAO (2013) International code of conduct on the distribution and use of pesticides. Guidelines on data requirements for the registration of pesticides. http://www.fao.org/3/a-bc870e.pdf. Accessed 03 Oct 2019, 36 p

FAO WHO (2014) The international code of conduct on pesticide management. Organization UNFaAOaWH. http://www.fao.org/fileadmin/templates/agphome/documents/Pests_Pesticides/Code/Code_ENG_2017updated.pdf.   Accessed 05 Oct 2019

FAO WHO (2019) Pesticide residues in food 2018 – report 2018 – Joint FAO/WHO meeting on pesticide residues. FAO Plant Prod Protect Paper 234:668 p

FAOSTAT (2019a) Pesticides use – Argentina. http://www.fao.org/faostat/en/#data/RP. Accessed 14 Dec 2019

FAOSTAT (2019b) Pesticides use – Australia. http://www.fao.org/faostat/en/#data/RP. Accessed 14 Dec 2019

FAOSTAT (2019c) Pesticides use – Brazil. http://www.fao.org/faostat/en/#search/pesticides. Accessed 20 Dec 2019

FAOSTAT (2019d) Pesticides use – Chile. http://www.fao.org/faostat/en/#search/pesticides. Accessed 14 Dec 2019

FAOSTAT (2019e) Pesticides use – Ethiopia. http://www.fao.org/faostat/en/#search/pesticides. Accessed 14 Dec 2019

FAOSTAT (2019f) Pesticides use – India. http://www.fao.org/faostat/en/#data/RP. Accessed 14 Dec 2019

FAOSTAT (2019g) Pesticides use – Nepal. http://www.fao.org/faostat/en/#data/ RP. Accessed 14 Dec 2019

FAOSTAT (2019h) Pesticides use – USA. http://www.fao.org/faostat/en/#search/ pesticides. Accessed 15 Dec 2019

Feller C, Brown GG, Blanchart E, Deleporte P, Chernyanskii SS (2003) Charles Darwin, earthworms and the natural sciences: various lessons from past to future. Agric Ecosyst Environ 99:29–49

FERA (2020) Pesticide usage statistics UK. https://secure.fera.defra.gov.uk/pusstats/ index.cfm. Accessed 11 Jan 2020

Fletcher W (1974) The Pest War. Basil Blackwell, Oxford

Fletcher M, Barnett L (2003) Bee pesticide poisoning incidents in the United Kingdom. Bull Insectol 56:141–145

Focus (2017) EU-Studie. Gefälschte Pestizide kosten Wirtschaft 1,3 Milliarden Euro. Focus Money Online, 08 February 2017

FoE (2019) Australian pesticides map. https://pesticides.australianmap.net. Accessed 24 Nov 2019

Gehrmann A-K (2016) Insektenbekämpfung mit den Waffen des Weltkriegs. Frankfurter Allgemeine. https://www.faz.net/aktuell/feuilleton/medien/makabre-werbekampagne-gegen-insektenbekaempfung-14227602.html;   Accessed 05 Oct 2019

Geisz HN, Dickhut RM, Cochran MA, Fraser WR, Ducklow HW (2008) Melting glaciers: a probable source of DDT to the Antarctic marine ecosystem. Environ Sci Technol 42:3958–3962

Gibbons D, Morrissey C, Mineau P (2015) A review of the direct and indirect effects of neonicotinoids and fipronil on vertebrate wildlife. Environ Sci Pollut Res 22:103–118

Gillam C (2017) White wash. The story of a week killer, cancer and the corruption of science. Island Press, Washington, DC

Global2000 (2016a) Osterhasen-check 2016. http://www.salzburg.com/download/2016-03/osterhasen_check_2016%2Bpz_sc_web.pdf. Accessed 30 Sept 2019

Global2000 (2016b) Schokoladen-check 2016. https://www.global2000.at/sites/ global/files/Schokolade_Check_2016.pdf. Accessed 30 Sept 2019

Goodson WI, Lowe L, Carpenter D, Gilbertson M, Manaf AA, Lopez de Cerain Salsamendi A (2015) Assessing the carcinogenic potential of lowdose exposures to chemical mixtures in the environment: the challenge ahead. Carcinogenesis 36:S254–S296

Goulson D (2013) An overview of the environmental risks posed by neonicotinoid insecticides. J Appl Ecol 50:977–987

Gowen A (2018) An Indian state banned pesticides. Tourism and wildlife flourished. Will others follow? Washington Post, Washington, DC

Grabmaier A, Heigl F, Eisenhauer N, van der Heijden MGA, Zaller JG (2014) Stable isotope labelling of earthworms can help deciphering belowground–aboveground interactions involving earthworms, mycorrhizal fungi, plants and aphids. Pedobiol 57:197–203

Grant PBC, Woudneh MB, Ross PS (2013) Pesticides in blood from spectacled caiman (Caiman crocodilus) downstream of banana plantations in Costa Rica. Environ Toxicol Chem 32:2576–2583

Gräslund S, Bengtsson BE (2001) Chemicals and biological products used in southeast Asian shrimp farming, and their potential impact on the environment – a review. Sci Total Environ 280:93–131

Graupner H (2019) Defending glyphosate: a 'Roundup' of German agribusiness sentiments. Deutsche Welle, 27 May 2019. https://www.dw.com/en/defending-glyphosate-a-roundup-of-german-agribusiness-sentiments/a-48841453. Accessed 17 October 2019

Gruber L (2017) Zankapfel Glyhosat. In: BR5 Aus Landwirtschaft und Umwelt. Bayerischer Rundfunk. https://www.podcast.de/episode/364161749/Zankapfel+Glyphosat. Accessed 05 Oct 2019

Gunatilake S, Seneff S, Orlando L (2019) Glyphosate's synergistic toxicity in combination with other factors as a cause of chronic kidney disease of unknown origin. Int J Environ Res 16:2734

Gunnarsson LG, Bodin L (2017) Parkinson's disease and occupational exposures: a systematic literature review and meta-analyses. Scand J Work Environ Health 43:197–209

Gurr GM et al (2016) Multi-country evidence that crop diversification promotes ecological intensification of agriculture. Nat Plants 2:16014

Haas G (2010) Wasserschutz im Ökologischen Landbau. Leitfaden für Land- und Wasserwirtschaft, vol. BÖL-Bericht-ID 16897. http://orgprints.org/16897/1/16897-06OE175-agraringenieurbuero-haas-2010-wasserschutz.pdf

Hallman T (2018) Experts warn of post-storm pesticide danger. https://newsstand.clemson.edu/mediarelations/experts-warn-of-post-storm-pesticide-danger/. Accessed 20 Dec 2019

Hallmann J, Quadt-Hallmann A, Tiedemann AV (2009) Phytomedizin. Grundwissen Bachelor, 2. Auflage edn. Verlag Eugen Ulmer, Stuttgart

Heids Mist (2019) Hurra! Blackenbekämpfung am Steilhang mit Drohne. Hurra? https://heidismist.wordpress.com, vol 2019, https://heidismist.wordpress.com/2019/09/18/hurra-blackenbekaempfung-am-steilhang-mit-drohne-hurra/. Accessed 24 November 2019

Heller M (2020) EPA proposes no new restrictions for bee-killing chemicals. E&E News. https://www.eenews.net/eenewspm/stories/1062221105. Accessed 01 Feb 2020

Heong KL, Escalada MM (1997) A comparative analysis of pest management practices of rice farmers in Asia. In: Heong K, Escalada M (eds) Pest management of rice farmers in Asia. International Rice Research Institute, Manila, Philippines, pp 227–245

Heong KL, Escalada MM, Sengsoulivong V, Schiller J (2002) Insect management beliefs and practices of rice farmers in Laos. Agric Ecosyst Environ 92:137–145

Heong K, Wong L, Delos Reyes J (2015a) addressing planthopper threats to Asian rice farming and food security: fixing insecticide misuse. In: Heong KL, Cheng JA, Escalada MM (eds) Rice planthoppers: ecology, management, socio economics and policy. Zhejiang University Press; Springer Science+Business Media, Hangzhou; Dordrecht, pp 69–80

Heong KL, Escalada MM, Chien HV, Reyes JHD (2015b) Are there productivity gains from insecticide applications in rice production? In: Heong KL, Cheng JA, Escalada MM (eds) Rice planthoppers: ecology, management, socio economics and policy. Zhejiang University Press; Springer Science+Business Media, Hangzhou; Dordrecht, pp 181–192

Hertz-Picciotto I et al (2018) Organophosphate exposures during pregnancy and child neurodevelopment: recommendations for essential policy reforms. PLoS Med 15:e1002671. https://doi.org/10.1371/journal.pmed.1002671

Hoeffner MK (2019) There are 2,000 untested chemicals in packaged foods – and it's legal. Troughout. https://truthout.org/articles/there-are-2000-chemicals-in-packaged-foods-and-its-legal/. Accessed 21 Nov 2019

Hollert H, Backhaus T (2019) Some food for thought: a short comment on Charles Benbrook's paper "How did the US EPA and IARC reach diametrically opposed conclusions on the genotoxicity of glyphosate-based herbicides?" and its implications. Env Sci Eur 31:3

IAASTD (2009) Agriculture at a crossroads. In. IAASTD – International Assessment of Agricultural Knowledge, Science and Technology for Development, Washington, DC

INKOTA (2019) Massensterben in Zentralamerika durch Pestizideinsatz. In INKOTA Berlin, 20 Sept 2019. https://www.inkota.de/aktuell/news/vom/27/aug/2019/massensterben-in-zentralamerika-durch-pestizideinsatz/. Accessed 05 Oct 2019

Isenring R (2010) Pesticides and the loss of biodiversity. How intensive pesticide use affects wildlife populations and species diversity. Pesticide Action Network. https://www.pan-europe.info/old/Resources/Briefings/Pesticides_and_the_loss_of_biodiversity.pdf. Accessed 01 Nov 2019

Jeffery S et al. (eds) (2010) European atlas of soil biodiversity. European Commission, Publications Office of the European Union, Luxembourg

Jeschke P, Nauen R, Schindler M, Elbert A (2011) Overview of the status and global strategy for neonicotinoids. J Agric Food Chem 59:2897–2908

Jezussek M (2012) Rückstandshöchstgehalte für Pflanzenschutzmittel-Rückstände. Lebensmittelsicherheit BLfGu. https://www.vis.bayern.de/global/script/drucken.php?www.vis.bayern.de/ernaehrung/lebensmittelsicherheit/unerwuenschte_stoffe/hoechstmengen.htm? Accessed 05 Oct 2019

JKI (2019) Statistische Erhebungen zur Anwendung von Pflanzenschutzmitteln in der Praxis. In. Julius Kühn Institut, Bundesforschungsinstitut für Kulturpflanzen. https://papa.julius-kuehn.de. Accessed 30 Nov 2019

Kainrath V (2019) Erdäpfelbauern wollen mehr Spritzmittel einsetzen. Der Standard, https://www.derstandard.at/story/2000101987312/gift-gegen-schwarze-loecher-erdaepfelbauern-wollen-mehr-spritzmittel-einsetzen. Accessed 18 Jan 2020

Kästel A, Allgeier S, Brühl CA (2017) Decreasing *Bacillus thuringiensis israelensis* sensitivity of *Chironomus riparius* larvae with age indicates potential environmental risk for mosquito control. Sci Rep 7:13565

Klátyik S, Bohus P, Darvas B, Székács A (2017) Authorization and toxicity of veterinary drugs and plant protection products: residues of the active ingredients in food and feed and toxicity problems related to adjuvants. Front Vet Sci 4:146

Klingenschmitt E (2016) Gifte belasten auch Bio-Äcker. SWR2 Impuls. http://www.swr.de/swr2/wissen/impuls-pestizide-protest/-/id=661224/did=17849754/nid=661224/eshk2f/index.html. Accessed 04 Oct 2019

Kortenkamp A (2014) Low dose mixture effects of endocrine disrupters and their implications for regulatory thresholds in chemical risk assessment. Curr Opin Pharmacol 19:105–111

Krausmann F (2001) Land use and industrial modernization: an empirical analysis of human influence on the functioning of ecosystems in Austria 1830-1995. Land Use Policy 18:17–26

Kretzschmar A, Aries F (1990) 3D images of natural and experimental earthworm burrow systems. Rev Ecol Biol Sol 27:407–414

Krol WJ (2000) Removal of trace pesticide residues from produce. https://portal.ct.gov/CAES/Fact-Sheets/Analytical-Chemistry/Removal-of-Trace-Pesticide-Residues-from-Produce. Accessed 22 Dec 2019

Krupke CH, Holland JD, Long EY, Eitzer BD (2017) Planting of neonicotinoid-treated maize poses risks for honey bees and other non-target organisms over a wide area without consistent crop yield benefit. J Appl Ecol 54:1449–1458

Kubsad D, Nilsson EE, King SE, Sadler-Riggleman I, Beck D, Skinner MK (2019) Assessment of glyphosate induced epigenetic transgenerational inheritance of pathologies and sperm epimutations: generational toxicology. Sci Rep 9:6372

Kumar DK (2017) Texas calls in U.S. Air Force to counter post-storm surge in mosquitoes. Reuters Health News, https://www.reuters.com/article/us-storm-harvey-mosquitoes/texas-calls-in-u-s-air-force-to-counter-post-storm-surge-in-mosquitoes-idUSKCN1BN2JR. Accessed 17 September 2019

Laetz CA, Baldwin DH, Collier TK, Hebert V, Stark JD, Scholz1 NL (2009) The synergistic toxicity of pesticide mixtures: implications for risk assessment and the conservation of endangered Pacific Salmon. Environ Health Perspect 117:348-353

Lal R (2004) Carbon emission from farm operations. Environ Int 30:981–990

Lechenet M, Dessaint F, Py G, Makowski D, Munier-Jolain N (2017) Reducing pesticide use while preserving crop productivity and profitability on arable farms. Nat Plants 3:17008

Leu A (2014) The myths of safe pesticides. Acres, Austin, TX

Levin S (2019a) California defies Trump to ban pesticide linked to childhood brain damage. The Guardian. https://www.theguardian.com/us-news/2019/may/08/california-pesticide-ban-chlorpyrifos-agriculture. Accessed 06 Dec 2019

Levin S (2019b) Trump administration won't ban pesticide tied to childhood brain damage. In: The Guardian, 18 July 2019. https://www.theguardian.com/us-news/2019/jul/18/epa-chlorpyrifos-ban-children-brain-damage-trump. Accessed 05 October 2019

Liebrich S (2012) Gift im Getreide. In: Süddeutsche Zeitung, 09 July 2012. https://www.sueddeutsche.de/gesundheit/herbizide-in-der-landwirtschaft-gift-im-getreide-1.1406344. Accessed 30 Sept 2019

Liu P, Guo Y (2019) Current situation of pesticide residues and their impact on exports in China. IOP Conf Ser Earth Environ Sci 227:052027. https://doi.org/10.1088/1755-1315/227/5/052027

Lumetzberger S (2016) Was steckt wirklich in meiner Banane? Kurier.at. https://kurier.at/wissen/ecuador-das-miese-geschaeft-mit-den-bananen/226.644.329. Accessed 06 April 2017

MacBean C (ed) (2013) The pesticide manual: a world compendium, 16th edn. British Crop Protection Council, London

Maggioni DA, Signorini ML, Michlig NA, Repetti MR, Sigrist ME, Beldomenico HR (2017) Comprehensive estimate of the theoretical maximum daily intake of pesticide residues for chronic dietary risk assessment in Argentina. J Environ Sci Health B:1–11

Mandavilli A (2018) The world's worst industrial disaster is still unfolding. The Atlantic July 10, 2018. www.theatlantic.com/science/archive/2018/07/the-worlds-worst-industrial-disaster-is-still-unfolding/560726/. Accessed 04 October 2019

Marquardt H, Schäfer SG (2004) Lehrbuch der Toxikologie, 2. Auflage. Wissenschaftliche Verlagsgesellschaft, Stuttgart

Mart M (2015) Pesticides, a love story. America's enduring embrace of dangerous chemicals. University Press of Kansas, Lawrence, KS

MAVRD (2020) Luxembourg, the first EU country to ban the use of glyphosate. Press release Luxembourg Minister of Agriculture, Viticulture and Rural Development. https://gouvernement.lu/en/actualites/toutes_actualites/communiques/2020/01-janvier/16-interdiction-glyphosate.html. Accessed 02 Feb 2020

McGrath MT (2020) What are fungicides? In: (APS) TAPS. https://www.apsnet.org/edcenter/disimpactmngmnt/topc/Pages/Fungicides.aspx. Accessed 18 Jan 2020

MDR (2017) Fakt exklusiv: Vorwurf der Industrienähe an die europäische Chemikalienagentur ECHA. In: Rundfunk M (ed) http://www.mdr.de/fakt/fakt-exklusiv-glyphosat-100.html. Accessed 05 Oct 2019

Mengistie BT, Mol APJ, Oosterveer P, Simane B (2015) Information, motivation and resources: the missing elements in agricultural pesticide policy implementation in Ethiopia. Int J Agric Sustain 13:240–256

Mengistie BT, Mol APJ, Oosterveer P (2016) Private environmental governance in the ethiopian pesticide supply chain: importation, distribution and use. NJAS - Wageningen J Life Sci 76:65–73

Mengistie BT, Mol APJ, Oosterveer P (2017) Pesticide use practices among smallholder vegetable farmers in Ethiopian Central Rift Valley. Environ Dev Sustain 19:301–324

Mesnage R, Antoniou MN (2018) Ignoring adjuvant toxicity falsifies the safety profile of commercial pesticides. Front Public Health 5:361

Mesnage R, Defarge N, Vendômois JSD, Séralini G-E (2014) Major pesticides are more toxic to human cells than their declared active principles. BioMed Res Int. Article ID 179691:8 pages

Mesnage R, Benbrook C, Antoniou MN (2019) Insight into the confusion over surfactant co-formulants in glyphosate-based herbicides. Food Chem Toxicol 128:137–145

Mie A, Rudén C, Grandjean P (2018) Safety of safety evaluation of pesticides: developmental neurotoxicity of chlorpyrifos and chlorpyrifos-methyl. Environ Health 17:77

Milner AM, Boyd IL (2017) Toward pesticidovigilance. Science 357:1232–1234

Mimkes P, Pehrke J (2016) 100 Jahre Giftgas-Tradition bei BAYER. In: Neue Rheinische Zeitung; http://www.nrhz.de/flyer/beitrag.php?id=20151. Accessed 05 Oct 2019, Köln

Mitchell EAD, Mulhauser B, Mulot M, Mutabazi A, Glauser G, Aebi A (2017) A worldwide survey of neonicotinoids in honey. Science 358:109–111

Moitzi G (2005) Kraftstoffeinsatz in der Pflanzenproduktion. In: ÖKL-Kolloquium 2005 "Kraftstoffkostensparen in der Landwirtschaft", 24 Nov 2005

Möseneder M (2008) Gift in Tadschikistans Erde: Kampf um die Quelle Asiens. Der Standard. http://derstandard.at/3353627/Gift-in-Tadschikistans-Erde-Kampf-um-die-Quelle-Asiens. Accessed 04 Oct 2019

Mourtzinis S et al (2019) Neonicotinoid seed treatments of soybean provide negligible benefits to US farmers. Sci Rep 9:11207

Müller W (2001) Pestizidverbrauch steigt drastisch an. Der Standard. http://derstandard.at/734378/Pestizidverbrauch-steigt-drastisch-an. Accessed 09 June 2017

Müller SK (2011) Airline zahlt Passagier 50.000€ Schadensersatz wegen Pestiziden an Bord. In: Chemical Sensitivity Network. http://www.csn-deutschland.de/blog/2011/08/22/airline-zahlt-passagier-50-000e-schadensersatz-wegen-pestiziden-an-bord/. Accessed 29 Sept 2019

Mullin CA, Fine JD, Reynolds RD, Frazier MT (2016) Toxicological risks of agrochemical spray adjuvants: organosilicone surfactants may not be safe. Front Public Health 4:92

MultiWatch (ed) (2016) Schwarzbuch Syngenta. Dem Basler Agromulti auf der Spur, edition 8, Liebefeld, Schweiz

Myers N, Mittermeier RA, Mittermeier CG, da Fonseca GAB, Kent J (2000) Biodiversity hotspots for conservation priorities. Nature 403:853–858

Myers JP et al (2016) Concerns over use of glyphosate-based herbicides and risks associated with exposures: a consensus statement. Environ Health 15:1–13

Nazarewska B (2013) Mückenplage: Umstrittenes Insektizid als Lösung? Münchner Merkur, 12 July 2013. https://www.merkur.de/bayern/insektizid-bti-eine-umstrittene-bazille-gegen-muecken-3001548.html. Accessed 06 October 2017

Nelsen A (2018) Ban entire pesticide class to protect children's health, experts say. The Guardian. https://www.theguardian.com/environment/2018/oct/24/entire-pesticide-class-should-be-banned-for-effect-on-childrens-health. Accessed 18 Jan 2020

Nentwig W (2005) Humanökologie: Fakten – Argumente – Ausblicke, 2. Auflage, Springer, Heidelberg

Neumeister L (2016) Von Menschen und Mäusen – Mythos Sicherheitsfaktoren. In: Essen ohne Chemie. https://www.essen-ohne-chemie.info/mythos-sicherheitsfaktoren/. Accessed 05 Oct 2019

Nischwitz G, Chojnowski P, Eller A (2019) Verflechtungen und Interessen des Deutschen Bauernverbandes (DBV), p 64

NPIC (2019) How can I wash pesticides from fruit and veggies? In: Center NPI (ed) http://npic.orst.edu/capro/fruitwash.html. Accessed 05 Oct 2019

NRC (1986) In: Press NA (ed) Drinking water and health, vol 6. National Research Council, Washington, DC

NRC (1993) Committee on pesticides in the diets of infants and children. In: Council NR (ed) National Academies Press: 8. https://www.ncbi.nlm.nih.gov/books/NBK236276/. Accessed 02 Nov 2019

Nyffeler M, Birkhofer K (2017) An estimated 400–800 million tons of prey are annually killed by the global spider community. Sci Nat – Naturwiss 104:30

Ocho FL, Abdissa FM, Yadessa GB, Bekele AE (2016) Smallholder farmers' knowledge, perception and practice in pesticide use in South Western Ethiopia. J Agric Environ Int Dev 110:307–323

OESF (2016) Soja – kleine Bohne, große Bedeutung. In: Forum Ö (ed) Factsheet. https://ökosozial.at/wp-content/uploads/2016/12/20151013_Factsheet_Soja_final.pdf. Accessed 05 Oct 2019

Ökotest (2013) Glyphosat in Getreideprodukten. https://www.oekotest.de/essen-trinken/20-Getreideprodukten-mit-Glyphosat-im-Test_102072_1.html. Accessed 30 Sept 2019

Ökotest (2017) Feldsalat. Rapunzels Geheimnis. Öko-Test 1:39–49

Oreskes N, Conway EM (2010) Merchants of doubt. How a handful of scientists obscured the truth on issues from tobacco smoke to global warming. Bloomsbury Publishing, London

ORF (2010) Pestizidgefahr im Flugzeug? https://sciencev1.orf.at/science/news/44685. Accessed 29 Sept 2019

ORF (2014) HCB nicht nur im Görtschitztal. kärntenorfat. http://kaernten.orf.at/news/stories/2683554/. Accessed 04 Oct 2019

ORF (2016a) Hohe Glyphosatwerte in Waldfrüchten. kaernten.orf.at, 21 Nov 2016. https://kaernten.orf.at/v2/news/stories/2810156/. Accessed 30 Sept 2019.

ORF (2016b) Millionen-Entschädigung für Südtiroler Bauern. tirol.orf.at. Accessed 27 September 2019

ORF (2018) Eingeschleppte Baumwanzen werden zur Plage. https://noe.orf.at/v2/news/stories/2941820/. Accessed 05 Oct 2019

Ornstein L (2010) Poisonous heritage: pesticides in museum collections. Master thesis Seton Hall University https://scholarship.shu.edu/theses/253. Accessed 28 Sept 2019

PAN (2019) PAN pesticide database. http://wwwpesticideinfoorg. Accessed 21 Dec 2019

PAN Europe (2019) Pesticide cocktails in European food. Pesticide Action Network Europe. https://www.pan-europe.info/press-releases/2019/07/pesticide-cocktails-european-food. Accessed 05 Oct 2019

PAN Germany (2019) Giftige Exporte. Die Ausfuhr hochgefährlicher Pestizide aus Deutschland in die Welt. Pestzid Aktions-Netzwerk e.V., Hamburg, p 20

PAN UK (2017) The truth about pesticide use in the UK. https://www.pan-uk.org/pesticides-agriculture-uk/. Accessed 11 Jan 2020

PCP (2010) Reducing environmental cancer risk. What we can do now? In: President's Cancer Panel (ed) https://deainfo.nci.nih.gov/advisory/pcp/annualReports/pcp08-09rpt/PCP_Report_08-09_508.pdf, 240 pp. Accessed 17 September 2017

Pelaez V, da Silva LR, Araújo EB (2013) Regulation of pesticides: a comparative analysis. Sci Publ Pol 40:644–656

Pesatori AC, Consonni D, Rubagotti M, Grillo P, Bertazzi PA (2009) Cancer incidence in the population exposed to dioxin after the "Seveso accident": twenty years of follow-up. Environ Health 8:39

Pettis JS, Lichtenberg EM, Andree M, Stitzinger J, Rose R, vanEngelsdorp D (2013) Crop pollination exposes honey bees to pesticides which alters their susceptibility to the gut pathogen Nosema ceranae. PLoS One 8:e70182

Phillips D (2019) Hundreds of new pesticides approved in Brazil under Bolsonaro. The Guardian. https://www.theguardian.com/environment/2019/jun/12/hundreds-new-pesticides-approved-brazil-under-bolsonaro, 12 June 2019. Accessed 05 Oct 2019

Philpott T (2011) Independent panel: EPA underestimates Atrazine's cancer risk. Mother Jones. http://www.motherjones.com/tom-philpott/2011/11/atrazine-cancer-epa, 07 Nov 2011. Accessed 03 Oct 2019

Pilling E, Jepson P (2006) Synergism between EBI fungicides and a pyrethroid insecticide in the honeybee (*Apis mellifera*). Pestic Sci 39:293–297

Pimentel D, Greiner A, Bashore T (1998) Economic and environmental costs of pesticide use. In: Rose J (ed) Environmental toxicology: current developments. Gordon and Breach Science Publishers, Amsterdam, pp 121–150

Pimentel D, Hepperly P, Hanson J, Douds D, Seidel R (2005) Environmental, energetic, and economic comparisons of organic and conventional farming systems. Bioscience 55:573–582

Pinto C (2018) Düngemittel, Unkrautvernichter, Insektizide. Gift im Schweizer Wein. Blick, 01 October 2018. http://www.blick.ch/news/schweiz/duengemittel-unkrautvernichter-insektizide-gift-im-schweizer-wein-id5582578.html. Accessed 30 Sept 2019

Pisa LW et al (2015) Effects of neonicotinoids and fipronil on non-target invertebrates. Environ Sci Pollut Res 22:68–102

Pistorius J, Bischoff G, Heimbach U (2009) Bienenvergiftung durch Wirkstoffabrieb von Saatgutbehandlungsmitteln während der Maisaussaat im Frühjahr 2008, J Kulturpflanzen 61:9–14

Ploetz RC (2001) Black sigatoka of banana. The Plant Health Instructor. doi:https://doi.org/10.1094/PHI-I-2001-0126-02:https://www.apsnet.org/edcenter/apsnet-features/Pages/BlackSigatoka.aspx. Accessed 07 Dec 2019

Primost JE, Marino DJG, Aparicio VC, Costa JL, Carriquiriborde P (2017) Glyphosate and AMPA, "pseudo-persistent" pollutants under real-world agricultural management practices in the Mesopotamic Pampas agroecosystem, Argentina. Environ Pollut 229:771–779

Quinones RA, Fuentes M, Montes RM, Soto D, Leon-Munoz J (2019) Environmental issues in Chilean salmon farming: a review. Rev Aquac 11:375–402

Rauh VA (2018) Polluting developing brains – EPA failure on chlorpyrifos. N Engl J Med 378:1171–1174

Rauh VA et al (2012) Brain anomalies in children exposed prenatally to a common organophosphate pesticide. Proc Natl Acad Sci USA 109:7871–7876

RC (2019) UNEP Rotterdam convention. http://www.pic.int/TheConvention/Overview/tabid/1044/language/en-US/Default.aspx. Accessed 06 Dec 2019

Ríos JM et al (2019) Occurrence of organochlorine compounds in fish from freshwater environments of the central Andes, Argentina. Sci Total Environ 693:133389

Robin M-M (2014) Our daily poison. From pesticides to packaging, how chemicals have contaminated the food chain and are making us sick. The New Press, New York

Sanderson K (2006) Fake pesticides pose threat. Flood of counterfeit chemicals is harming people and industry. Nature 5 November 2006. https://doi.org/10.1038/news061030-14

Sattelberger R (2001) Einsatz von Pflanzenschutzmitteln und Biozid-Produkten im Nicht-Land- und Forstwirtschaftlichen Bereich. In: Monographien, vol. Band 146. Umweltbundesamt Wien, Wien, p 106

Sattler C, Schrader J, Farkas VM, Settele J, Franzén M (2018) Pesticide diversity in rice growing areas of Northern Vietnam. Paddy Water Environ 16:339–352

Scherbaum E, Marks H (2019) Insect spray as contaminant in food: incidence and legal assessment. In: Stuttgart C (ed) https://www.cvuas.de/pub/beitrag_printversion.asp?subid=1&Thema_ID=5&ID=3060&Pdf=No&lang=EN.   Accessed 17 Jan 2020

Schmider F (2016) Giftwolke über dem Dreiländereck. Badische Zeitung. http://www.badische-zeitung.de/suedwest-1/giftwolke-ueber-dem-dreilaendereck%2D%2D129251349.html. Accessed 04 Oct 2019

Schmitz PM, Garvert H (2012) Die Ökonomische Bedeutung des Wirkstoffes Glyphosat für den Ackerbau in Deutschland. J Kult 64:150–162

Schnepf R (2004) Energy use in agriculture: background and issues. CRS report for congress, https://nationalaglawcenter.org/wp-content/uploads/assets/crs/RL32677.pdf. Accessed 30 September 2017

Schuller J (2020) Chlorothalonil: so begründet Syngenta seine Beschwerde. Bauernzeitung. https://www.bauernzeitung.ch/artikel/chlorothalonil-so-begruendet-syngenta-seine-beschwerde. Accessed 01 Feb 2020

Sda (2016) Boden ist immer noch belastet. Badische Zeitung. http://www.badische-zeitung.de/baselland/boden-ist-immer-noch-belastet. Accessed 04 Oct 2018

Seiwald RV (2019) Pesticide practices and pesticide-related health effects among female-headed smallholders in the Amhara region (Ethiopia). Master thesis, Institute of Organic Farming, University of Natural Resources and Life Sciences Vienna, Austria, 96 pp

Seok J et al (2013) Genomic responses in mouse models poorly mimic human inflammatory diseases. PNAS 110:3507–3512

Settele J et al (2018) Rice ecosystem services in South-east Asia. Paddy Water Environ 16:211–224

Shaner DL, Lindenmeyer RB, Ostlie MH (2012) What have the mechanisms of resistance to glyphosate taught us? Pest Manag Sci 68:3–9

Shiva V (2013) Making peace with the earth. Pluto Press, New Delhi

Shiva V, Shiva M, Shiva V (2013) Poison in our foods: links between pesticides and diseases. Natraj Publishers, New Delhi

Simon-Delso N et al (2015) Systemic insecticides (neonicotinoids and fipronil): trends, uses, mode of action and metabolites. Environ Sci Pollut Res 22:5–34

SNH (2017) Use of pesticides in nature conservation. Scottish Natural Heritage, Scientific Advisory Committee SAC/2017/03/04:11p

Snyder F, Ni L (2017) A tale of eight pesticides: risk regulation and public health in China. Eur J Risk Regul 8:469–505

Spangenberg JH, Douguet JM, Settele J, Heong KL (2015) Escaping the lock-in of continuous insecticide spraying in rice: developing an integrated ecological and socio-political DPSIR analysis. Ecol Model 295:188–195

Statesman Journal (2014) Bumblebee die-off in Eugene under investigation. www.statesmanjournal.com/story/tech/science/environment/2014/06/18/bumblebee-die-eugene-investigation/10804135/. Accessed 28 September 2017

Strassheim I (2019) Selbst Syngenta rechnet mit Verbot der meisten Pestizide. Tagesanzeiger. https://www.tagesanzeiger.ch/wirtschaft/selbst-syngenta-rechnet-mit-verbot-der-meisten-pestizide/story/26294606. Accessed 03 Jan 2020

Strubelt O (1996) Gifte in Natur und Umwelt. Pestizide und Schwermetalle, Arzneimittel und Drogen. Spektrum Akademischer Verlag, Heidelberg

Stykel MG et al (2018) Nitration of microtubules blocks axonal mitochondrial transport in a human pluripotent stem cell model of Parkinson's disease. FASEB J 32:5350–5364

Sur R, Stork A (2003) Uptake, translocation and metabolism of imidacloprid in plants. Bull Insect 56:35–40

SustainablePulse (2017) US Court Documents Show Monsanto Manager Led Cancer Cover Up for Glyphosate and PCBs. https://sustainablepulse.com/2017/05/19/us-court-documents-show-monsanto-manager-led-cancer-cover-up-for-glypho-sate-and-pcbs/?utm_source=newsletter&utm_medium=email&utm_campaign=breaking_us_epa_continues_glyphosate_cancer_cover_up_with_regu-latory_review_publication&utm_term=2020-01-31#.XjPYrS2X9QJ. Accessed 31 Jan 2020

Syngenta (2017) So sauber war Ihr Feld noch nie! Die breite Komplettlösung gegen alle Unkräuter und Hirsen in Mais. Die Landwirtschaft, noe.lko.at/die-landwirtschaft. Accessed 17 October 2017

SZ (2019) Kritik an Monsanto wegen Finanzierung von Glyphosat-Studien. Süddeutsche Zeitung. https://www.sueddeutsche.de/wirtschaft/agrar-kritik-an-monsanto-wegen-finanzierung-von-glyphosat-studien-dpa.urn-newsml-dpa-com-20090101-191205-99-18346. Accessed 05 Dec 2019

Székács A, Darvas B (2018) Re-registration challenges of glyphosate in the European Union. Front Environ Sci 6:78

Tanabe S, Hidaka H, Tatsukawa R (1983) PCBs and chlorinated hydrocarbon pesticides in Antarctic atmosphere and hydrosphere. Chemosphere 12:277–288

Teklu BM (2016) Environmental risk assessment of pesticides in Ethiopia: a case of surface water systems. Chemical Stress Ecology. Wageningen University. https://library.wur.nl/WebQuery/wurpubs/fulltext/380652. Accessed 14 Dec 2019

Thakur JS, Rao BT, Rajwanshi A, Parwana HK, Kumar R (2008) Epidemiological study of high cancer among rural agricultural community of Punjab in Northern India. Int J Environ Res 5:399–407

Thongprakaisang S, Thiantanawat A, Rangkadilok N, Suriyo T, Satayavivad J (2013) Glyphosate induces human breast cancer cells growth via estrogen receptors. Food Chem Toxicol 59:129–136

Tilman D, Cassman KG, Matson PA, Naylor R, Polasky S (2002) Agricultural sustainability and intensive production practices. Nature 418:671–677

TNS Emnid (2012) Das Image der deutschen Landwirtschaft. Ergebnisse einer Repräsentativbefragung in Deutschland, www.yumpu.com/de/document/read/22616473/das-image-der-deutschen-landwirtschaft-information medienagrar-. Accessed 30 September 2017

Tubiello FN (2019) Greenhouse gas emissions due to agriculture. In: Ferranti P, Berry EM, Anderson JR (eds) Encyclopedia of food security and sustainability. Elsevier, Amsterdam, NL, pp 1–11

Tweedale AC (2017) The inadequacies of pre-market chemical risk assessment's toxicity studies—the implications. J Appl Toxicol 37:92–104

UBA Berlin (2019) Pflanzenschutzmittelverwendung in der Landwirtschaft. www.umweltbundesamt.de/daten/land-forstwirtschaft/landwirtschaft/pflanzen-schutzmittelverwendung-in-der#textpart-1. Accessed 04 Oct 2019

UCS (2013) "Superweeds" resulting from Monsanto's products overrun U.S. farm landscape. Union of Concerned Scientists, www.ucsusa.org/news/press_release/superweeds-overrun-farmlands-0384.html#.WKcKCxiX_MU. Accessed 17 Sept 2017

UCS (2018) Betrayal at the USDA. How the Trump Administration Is Sidelining Science and Favoring Industry over Farmers and the Public. Union of Concerned Scientists, www.ucsusa.org/USDAbetrayal, 27pp. Accessed 04 Oct 2019

Umweltinstitut München (2016) Umweltinstitut findet Glyphosat in deutschem Bier. http://www.umweltinstitut.org/aktuelle-meldungen/meldungen/umweltinstitut-findet-glyphosat-in-deutschem-bier.html. Accessed 30 Sept 2019

UNEP (2017) Stockholm Convention on Persistent Organic Pollutants (POPS). Texts and Annexes revised in 2017. United Nations Environment Programme (UNEP) SotSCS (ed) www.pops.int/TheConvention/Overview/Textofthe Convention/tabid/2232/Default.aspx, 78pp. Accessed 04 Oct 2019

UNEP (2018) Fake pesticides, real problems: addressing Ukraine's illegal and counterfeit pesticides problem. United Nations Environmental Programme. https://www.unenvironment.org/news-and-stories/story/fake-pesticides-real-problems-addressing-ukraines-illegal-and-counterfeit. Accessed 04 Oct 2019

UNHRC (2017) Report of the Special Rapporteur on the right to food. In vol. A/HRC/34/48. United Nations Human Rights Council. https://documents-dds-ny.un.org/doc/UNDOC/GEN/G17/017/85/PDF/G1701785.pdf?OpenElement, 24 Jan 2017. Accessed 29 Sept 2019

US EPA (2019) Summary of the Federal Insecticide, Fungicide, and Rodenticide Act – 7 U.S.C. §136 et seq. (1996). Laws & Regulations 2017. https://www.epa.gov/laws-regulations/summary-federal-insecticide-fungicide-and-rodenticide-act. Accessed 21 Dec 2019

USPA (1964) Amenomethylenephosphinic acids, salts thereof, and process for their production. United States Patent Office 3,160,632; patented Dec 8, 1964. Accessed 04 Oct 2019

USPA (1974) Phosphonomethylglycine phytotoxicant compositions. United States Patent Office 3,79,758. N. Phosphonomethylglycine phytotoxicant compositions. Accessed 04 Oct 2019

USPA (2010) Glyphosate formulations and their use for the inhibition of s-enolpyruvylshikimate-3-phosphate synthase. United States Patent Office US7,771,736B2. Accessed 04 Oct 2019

Vaccari C, El Dib R, de Camargo JLV (2017) Paraquat and Parkinson's disease: a systematic review protocol according to the OHAT approach for hazard identification. Syst Rev 6:98–98

van der Putten WH et al (2004) The sustainable delivery of goods and services provided by soil biota. In: Wall DH (ed) Sustaining biodiversity and ecosystem services in soils and sediments. Island Press, San Francisco, CA, pp 15–43

van Groenigen JW, Lubbers IM, Vos HMJ, Brown GG, De Deyn GB, van Groenigen KJ (2014) Earthworms increase plant production: a meta-analysis. Sci Rep 4:6365. https://doi.org/10.1038/srep06365

Van Nguyen N, Ferrero A (2006) Meeting the challenges of global rice production. Paddy Water Environ 4:1–9

Vandenberg LN, Blumberg B, Antoniou MN, Benbrook CM, Carroll L, Colborn T, Everett LG, Hansen M, Landrigan PJ, Lanphear BP, Mesnage R, vom Saal FS, Welshons WV, Myers JP (2017) Is it time to reassess current safety standards for glyphosate-based herbicides? J Epidemiol Commun Health 71:613–618

Vandenberg LN, Colborn T, Hayes TB, Heindel JJ, Jacobs DR, Lee DH (2012) Hormones and endocrine-disrupting chemicals: low-dose effects and nonmonotonic dose responses. Endocr Rev 33:378–455

Vandenberg LN, Colborn T, Hayes TB, Heindel JJ, Jacobs DR, Lee DH (2013) Regulatory decisions on endocrine disrupting chemicals should be based on the principles of endocrinology. Reprod Toxicol 38C:1–15

Walker G, Hostrup O, Hoffmann W, Butte W (1999) Biozide im Hausstaub. Gefahrstoffe - Reinhaltung der Luft 59:33–41

Wang A, Costello S, Cockburn M, Zhang XB, Bronstein J, Ritz B (2011) Parkinson's disease risk from ambient exposure to pesticides. Eur J Epidemiol 26:547–555

Wang J, Tao J, Yang C, Chu M, Lam H (2017) A general framework incorporating knowledge, risk perception and practices to eliminate pesticide residues in food: a Structural Equation Modelling analysis based on survey data of 986 Chinese farmers. Food Control 80:143–150

Webb W (2017) US gov't pesticide spraying in wake of Harvey A toxic boon to agrochemical giants. MintPress News, September 12, 2017

Weiland M (2016) Greenpeace-Untersuchung: Pestizid Ethoxyquin in Speisefischen. Vorsorge ist besser. www.greenpeace.de/themen/meere/vorsorge-ist-besser. Accessed 08 Oct 2017

Weiss S (2019) Interview: "Die Europäer haben keine Ahnung, was sie da konsumieren". Blickpunkt Lateinamerika, 02 July 2019. https://blickpunkt-lateinamerika.de/artikel/interview-die-europaeer-haben-keine-ahnung-was-sie-da-konsumieren/. Accessed 05 Oct 2019

Wetzenkircher M, Llubic Tobisch V (eds) (2014) Gefahrstoffe in Museumsobjekten. Erhaltung oder Entsorgung? Technisches Museum Wien, Wien, Österreich

WHO (2019) World hunger is still not going down after three years and obesity is still growing – UN report. https://www.who.int/news-room/detail/15-

07-2019-world-hunger-is-still-not-going-down-after-three-years-and-obesity-is-still-growing-un-report. Accessed 17 Jan 2020

Winkler AJ, Cook JA, Kliewer WM, Lider LA (1974) General viticulture. University of California Press, Berkeley, CA

Wolfarth F, Schrader S, Oldenburg E, Brunotte J (2016) Mycotoxin contamination and its regulation by the earthworm species *Lumbricus terrestris* in presence of other soil fauna in an agroecosystem. Plant Soil 402:331–342

Woods J, Williams A, Hughes JK, Black M, Murphy R (2010) Energy and the food system. Philos Trans R Soc Lond Ser B Biol Sci 365:2991–3006

Yamamuro M, Komuro T, Kamiya H, Kato T, Hasegawa H, Kameda Y (2019) Neonicotinoids disrupt aquatic food webs and decrease fishery yields. Science 366:620–623

Yang T, Doherty J, Zhao B, Kinchla AJ, Clark JM, He L (2017) Effectiveness of commercial and homemade washing agents in removing pesticide residues on and in apples. J Agric Food Chem 65:9744–9752

Zaller JG (2004) Ecology and non-chemical control of Rumex crispus and R-obtusifolius (Polygonaceae): a review. Weed Res 44:414–432

Zaller JG (2018) Unser täglich Gift. Pestizide – die unterschätzte Gefahr, 3. Auflage, Deuticke Verlag, Vienna

Zaller JG et al (2013) Herbivory of an invasive slug is affected by earthworms and the composition of plant communities. BMC Ecol 13:20

Zand-Vakili A (2014) Hamburg ist die Drehscheibe für die Pestizid-Mafia. https://www.welt.de/regionales/hamburg/article135436727/Hamburg-ist-die-Drehscheibe-fuer-die-Pestizid-Mafia.html. Accessed 04 Oct 2019

Zhang W (2018) Global pesticide use: profile, trend, cost/benefit and more. Proc Int Acad Ecol Environ Sci 8:1–27

Zhang Z-Y, Liu X-J, Hong X-Y (2007) Effects of home preparation on pesticide residues in cabbage. Food Control 18:1484–1487

Zhang W, Jiang F, Ou J (2011) Global pesticide consumption and pollution: with China as a focus. Proc Int Acad Ecol Environ Sci 1:125–144

Zhang C, Hu R, Shi G, Jin Y, Robson MG, Huang X (2015) Overuse or underuse? An observation of pesticide use in China. Sci Tot Environ 538:1–6

Zhang L, Rana I, Shaffer RM, Taioli E, Sheppard L (2019) Exposure to glyphosate-based herbicides and risk for non-Hodgkin lymphoma: A meta-analysis and supporting evidence. Mutat Res 781:186–206

Zimmer K (2018) We're asking the wrong questions about glyphosate. The New Food Economy, newfoodeconomy.org/glyphosate-safety-debate/. Accessed 24 Nov 2019

2

Pesticide Impacts on the Environment and Humans

There is mounting evidence that pesticides do not only control pests or pathogens but have also various side effects on nontarget organisms, the environment, and us humans. Depending on where they are applied, pesticide active substances are usually degraded in soil or water or converted into degradation products, so-called metabolites. In pesticide approval studies, particular attention is paid to the degradation behavior of active substances or its metabolites in soil and water, transport in the air.

In the field of ecotoxicology, the effects on so-called surrogate species, some bird and mammal species, as well as some species of water and soil organisms, insects, arachnids, and plants are examined. As a rule, simple laboratory tests are initially carried out on a few, mostly lab-cultured animals. Standard animal species as surrogates for wildlife species are for example Japanese quails (*Coturnix japonica*) for birds, rainbow trout (*Oncorhynchus mykiss*) for fish, or compost worms (*Eisenia fetida*) for earthworms. When it comes to assessing the risks for humans, experiments are carried out with lab mice (*Mus musculus*) or lab rats (*Rattus norvegicus domestica*). In order to avoid ethically questionable animal experiments, human cell cultures from organs or embryos are also used.

In the risk assessment, little consideration is given to epidemiological data from poisoning centers or general medical practitioners. This could be used, for example, to determine the risks for certain pesticide user groups (arable farmers, winegrowers, animal breeders, gardeners, etc.) on the basis of human findings. When it is not possible to evaluate the substances on the basis of laboratory results, tests are carried out under more realistic semi-field conditions or even in field trials. If these tests result in an unacceptable risk, appropriate risk

© Springer Nature Switzerland AG 2020
J. G. Zaller, *Daily Poison*, https://doi.org/10.1007/978-3-030-50530-1_2

reduction measures must be examined. This could mean that special equipment is required for the application of pesticides for instance to reduce drift. Or as a consequence an increased distance to water bodies or field margins is prescribed, or the application rate is limited.

Before I started researching pesticide effects, I was not so attracted by the scientific discussions on pesticides and their nontarget effects. Moreover, I naively assumed that these products were anyway being investigated by good and well-paid scientists in the research departments of agrochemical companies and that there was not much to contribute scientifically. As many other ecologists, I was and still am an advocate of organic farming where synthetic pesticides are not used anyway. Despite all the enthusiasm for organic farming and although my home country Austria is doing pretty good with 22% of the agricultural area under organic farming in 2017, it must not be ignored that even there 78% of the agricultural land is managed conventionally, i.e., with more or less use of pesticides. Worldwide, only 1.4% of the farmland is (certified) organically farmed; thus, pesticides are theoretically applied to the vast majority of croplands (Willer and Lernoud 2019).

Even if it is rarely admitted, many scientists inform themselves in Wikipedia, especially when one needs to get a quick overview of a new field. After all, Wikipedia is great and the world's largest encyclopedia and has driven such renowned reference works as the German Brockhaus or the Encyclopedia Britannica out of the market. Overall, articles on pesticides in Wikipedia are quite relaxed regarding their side effects on humans and the environment. One gets the feeling that there seem to be a few studies that are controversially discussed, but those appear to have methodological flaws and are therefore not really relevant. Back then this also confirmed me that pesticides are among the best investigated chemicals. I already tried to elaborate in Chap. 1 that this is definitely not the case. Only later did I find out that the pesticide manufacturers and their allies of course also edit entries in Wikipedia in order to spread their narrative of the issue.

The next step for me was to consult scientific databases on the subject. With appropriate access via university libraries, you will find a vast amount of information published in tens of thousands of scientific journals worldwide. The results of the literature search were impressive: since the year 1928, well over 100,000 studies on pesticides have been published. However, only about 1200 studies have dealt with the effect on soil organisms, just about 1.2% of all studies! And this despite the fact that practically all pesticides get onto or into the soil. A closer look at the studies on soil organisms also revealed that many aspects an ecologist would consider important were apparently ignored. Only a few studies dealt with the effect of pesticides on wild animal species and their interaction with different organisms such as plants, animals, fungi, or microorganisms; even fewer considered effects on populations and communities or addressed climate change issues (Köhler and Triebskorn 2013).

Since the focus of our research group in Vienna is in soil ecology and ecological interactions between different organisms, this was a very good starting point for our own studies in this field.

2.1 From "Silent Spring" to Own Experiments

When reading *Silent Spring* authored by Rachel Carson in its German edition during my college time I was quite shocked about the revelations regarding the widespread pesticide use in the USA at that time (Carson 1962). The book is still worth reading today and deals with the nontarget effects of pesticides and especially the insecticide DDT on the environment, fish, birds, and humans. In 1943, the production of DDT was low. Given its importance as a control agent for diseases such as malaria, where insects are the primary vectors, DDT production was rapidly ramped up during the final years of the World War II. By 1945, production increased to over two million pounds of DDT synthesized per month. When the war ended, increases in the production of organochloride pesticides continued virtually unabated. Carson has shown that DDT and the other organochloride insecticides aldrin and dieldrin, which were very popular at the time, accumulate in human and animal tissues. These pesticides are part of a larger group of toxic compounds known as persistent organic pollutants or POPs. POPs are lipophilic tending to accumulate in fat cells, where they remain essentially unchanged for long periods of time (Kolok 2016).

In natural food nets biomagnification takes place leading to high concentrations in predators such as eagles or ospreys and ultimately to their decimation. Over time, DDT and some of its degradation products have also been found to have hormone-like effects. Birds of prey laid eggs with thinner shells, which led to considerable population losses. The stories in the book impressed me so much that I decided to start studying biology and ecology after completing my training as an electrical engineer at a technical college. Growing up in a rural area in the Austrian Alps, pesticide application in agriculture was new to me as no grain or potato field in the Alpine region was treated with pesticides at that time.

In the meantime, the situation has changed a bit but is still quite low in Alpine compared to lowland agriculture. In retrospect the environment was not completely pesticide free either. Insecticides against annoying flies and mosquitos were used in households and stables. DDT was used in these

Alpine areas by the military during World War II and that schoolchildren during the 1950s were regularly treated with a powder against head lice. Most likely this was the insecticide DDT introduced to Austrian provinces by the US-American occupation authorities.

Reading Carson's book has opened my eyes to the widespread use of pesticides and their consequences for people and the environment. I assumed, however, that this would only apply to the USA and that we in Europe would be little affected. Carson, for example, observed that the stock of juvenile salmon in a river in the north-west of the USA declined after DDT was used to control the eastern spruce budworm (*Choristoneura fumiferana*), a moth species. She found that while the pesticide treatment destroyed the spruce moth it also reduced aquatic insects of which young salmons lived. Other cases of pesticide poisoning were also a serious problem very early on. In the USA in the 1950s, the large-scale treatment of elm bark beetles (Scolytidae, a group with 6000 species) with DDT led to the starvation of blackbirds and other bird species. Soon afterward, DDT came under suspicion of causing cancer in humans. Carson's book sold more than a million copies and even prompted President John F. Kennedy to appoint a science advisory committee which ultimately confirmed the findings described in *Silent Spring*. Finally, DDT was banned for use in the USA in 1972, and both aldrin and dieldrin were banned in 1974. *Silent Spring* is now considered a foundational text of the environmental movement worldwide and also led to the passing of environmental laws and the establishment of the US Environmental Protection Agency in 1970. Five years after the USA banning DDT, Germany followed and it took another 15 years for Austria to be convinced that DDT was harmful also in this country. Nowadays, the use of DDT is banned worldwide and only permitted in exceptional cases for in-door use in order to control disease-transmitting insects, in particular to fight the carriers of malaria.

In an interview Rachel Carson gave for a radio station, it became clear that she was also interested in ecotoxicological topics such as potential effects that might be passed on to the next generation—topics that are still not well investigated today. In this interview Carson stated "We have to remember that children born today are exposed to these chemicals from birth perhaps even before birth. We simply don't know how these chemicals will affect their lives." Although not using the term ecology she was thinking in a way every modern ecologist would do by stating: "Balance of nature is built by a series of interrelationships between organisms and their environment. You can't just change one thing without changing many others."

The publication of *Silent Spring* also corresponded with a profound paradigm shift in ecotoxicology (Kolok 2016) and US culture and environmentalism (Mart 2015). Humans were shown to have the capacity to alter entire ecosystems through the excessive release of industrial chemicals into the environment.

Testing chemicals for toxicity is straightforward, but detecting their effects at the population, community, or ecosystem level is exceedingly difficult. As one moves to higher levels of ecological organization, the number of interacting factors and compensatory mechanisms increases. Standardized, laboratory studies of pesticides, required by regulatory agencies, typically focus on the short-term effects of acute exposure to individual model organisms with the results mathematically scaled up to estimate long-term and indirect effects. However, long-term and ecosystem-scale studies frequently show surprises and emergent phenomena that could not be predicted by extrapolating from results at smaller temporal, spatial, and organizational scales (Jensen 2019).

Well, I finally decided with my working group to take on the endeavor to scientifically investigate the effects of pesticides on soil life. However, we did not want to proceed according to strict test protocols as required for ecotoxicological pesticide registration studies, because this seemed rather far away from what is happening in the field.

For the investigation of pesticide effects on earthworms, two methods are usually used in these standard tests: the avoidance test (ISO 2008) and the reproduction test (OECD 2004). For the avoidance test, soil substrate is filled into small plastic trays, with half of the soil surface sprayed or mixed with a given amount of pesticide, and the other half serving as the control treatment which is sprayed or mixed with the same amount of water only. Then earthworms are added in the border between the two treatments and over 2 days it is recorded whether earthworms avoid pesticide-treated areas or not. Reproduction tests run for 8 weeks also according to a fixed protocol. Usually, longer tests are not common, although earthworms can live for several years and are exposed to pesticides over this period. The aim of this test is to investigate how soil treated with a pesticide affects the propagation rate of earthworms.

These tests are carried out under controlled laboratory conditions under standard temperature and moisture conditions and have not much to do with the situation in the field. Another shortcoming of these tests is that they are usually carried out with compost worms (*Eisenia fetida*). As its name implies, this earthworm species actually occurs in habitats where there is a layer of dead plant material—on the forest floor with a thick leaf layer or in the compost heap—but not on arable land. Therefore, the question arises how relevant these tests are if the test species actually lives in a different habitat and would never come into contact with agrochemicals. Moreover, studies have also shown that earthworm species differ in their sensitivity to pesticides and that compost worms are among the least sensitive ones (Pelosi et al. 2013).

Interestingly, earthworms were also suggested by Rachel Carson as being responsible for the dying of robins (Carson 1962). In *Silent Spring*, she suggests that robins were not being directly poisoned by insecticides but indirectly by eating poisoned earthworms. She also mentions feeding studies with crayfish or snakes kept in a laboratory that had gone into violent tremors after being fed with poisoned earthworms. Earthworms came in contact with the insecticide via feeding on elm tree leaves that had been sprayed with DDT. Obviously as few as eleven large earthworms, actually a small part of a robin's day ration, can transfer a lethal dose of DDT to a robin.

Carson did not only observe devastating effects of insecticide use but also addressed effects of widespread herbicide use. She mentions that already back in the 1950s millions of hectares of sagebrush country and mesquite lands in the USA were sprayed with herbicides. Treatment of agricultural lands with herbicides already totaled 21 million hectares in 1959. Herbicides also seemed to be already used in private lawns, parks, and golf courses. The most widely used herbicides were 2,4-D, 2,4,5-T, and related compounds. It is fascinating to see how prospective her observations were on complex interactions between herbicides and topics that are still not well investigated today such as the changed nutrient content of sprayed plants, the attraction of herbicide-treated plants to herbivores, and their impact on rumen bacteria in cattle.

Another aspect of ecotoxicological protocols prompted us to focus on more ecologically oriented experiments. Ecotoxicology mainly tests pure active substances of pesticides. For instance, from a glyphosate-based herbicide mainly the active substance glyphosate is tested. However, we have seen before that a pesticide consists of one or more active substances and many co-formulants or adjuvants. It is hard to believe, but only the amount of the active substance in a pesticide formulation is indicated on the package. The exact composition of the pesticide is only known to the manufacturer and the registration authorities but is concealed to the public for reasons of trade secrecy. It is a scandal that this intransparency is still tolerated by the legislators given the fact that these pesticides are sprayed in our landscape to the tens of thousands of tons per year. Moreover, the active substance often makes up only a small fraction of the pesticide. For example, the label of a very frequently used broadband Roundup herbicide only states that the active substance glyphosate is present at 480 g/liter in a weight proportion of 35.75%. This means we are not informed about the remaining 64.25% of the ingredients. And this is not an exception but the rule for all pesticides used worldwide. Tolerating this secrecy by our authorities is even more astonishing as there is plenty of evidence that the commercial pesticides on the market can be more toxic than the pure active ingredients (Mesnage and Antoniou 2018).

For our working group it was clear that we should first focus on testing nontarget effects of pesticide products that are actually used in agriculture or private gardens and, secondly, to use as model organisms those species that could potentially come into contact with pesticides in the fields. Thirdly, we also strived to combine the factor pesticide with at least one additional stressor to get away from the simplified experimental approaches used in ecotoxicological registration tests.

Figure 2.1 illustrates the scenery in an agricultural landscape including aspects that will be elaborated in more details in the following chapters.

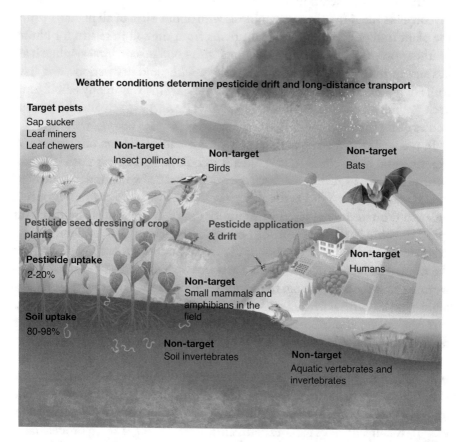

Fig. 2.1 The fate of pesticides in our environment. When systemic pesticides are used for seed dressings, only a small fraction is actually taken up by the crop plant. Besides protecting the crop plant of pest insects and fungi, many nontarget organisms living off the crop plant, in the soil, or in neighboring aquatic ecosystems will be affected. Regular pesticide applications can lead to pesticide drift to neighboring cropping systems, non-crop areas, or private land. Illustration inspired by Sánchez-Bayo (2014) and drawn by A. Neubauer

2.1.1   Earthworms Become Fat, Lazy, and Have Less Offspring

Ecological experiments require careful planning. There is also the ethical responsibility when working with living animals. So, different options should be discussed before starting; if possible also emergency plans about the fate of test organisms should be made in case something goes wrong. In order to reproduce the situation in nature as well as possible we established a weed community and asked ourselves whether the killing of weeds with herbicides influences soil organisms. Specifically, we were interested in the effects on earthworms and on mycorrhizal fungi. As already mentioned, earthworms are important for the aeration and fertility of the soil. Not less important are mycorrhizal fungi that form a symbiosis with about 80% of all plants, helping the plant to digest hardly soluble nutrients from the soil in return for photosynthesis products from the plant. Plants with good mycorrhizal symbiosis are less susceptible to diseases, are better able to withstand dry periods, and have fruits of better quality (Smith and Read 2008). By the way, many edible fungi such as the golden chanterelle (*Cantharellus cibarius*) are actually the fruiting bodies of mycorrhizal fungi associated with trees.

In preparation for the experiments, we took arable soil from our university's experimental farm, sifted it, and filled it into large 50-l flower pots. The pots were placed in a greenhouse because we wanted to work under controlled weather conditions. Wild herbs and grasses were sown into the flower pots in order to create a model weed community that was then killed by herbicide spraying. This is the situation of herbicide applications in arable fields for example before sowing new crops or in private gardens. In order to investigate the mutual influence of earthworms and mycorrhizal fungi, we added earthworms and mycorrhizal fungi to half of the pots. I excavated the experimental night crawler earthworms (*Lumbricus terrestris)* in my kitchen garden. These earthworms are about 25-cm-long vertical burrowers and quite common in fertile garden soil. The mycorrhizal fungi can be bought as so-called inoculum from special companies. This kind of experiment is called a full-factorial experiment; they get bigger and more complex with each factor added. With two earthworm levels (earthworm addition vs. no earthworm addition), two mycorrhizae levels (addiction vs. no addition), and two herbicide levels (yes vs. no), we already have eight experimental units. In order to do proper statistical analyses of the data and assure that the findings are not just random, one has to replicate every specific setting and factor combination. With six replicates one ends up with 48 experimental units in our case.

At the end of the experiment we additionally wanted to know what happens to the herbicides when heavy rain suddenly falls. Therefore, we simulated strong rain by sprinkling a defined amount of water onto the pots. In principle, we wanted to test claims that glyphosate-based herbicides are rapidly bound to soil particles and therefore hardly washed out.

After the seeds had germinated and the vegetation well developed, we purchased a ready-to-use glyphosate-based herbicide in a garden store and sprayed the plants with it. The effect was as planned: After a few days, the plants withered and died. In the farmland, the field would now be ready for sowing the following crop, even without previous mechanical tillage. We assessed the activity and growth of earthworms and the colonization of plant roots and soil with mycorrhizal fungi. Our results showed that the herbicide significantly affected both earthworms and mycorrhizal fungi. Earthworms were thicker and less active. Mycorrhization of plant roots and soil was reduced by almost half. After the simulated heavy rainfall of 40 l per square meter, we found also that a large amount of the herbicide was actually leached out of the study system. Considering that climate change has been shown to lead to an increase in heavy rainfall events, this would also mean an increased pesticide leaching into neighboring waters or the groundwater. The results of the experiment were published in a scientific journal (Zaller et al. 2014a) and also widely reported by several newspapers and agricultural journals.

Others also found that glyphosate herbicide can reduce mycorrhizal symbiosis of weed and crop species, even 10 months after its application (Helander et al. 2018). These researchers also detected glyphosate residues in weeds and crop plants in the growing season following the glyphosate treatment.

Encouraged by the results of our first experiment, we planned a follow-up study. We were now interested in herbicide effects not only in the activity and growth of earthworms but also in their reproduction. Since several earthworm species live in the soil, we added a second species, the grey worm (*Aporrectodea caliginosa*), a horizontal burrower or endogeic species. Earthworms are commonly grouped according to their feeding behavior: epigeics (as the former mentioned compoest worm) live in the uppermost soil layers and feed mainly in the litter layer, endogeics live mainly in the soil and create horizontal channels, and anecics such as *Lumbricus terrestris* in the other previous experiment, which form vertical channels, come to the surface and drag leaf litter into their tunnels. The two earthworm species also differ in their behavior and function in the soil system.

The experimental setup was similar to the previous study, but now only earthworms and no mycorrhizal fungi were considered. The result was that after the herbicide treatment the activity of earthworms was drastically reduced and that earthworms had only half as much offspring (Gaupp-Berghausen et al. 2015). This was really a finding with potentially far-reaching consequences for soil health and soil fertility if one recalls the positive effects of earthworms outlined before.

Furthermore, we also found that the amount of nitrate and phosphate in the soil was much higher after herbicide application in comparison to no herbicide use. This is actually not surprising but rarely addressed: after killing plants with herbicides, the nutrients normally absorbed by plants suddenly remained unused in the soil. As we showed in the former experiment, rain or soil erosion can transfer these nutrients into neighboring water bodies or the groundwater. The influence of a widespread glyphosate addition on the accumulation and cycling of phosphorus to agricultural landscapes is largely ignored. A study showed that alone across the USA, mean inputs of glyphosate-derived phosphorus increased from 1.6 kg/km^2 phosphorus in 1993 to 9.4 kg/km^2 phosphorus in 2014, with values frequently exceeding 20 kg/km^2 phosphorus in areas planted with glyphosate-resistant crops (Hébert et al. 2019). While these amounts are still much lower than those added by fertilizers, phosphorus inputs via glyphosate now reach levels for which environmental regulations for water protection were initiated in the past. After all, pesticides are one of the most common nonpoint sources of water pollution, meaning that no identifiable source like a pipe or drain is known.

These strong effects in our experiments were quite surprising, and we wondered why such simple investigations have not been carried out long ago in the course of the registration of glyphosate. It is important to note that these findings apply only to one glyphosate herbicide, while normally dozens of different pesticide products—insecticides, fungicides, herbicides, or acaricides—are used simultaneously or during the course of the season. The combined effects of these pesticide cocktails are unknown.

After its publication in an international scientific journal (Gaupp-Berghausen et al. 2015), this study caused much more uproar than the previous one. An Austrian NGO, which works a great deal on pesticides since decades, convened a press conference after the publication of the study in which our study and corresponding political demands were formulated. I refused an invitation to attend the press conference at that time because I did not want to expose myself politically with my work, but actually only wanted to provide facts for the debate.

Following this press conference, an agrochemical lobby group in Austria responded immediately. This group describes itself as a community of interests of pesticide-producing companies in Austria that advocates "open and objective information on all aspects of plant protection." This formulation already implies that everything else is not open and objective. This industry lobby group immediately assigned our study to the NGO that organized the press conference. Since then I am often considered being an NGO-scientist, although I never received any funding from NGOs or organic farming institutions (and of course also not from agrochemical companies).

We were also accused of working with excessively high dosages. The results would therefore have nothing to do with the reality of agricultural practice (IGP 2015). As a newcomer in this field, I was quite surprised by these attacks. Usually, only colleagues within the safe academic glasshouse would respond or refer to a scientific publication, but not mighty lobby organizations. Meanwhile, after being in pesticide research for several years, I know that these attacks are part of the strategy of the agrochemical industry and their lobbyists to discriminate and unsettle critical scientists. This thread will be taken up in a later chapter.

Let us come back to the accusation of using incorrect dosages in our experiment. The product user manuals of glyphosate herbicides give very contradictory dosage information. There is the option to use either an area-based dosage or a plant-based dosage. The area-related dosage is 33 ml/m^2. That is not much. Of course, we have scientific balances in our labs to exactly measure the dosage needed per pot surface. However, we wanted to look what home gardener would do and it is impossible for her/him to fill up this tiny amount with kitchen utensils. Also, the herbicide comes in a spray bottle without any measuring aid. The other dosage recommendation states that 1 liter of the herbicide is good killing for up to 1000 plants. Here, the question arises who counts the plants that are to be sprayed before the treatment? Would not the efficacy of the herbicide also depend on the size of the plants? Anyway, the manual does not differentiate between plant sizes. Thankfully, we knew exactly how many plants we had in our pots and therefore applied the plant-related application rate (which was about five times higher than the one on an area basis). In conclusion, we applied herbicides according to recommended dosages by manufacturers.

Well, I learned from this dosage episode that herbicides sold in spray bottles cannot be dosed properly and are most likely significantly overdosed. For this reason, I strongly advocate for a complete ban of pesticides for non-professional users. In addition, it is safe to assume that only a few private gardener will read the instructions for its use (Sattelberger 2001). In comparison to professional use by farmers wrong dosages in the private garden also have no consequences, because no one controls them. If it is stated in the manual that the product should be used in such a way that the weeds are evenly moistened and the application should be repeated if necessary, then any legally stipulated maximum quantity becomes invalid anyway. On top of it often the motto prevails that "more helps more."

Regarding dosage, it is also interesting to note that the dosage recommendations for private users (330 liters/ha) are up to 100 times higher than that for professional users (3–4 liters/ha). Unfortunately, using wrong dosages is not only an issue for private users. It has already been mentioned that herbicides are applied with insufficient equipment in Romanian vineyards. Colleagues and farmers also told me that in some tropical regions in Central America often two–ten times the recommended dosages of glyphosate herbicides are applied from the outset, since the tropical soils are rich in iron (red color) that decreases the efficacy of these herbicides. From an African country I was told that principally twice the recommended dosage is used for glyphosate herbicides because pesticide containers often get tapped and filled up with water. To avoid that the diluted herbicide is ineffective, farmers use a higher than recommended dose right away. A colleague also told me that even some farmers in Central Europe fill their spray tanks without using measuring vessels but rather prefer counting the gulps when pouring the pesticides into the tank.

Following the criticism of the pesticide lobby group, the Austrian pesticide registration authority also responded to our study with a four-page statement (AGES 2015). The main point of criticism was that our study tells nothing about the effect of glyphosate on earthworms. That is actually correct, but we never said it would. We always emphasized that we investigated the effect of a glyphosate-based herbicides with all possible co-formulants not just the active substance glyphosate alone. We were interested in practical effects and have therefore investigated a product that is actually used in the field. It is very important to keep this separated. Further criticism concerned the experimental conditions that dead plant material after herbicide killing alone would lead to a reduction in earthworm activity. Interestingly, this was exactly the same criticism raised in an US internet blog (Kniss 2015). In this blog, our study was cited as a prime example of bad science. The blog entry was written by a scientist who is quite open with his receiving of broad funding from various agrochemical companies (Kniss 2019). By the way, earthworms feed primarily on dead plants; hence, the presence of dead plant material would actually substantially increase earthworm activity. The Austrian authority at least concluded in their statement that after all our study showed that there definitely seemed to be nontarget effects of glyphosate herbicides on earthworms.

The coverage of our studies in German-language media has already been mentioned. A few days after the publication of our study I returned to my office from the lecture hall. The phone rang and a call came in from the university's public relations department asking where I was because the BBC wanted to put me on a live show on Radio 4. They wanted to talk with me about our earthworm study. Of course, I was excited (and nervous) of this call. When I then contacted the BBC again to confirm that I would be delighted to appear on a show the next day, they gave me to understand that one gets usually only one call from the BBC, and other topics were already planned for the next few days. Apart from that, national and international media interest has continued for some time. The study was mentioned in dozens of media articles; on top of this I gave lectures to farmers and talked about our experiments at scientific meetings. Finally, there was also a parliamentary question about the study in the Austrian parliament (PA 2015). However, no concrete political reactions on the subject were taken by the responsible ministry since then.

Half a year later, also the Bavarian Broadcasting company (BR) called in. They wanted to make a documentary about glyphosate, but all German scientists they contacted refused to participate. Because we just had a field trial running I accepted this invitation. The aim of this study was to test if our findings from the greenhouse will also be seen in the field. Usually, the situation in the field is infinitely more complex than in the greenhouse because many more uncontrollable influencing factors are in place. The film shooting with the Bavarian film team was a very nice and interesting experience—one has no idea how tedious it is to shoot a documentary. For the 6-min clip, the shooting with different settings and locations lasted a full day. The result is quite respectable, the contribution is objective and not too sensational, and the movie has already been viewed several thousand times on YouTube (BR 2016).

Later we continued our research investigating potential side effects of herbicides applied in vineyards under the grapevines (Zaller et al. 2018). We examined the impacts of three different herbicides with different active ingredients (flazasulfuron, glufosinate, and glyphosate) and found that all three herbicides reduced grapevine root mycorrhization on average by 53% compared to mechanical weeding. Soil microorganisms were also significantly affected by herbicides. Herbicides also altered nutrient composition in grapevine roots, leaves, grape juice, and xylem sap that was collected 11 months after herbicide application. This was among the first studies using an interdisciplinary approach to assess pesticide effects in the field. The herbicide treatments also had complex effects on soil microorganisms (Mandl et al. 2018). We found about 160 different bacteria species; total bacterial counts under herbicides were on average 260% higher than under mechanical weeding. The significance of these herbicide-induced alterations of soil microorganisms is not clear yet. However, evidence is increasing that the terroir of wine, that is the specific taste of wine, is not just a result of grape varieties and soil minerals but also of biological soil parameters (Belda et al. 2017). Soil microorganisms have been described as responsible for the chemistry and nutritional properties of soils, but also for health, yield, and the quality of grapes.

In this experiment, we did not find herbicide-induced reductions in earthworm activity or biomass and were up to further investigate the reasons for this in another project. But unfortunately, our research was abruptly stopped after one of the biggest boulevard newspapers in Austria (Heute 2018) and the science section of the Austrian broadcasting station (ORF 2018) featured our study results with headlines warning that we found that Austrian wine is threatened by glyphosate. A manufacturer of a tested herbicide asked for our study data, reanalyzed everything, and finally had to admit that our conclusions were correct. I have never experienced such a massive intervention against our research activities before. The cowardly move was that none of the responsible parties talked to me personally but rather intimidated people working with me. I am sure they sensed themselves that their actions endangered the constitutional right to freedom of research.

After we had carried out some experiments on the effect of herbicides, we turned our attention to a group of substances that had attracted a lot of attention in connection with bee mortality: insecticides with the active ingredient group of neonicotinoids. I already mentioned that neonicotinoids are the world's most important insecticides and that they act systemically protecting the whole crop. Neonicotinoids are mainly used as seed dressings; their degradation is very slow and can take years (Goulson 2013).

It is evident that only about 2–20% of the neonicotinoids are taken up by the crop plant. The resulting neonicotinoid concentrations in the sap (5–10 µg per liter of sap) are sufficient to control sucking and chewing pest insects (Sánchez-Bayo 2014). Accordingly, 80–98% of the neonicotinoids remain in the soil of treated crops, might affect soil biota, and eventually move into surface waters or the groundwater because they are easily soluble in water.

A significant pressure for surface waters and groundwater in Europe comes from agriculture leaving only a third of the water bodies in good chemical quality (EEA 2018). Good chemical quality means that the concentration of one active ingredient is below 0.1 µg/liter (or parts per billion, ppb) water or a total maximum of 0.5 µg/liter when more than one chemical is found. Pesticide-contaminated insect pollinators, soil biota, or aquatic insect larvae are consumed by birds, fish, and other predators and can finally get into the food system of humans (Fig. 2.1). Surveys from various countries show 80% of surface waters contaminated with neonicotinoids at levels of 0.14–18 µg/liter, which are not lethal but sublethal to aquatic arthropods (Main et al.

2014). Micrograms per liter sound like nothing but experiments in aquatic model ecosystems treated with single or repeated dosages of the neonicotinoid imidacloprid confirm that midges, ostracods (seed shrimps), and mayflies disappear and their populations do not recover when neonic residues are above 1 μg/liter (Hayasaka et al. 2012). After 8 years of field monitoring, it was reported that imidacloprid concentrations as low as 0.01 μg/liter led to significant reduction of macro-invertebrates in surface waters (Van Dijk et al. 2013). These examples just demonstrate how toxic neonicotinoids are.

Anyway, our next experiment focused on the effect of pesticide seed dressings. In seed dressing or seed coating, the seed is uniformly surrounded by insecticides, fungicides, or both. This method is applied to seeds of different cereals such as maize or wheat, or in rapeseed and potatoes. The basic principle of these products does not lack a certain attractiveness, as the pickled seed is already equipped for the potential diseases and pests.

In our experiment, we were initially interested in the effects on earthworms and other soil organisms. Our working hypothesis was that the pesticide seed dressings will influence soil organisms—an aspect that has been little studied so far. We were really curious to see what we would find, as the pesticide quantities that adhere to the plant seeds are extremely low. However, we also know that some neonicotinoids (e.g., clothianidin) are extremely potent with a 10,000 times higher toxicity to honeybees than DDT (Pisa et al. 2015).

We set up our experiment again in the greenhouse. This time we also wanted to work with deep-digging earthworms and used 60-cm long sewer pipes as experimental units. Funnily enough, the pipes are smooth on the inside but corrugated on the outside and look like big segmented earthworms themselves. We filled the pipes with arable soil and seeded pesticide-treated wheat grains. The recommended seeding rate for this wheat variety was 367 seeds per square meter, converted to the diameter of our pots; this was a total of 18 seeds per pot. In order to make the experiment as practical as possible, we used the seed material that farmers also use. The commercial seeds are coated with a combination of different pesticides, neonicotinoid-based insecticides, plus fungicides or only fungicides. As a control, we used untreated seeds.

Another consideration was that earthworms or other soil animals like the tiny springtails could themselves influence the effect of the pesticides. Soil animals could, for example, nibble on the seeds or even move around the whole seed, thereby impairing the effect of the pesticides on other soil organisms. Therefore, we also set up experimental units with and without earthworms, with and without springtails. Springtails, scientifically called collembolans, are microscopically small insect-like animals (they have long been among a group of wingless insects but in modern taxonomy have been separated from them). They owe their common name to a jump-fork at their back, with which they can catapult themselves over long distances on the ground surface. Sir David Attenborough in his magnificent documentary "Live in the underground" states that a human would be able to jump over the Eiffel Tower in Paris with the jumping power of a springtail. There are about 3600 species of springtails worldwide, and they can occur in very large numbers in the soil—up to 100,000 individuals/m² in the upper 20 cm soil. Sometimes they can be seen with the naked eye as little white dots jumping around on compost heaps, or forming black carpets on snow as so-called glacier fleas (*Desoria saltans*) feeding on pollen or snow algae. Springtails are very important in soils as they decompose organic material and build up humus.

The results of our experiment showed that especially the fungicides in the seed dressings led to an increased activity of the springtails on the soil surface and reduced organic matter decomposition in the soil (Zaller et al. 2016). Our interpretation of this finding is that springtails seemed to avoid soil contaminated with pesticides. When soil biota are more active on the soil surface than in the soil they are also more prone to predation by other animals. Additionally, earthworm activity was reduced by pesticides used for seed dressing. Overall, we were quite surprised by these effects especially because we only investigated the influence of 18 pesticide treated seeds per experimental unit.

In a subsequent experiment, we were interested to see what happens when cereals with seed dressing were additionally sprayed with a herbicide either for ripening spraying or desiccation or later for weed control before sowing the next crop (van Hoesel et al. 2017). Without too much details, our results showed that the herbicides aggravated the side effects of the seed dressings. Therefore, it appears that different pesticide classes interact with each other showing synergistic effects. A further aspect that is commonly not considered during environmental risk assessments of pesticides.

By the way, wheat growth in our experiment was not influenced by seed dressing. This is not surprising at first, since the disease and pest pressure in the greenhouse was not great and consequently the pesticides did not have to unfold their effects. But this also reveals another dilemma using pesticide seed dressings: it is a preventive measure in the event that pests or plant diseases occur. However, the occurrence of pests and plant pathogens is strongly influenced by the climate, the surrounding landscape, and naturally occurring beneficial insect–pest interactions. If circumstances are favorable, pests and plant diseases may not occur at all and pesticide application would be unnecessary. This prophylactic application is also against the principles of integrated pest management after which a pesticide is only applied when certain pest population densities are reached (Goulson 2013).

Neonicotinoids are among the most toxic substances used as pesticides and have been shown to affect many other wild animals on land as well as in fresh and saltwater (Pisa et al. 2015). The lethal concentration, at which 50% of an earthworm population dies in an experiment to assess the environmental risk, varies depending on the type of earthworm and is between 2–4 mg/kg of soil for sensitive earthworms (Capowiez et al. 2005). Sublethal effects, i.e., those that do not directly lead to death, occur at concentrations of between 0.5–1 mg/kg of soil (Tu et al. 2011). These sublethal effects such as behavioral changes or other chronic effects are usually not taken into account in the tests for pesticide approval.

We usually get told that soil contamination is no issue with modern pesticides because they degrade rather readily. Studies showed that the half-life for some neonicotinoids in the soil is up to 8000 days, meaning that half of the original amount will be available after almost 20 years (Bonmatin et al. 2015). A study analyzing 317 agricultural topsoils from across Europe revealed that more than 80% of the tested soils contained pesticide residues (Silva et al. 2019). A quarter of the samples had one residue; more than half (58%) of samples had a mixture of two or more residues; in total 166 different pesticide combinations were found. Given the widespread use, it was not surprising that glyphosate and its metabolite AMPA were most often found. Rather alarming was the finding that DDT and its metabolites, which were banned in Europe for decades, were among the most frequently found pesticides in European soils in 2015. Quite often found were also several widely used broad-spectrum fungicides (boscalid, epoxiconazole, and tebuconazole). There is no regular pesticide monitoring performed in Europe (and nowhere else either) and nobody knows what these pesticide residues do in our environment. However, it is clear that they do not belong there.

Besides earthworms or collembolans, other nontarget soil organisms, aquatic organisms, or herbivorous insects feeding on non-crop plants in farmland will inevitably be exposed with neonicotinoids. Studies of food stores in honeybee colonies from across the globe demonstrate that colonies are routinely and chronically exposed to neonicotinoids in a range between 1–100 parts per billion (Bonmatin et al. 2015). We will later see that neonicotinoids also easily leach into groundwater or surface waters. Hence, although neonicotinoids have only been in use for about 30 years, they can already be found everywhere in our environment. Once again, scientific evidence disproves the myth that modern pesticides are less persistent and better degradable than the old ones. In most cases, the new products are only supposedly better, because fewer negative impacts are known, simply because there was less time to study them properly.

Another problem with ecotoxicological standard tests is the requirement that all experiments should be carried out at standard temperatures (usually 20 °C). In further experiments, we asked ourselves whether potential pesticide effects are altered by temperature. Everybody will affirm this question as temperature will alter chemical and physiological processes of the pesticides and in the organisms. For a change, we explored amphibians and algae.

2.1.2   Tadpoles with Crippled Tails

Farmers must maintain a distance of 5–20 m from the next water body when spraying pesticides in order to minimize the risk of pesticides being leached or drifted into the water. However, when the field plots are small, it can be difficult to maintain the required distances.

Official monitoring documents of German water bodies show that the number of pesticides in near-surface groundwater has continuously declined over the last few years (UBA Berlin 2019). Between 2013–2016 still about 4% of more than 10,000 near-surface groundwater samples exceeded the respective legal limit for good chemical quality of 0.1 µg/liter for at least one active substance. The decline in groundwater pollution is mainly due to decreasing discovery frequencies of the herbicide atrazine, desethylatrazine, and a few other active substances and their metabolites, the use of which has been banned for years or even decades. Modern pesticides occur much less frequently in groundwater than older pesticides. However, the report considers only larger groundwater bodies while excluding the numerous small groundwater bodies in the agricultural landscape with more or less intensive pesticide application. Across the EU, the following pesticides make problems for groundwater contamination: herbicides—trifluralin, atrazine, simazine, alachlor, and pentachlorophenol (used in herbicides, insecticides, fungicides, algaecides, and as disinfectant and in antifouling paints) (EEA 2018).

For the following experiments, we focused on pesticide effects on aquatic organisms: toads and algae. When seeing a frog or a toad in the garden or in nature, the heart of every nature lover rejoices. Not only do frogs and other amphibians look nice; they also consume lots of insects that could otherwise become a nuisance to us, like flies or mosquitoes. By consuming these insects, amphibians carry out classical biological pest control, which takes place unnoticed in all ecosystems. Amphibians are now among the most endangered vertebrates in the world because their habitats like swamps are being drained, ponds are being filled up with excavation material from construction sites or household trash, and they are threatened by new diseases and pesticides deliberately applied to destroy amphibians' food.

However, also herbicides can threaten amphibians by eliminating the plant food or vegetated habitat for insects that are eaten by amphibians. In addition, amphibians have a very sensitive skin. Unlike the human skin, it is not built up in three quite robust layers, but much thinner and better able to directly absorb water and pesticides dissolved in it. As a consequence of these threats, most species of toads, frogs, salamanders, and newts are protected in Europe and many other countries.

In two experiments, we examined what would happen when spawning ponds of toads are contaminated with glyphosate herbicides (Baier et al. 2016a, b). European toads are the most common amphibians in central Europe. In spring they migrate from their overwintering habitats (usually woody areas), to ponds and lakes in order to spawn.

Since toads are vertebrates, a lot of permits for animal protection and approval of animal experiments need to be obtained before starting with the experiments. This was not necessary for experimenting with earthworms and springtails as they are invertebrates. In possession of these permits, we took a few hundred toad eggs from the spawning ponds and distributed them evenly in plastic containers filled with active-coal filtered pond water. The populations of the European toads were not endangered by this sampling as a female toad lays up to 8000 eggs during spawning and dozens of female European toads lived in the pond.

Then we added realistically low amounts of glyphosate herbicides to the experimental units. We additionally wanted to find out how the pesticide effects would be altered by water temperature and maintained half of the containers in climate chambers at 15°C and the other half at 20°C. This approach was at least a first attempt to address whether temperature interacts with pesticide effects. One could also link this approach to man-made climate change with more extreme weather events such as stretches of unexpectedly cold or hot temperatures. However, this is a rather long shot. Our results showed that the herbicides influenced tadpole development mainly at 15°C and much less at 20°C. Under colder temperature, the tadpoles showed many tail malformations; in warmer temperatures there were no malformations at all. We observed that the tadpoles with the crippled tales were less skilled swimmers which would have handicapped them when escaping from predators in the water. The tadpoles also grew more slowly under cooler conditions and were therefore longer exposed to the herbicide, while at higher temperatures they were somehow able to outgrow the herbicide effect. In addition to aquatic animals, a large number of partially microscopically small algae species live in natural water bodies. Colleagues from hydrobiology found up to 43 algae species in our pond water; the diversity of algae communities was also altered by herbicides, regardless of the water temperature.

We were among the first studying these effects on European amphibian species, but of course much research was done on other species and found for instance reduced survival of juvenile Great Plains toads (*Anaxyrus cognatus*) and New Mexico spadefoot toads (*Spea multiplicata*) after exposure to certain formulations of the herbicides glufosinate and glyphosate (Dinehart et al. 2009).

The interesting and novel aspect of these findings was that temperature influenced herbicide effects. This is insofar important as the tests for pesticide approval are carried out at a standard temperature of 20 °C. Thus, our results gave rise to doubts as to whether these approval studies had much to do with the situation in nature (admittedly, our studies were also simple, but we had at least one additional factor included). Temperature dependence of pesticide toxicity was also found for other substances and amphibian species around the world (Lau et al. 2015; Rohr et al. 2011). It may come as no surprise that such a temperature dependence of fungicide toxicity has also been shown on soil animals, e.g., springtails (Bandow et al. 2014).

Glyphosate herbicide contamination of a pond has also been shown to kill 50–100% of the populations of leopard frogs (*Rana pipiens*), American toads (*Bufo americanus*), and gray tree frogs (*Hyla versicolor*) after only 20 days (Relyea 2005). When experimental ponds were contaminated with a pesticide cocktail consisting of single or combined contamination of five insecticides (malathion, carbaryl, chlorpyrifos, diazinon, and endosulfan) and five herbicides (glyphosate, atrazine, acetochlor, metolachlor, and 2,4-D) at low concentrations (2–16 μg/liter), a wide range of direct and indirect effects on zooplankton, phytoplankton, and larval amphibians were observed (Relyea 2009). Similar damaging effects to amphibians of the insecticides chlorpyrifos and endosulfan have also been found by others (Sparling and Feller 2009). Adverse effects of endosulfan on fish and invertebrates are a concern when this insecticide is used near aquatic ecosystems (Carriger and Rand 2008). In field tests, the insecticide carbaryl affected the composition of an aquatic community of amphibians and insects by changing colonization of pools and numbers of eggs laid (Vonesh and Kraus 2009). These and many other studies show that aquatic ecosystems can be seriously impacted even by low contamination of pesticides.

The situation with regard to pesticides and amphibians is of course more complicated than these simple experiments would suggest. In addition to pesticides and temperature fluctuations, predators such as dragonflies and water beetle larvae, fish, and ring snakes also lurk in spawning waters, and it has been shown that amphibians react differently to pesticides in the presence of predators (Relyea 2003a, b).

Amphibians migrate from their overwintering sites to the spawning ponds and spend the summer on land and then also roam through agricultural land where they run the risk of being sprayed directly by pesticides. Carsten Brühl and his lab in Germany were the first to show that grass frogs are in jeopardy when exposed to pesticides (Brühl et al. 2013). They tested how grass frogs are affected by direct oversprays of fungicides, herbicides, and insecticides and found that frogs died within a week after exposure to field-relevant concentrations. The astonishing aspect here is that these pesticides come on the market without an assessment of possible effects on amphibians. Up to now it is assumed in the official risk assessments that a pesticide that is safe for birds or mammals is also safe for amphibians. This is another example of rejecting the narrative that pesticides are rigorously tested before they are allowed. Overall, farmers should also have a keen interest in intact nature and the protection of amphibians, as these animals decimate many potential insect pests.

Frogs also live in shrubby coffee plantations where a great amount of pesticides is sprayed. Frogs collected in coffee plantations in the Western Ghats in India showed a worse health status when regular pesticides were used (including Bordeaux mixture, lindane, Di-Syston, chlorpyrifos, bavistin, phorate) compared to those in plantations without pesticide use (Hegde et al. 2019).

When introducing herbicides, I briefly mentioned the active ingredient atrazine. Atrazine is still one of the most commonly used pesticides worldwide and has been shown leading to infertility and even sex change in adult frogs. In one study, three-quarters of male frogs exposed to atrazine were "chemically castrated" (Hayes et al. 2010b); they were then no longer able to reproduce. The male frogs suffered from testosterone deficiency and lost all the abilities controlled by this hormone. Every tenth male frog even underwent a sex change; they were then able to mate successfully, but gave birth only to male offspring. In the wild, where there is competition for sex partners, these animals would probably have no chance. Although these experiments were carried out on African clawed frogs, it can be assumed that atrazine has similar effects as endocrine disruptors also in other species. Endocrine disruptors are substances that act like hormones and can disturb the sensitive balance of the hormone system of living beings. In the USA, many rivers across the country are so heavily polluted by endocrine disruptors that up to 70% of the fish examined already show intersexual phenomena (Hinck et al. 2009). The Pee Dee River at Bucksport in South Carolina even contained 91% of intersex fish. In a later chapter, the role of endocrine disruptors for humans will be also be addressed.

Pesticides and other environmental pollutants also arise as likely factors in worldwide amphibian declines (Hayes et al. 2010a). However, as for other species declines, not only one factor can be made responsible for this but factors like pathogens, climate change, habitat modification, and invasive species also play an important role. Later we will see that one can get into huge troubles with agrochemical companies after revealing such findings.

In Europe, atrazine has been banned since the 1990s; however, it is still omnipresent in our environment, especially in soil and water samples. In the USA, this herbicide is still used despite these frightening side effects, for example in maize, but also in asparagus, potato, and tomato cultivation. Despite a ban, we consumers can still come into contact with atrazine residues in these foods more so when free-trade agreements foster the import of these commodities.

2.1.3   Bees and Bumblebees Without Orientation

So far, we did not conduct our own experiments with bees. However, many other studies clearly demonstrated that neonicotinoids not only kill insect pests but also beneficial insects such as wild bees like bumblebees or the domestic honeybees (Fischer et al. 2014). When talking about bees, most people think only of honeybees and perhaps bumblebees. However, there are about 20,000 bee species worldwide, about 4000 alone in the USA; even in a small country like Austria, there are 700 different bee species known.

According to the assessment report on pollinators, pollination, and food production of IPBES, the Intergovernmental Science-Policy Platform on Biodiversity and Ecosystem Services (IPBES 2017), a growing number of pollinator species worldwide are being driven toward extinction by diverse pressures. Pollinators are important contributors to world food production and nutritional security, and their health is directly linked to our own well-being. Besides the many wild bees, many species of butterflies, flies, moths, wasps, beetles, birds, bats, and other animals contribute to pollination. Insect pollinated crops include those that provide fruit, vegetables, seeds, nuts, and oils. It is estimated that more than three-quarters of the world's food crops rely at least in part on pollination by insects and other animals, which sums up to between 212 and 520 billion € worth of annual global food production that relies on pollinators.

A fact that is often ignored is that nearly 90% of all wild flowering plants also depend at least to some extent on animal pollination. In addition to food crops, pollinators also contribute to crops that provide biofuels such as canola or palm oils, fibers such as cotton, medicines, forage for livestock, and construction materials. Some species also provide materials such as beeswax for candles and musical instruments and arts and crafts. Additionally, an estimated 16% of vertebrate pollinators are threatened with global extinction with a trend toward more extinctions (IPBES 2017).

Insect pollinators are threatened in different regions by a variety of factors. However, generally it can be said that their decline in certain regions is primarily due to changes in land use, intensive agricultural practices and pesticide use, alien invasive species, diseases and pests, and climate change (Sánchez-Bayo and Wyckhuys 2019).

Among the differenct pesticide classes, insecticides are of course the greatest threat to bees. We heard that neonics are extremely toxic to bees; the frequently used active substance clothianidin, for example, is 10,800 times more toxic for bees than the notorious DDT (Pisa et al. 2015). The side effects of neonics have been researched in much detail in the last few years, and a complete ban on this group of active substances is meanwhile in place in Europe and under evaluation in Canada. The discussion has focused primarily on honeybees. However, the findings show that the environmental damage is far more extensive than previously thought. A study found that US agriculture has become 48 times more toxic to insects over the past 25 years and pinned 92% of the toxicity increase on neonicotinoids (DiBartolomeis et al. 2019).

Bumblebees that came into contact with a field-relevant dose of a neonicotinoid remained smaller and produced up to 85% fewer queens (Whitehorn et al. 2012). Bumblebees have a 1-year life cycle: only the new queens survive the winter and can start new colonies in the following spring. Fewer queens simply means fewer bumblebee colonies that can be newly founded. Like other bee species, bumblebees feed on the pollen of wild flowers on the edges of fields and in particular on rape blossoms and sunflowers. Neonic concentrations in the range of one parts per billion (ppb) in food led to a 30% reduction in egg deposition (Stoner and Eitzer 2012). One ppb corresponds to an unimaginably small amount of one gram per thousand tons. The neonic concentrations found in pollen from cultivated fruits were typically between 1–10 ppb. Concentrations of 50 ppb were found in alfalfa pollen and over 100 ppb in melon pollen.

When studying effects of pesticides on insects, it is also important to bear in mind their life cycles. For instance, bumblebee queens forage for pollen and nectar and are thus exposed to more risk of direct pesticide exposure than honeybee queens and almost no research has been done on pesticide exposure to and effects on bumblebee queens (Stoner 2016). Many studies have investigated the effect of neonics on female honeybees or queens. But a study on male drones showed that neonics were effective as a contraceptive and drones produced less sperm (Straub et al. 2016). If the mated queen bee then lays fewer eggs, this has an effect on the vitality of the entire colony.

Field trials conducted in Sweden did find detrimental effects of neonics (clothianidin): reduced wild bee density, solitary bee nesting, and bumblebee colony growth and reproduction, but not on honeybees (Rundlöf et al. 2015). A proposed mechanism for these negative effects of neonicotinoids on bee foraging performance is an impairment to learning and memory (Siviter et al. 2018b). One of the first free-foraging study into the effects of acute exposure of a neonicotinoid (imidacloprid) on bumblebees (*Bombus impatiens*) uncovered dose-dependent detrimental effects on motivation to initiate foraging, amount of nectar collected, and initiation of subsequent foraging bouts, but no impairment to bees' ability to learn visual associations (Muth and Leonard 2019). These are interesting scientific questions to be further studied in order to understand the mechanisms responsible for the decline in insects.

Europe's largest (industry financed) field trial on neonics and their danger for bees also yielded mixed results, suggesting that neonicotinoids are most likely not the only threat (Woodcock et al. 2017). The study was conducted in the UK, Hungary, and Germany and provides the first real-world demonstration that agricultural use of neonicotinoids can hurt both domesticated honeybees and wild bees. However, in Germany, researcher saw no lasting effects in honeybee colonies near the treated crops, while in Hungary, colonies near oilseed rape treated with clothianidin had, on average, 24% fewer workers in the following spring. UK trends were similar, but not statistically significant. Researchers suspect that the German honeybees might have been generally healthier than those in the other two countries. Wildflowers growing near the German fields might also have provided extra resources that could have made the bees more resilient. The study also showed that in wild bees (*Bombus terrestris* and *Osmia bicornis*), reproduction was negatively correlated with neonicotinoid residues. Moreover, imidacloprid, a neonicotinoid that has not been used in Europe since several years, have been shown to persist in the environment and is taken up by wildflowers, exposing bees even years later. Another study of honeybees in Canadian maize fields identified up to 4 months of chronic exposure to neonicotinoids that also appear to have persisted from prior plantings which is much longer than admitted by the manufacturer (Tsvetkov et al. 2017).

A joint report by several European academies of sciences states that the widespread preventive use of neonics has strong side effects on nontarget organisms (EASAC 2015). This includes insects providing ecosystem services such as pollination and natural pest control. The additional effects on soil organisms have already been discussed in an earlier chapter. This prophylactic use of neonics currently being practiced is contrary to the basic principles of integrated pest management which was quite popular in the 1990s and is also laid down in actual EU directive for the sustainable use of pesticides (EPRS 2018). This prophylactic approach also endangers the restoration of biodiversity in agricultural land, which is actually also an objective of EU agricultural policy.

I already mentioned that in 2013, the EU prohibited applications of the three neonics on crops attractive to bees, such as sunflowers, oilseed rape, and maize. It was found that foraging bees are exposed to harmful levels of pesticide residues in pollen and nectar in treated fields and contaminated areas nearby, as well as in dust created when treated seeds are planted. It also concluded that neonicotinoids can sometimes persist and accumulate in the soil, and can so affect generations of planted crops and the bees that forage on them.

Then in early 2018 the European Union Food Safety Agency EFSA in a long-awaited assessment has concluded that three controversial neonicotinoid insecticides (clothianidin, imidacloprid, and thiamethoxam) pose a high risk to wild bees and honeybees (EFSA 2018b, c, d, e). The agency surveyed more than 1500 studies, including all the relevant published scientific literature, together with data from academia, chemical companies, national authorities, NGOs, and beekeepers' and farmers' associations. The assessment found that each of the three neonics posed at least one type of high risk to bees in all outdoor uses.

EFSA's advice was often criticized by interested parties such as NGOs or agrochemical companies, but this time the agency demonstrated that it can also make decisions purely based on scientific facts (Butler 2018a). Of course, the agrochemical industry criticized EFSA's conclusions on neonicotinoids as overly conservative. The industry states that their evidence clearly shows that neonics pose a minimum threat to bee health compared to a lack of food, diseases, and cold weather and that they of course stand by their products and science (Butler 2018b).

What will farmers do without neonics? Immediately after the 2013 moratorium on three neonics, the agrochemical industry turned to farmers recommending alternative products. A complete ban would certainly immediately bring several other chemical insecticides as substitutes. So, one should not be so naive and thinking that pesticide applications will be abandoned. Agrochemical industry of course has alternative products in their portfolios and neonicotinoids will be replaced by other substances.

One such substitute substance for neonics is sulfoxaflor, a substance with a similar mode of action. EFSA also found that sulfoxaflor may pose a high risk to honeybees and bumblebees in a treated crop scenario except after flowering period, weed scenario, and field margin scenario and long-term risks to small herbivorous mammals in certain applications (EFSA et al. 2019a). A low risk was concluded at low exposure in greenhouses. Nevertheless, mixtures of pesticides containing sulfoxaflor are now approved in 15 EU member states (EC 2019). Also a study showed serious effects of sulfoxaflor on the buff-tailed bumblebee (*Bombus terrestris*) with only half as many offspring when colonies were exposed to sulfoxaflor in field-realistic quantities compared to sulfoxaflor free fields (Siviter et al. 2018a).

Another replacement of neonicotinoids is cyantraniliprole with two insecticides also approved in Germany (BVL 2019b). Several other European countries approved this insecticide or are considering their approval. For viticulture, it has an emergency approval against the spotted wing drosophila (*Drosophila suzukii*); farmers are advised to take certain safety precautions because of its high toxicity for bees (BVL 2019a).

Flupyradifurone is another active substance similar to neonics. It also binds to the receptors of the nerve cells of insects, thereby disrupting the transmission of nerve stimuli. A study has shown that nonfatal doses of flupyradifurone after single administration to collecting honeybees have a negative effect on the taste perception, learning, and memory of insects (Hesselbach and Scheiner 2018).

So, I guess we should not be worried that the farmers have nothing to spray, when neonics are finally all banned. The more sensible consequence of this situation, however, would be to at least return to the principles of integrated crop protection. Spraying only when certain damage thresholds are exceeded would drastically reduce the quantities of pesticides used. But of course, also the sales for the agrochemical industry would drop. Experience tells that most likely everything will continue as before and that banned insecticides will be replaced by supposedly new, better ones; and the game starts over again.

The general insect decline has most likely not be solved by banning neonics. For decades there has been a worldwide decline in insects, and pesticides are not exclusively to blame for this. For honeybees, we know that besides pesticides also parasites (*Varroa* mite), various bee diseases, prolonged periods of bad weather, and poor flower resources affect bee health. In addition to neonics, herbicides can also indirectly damage bees, as they destroy native herbs and thus the food basis of bees. Even direct effects of glyphosate on honeybees have been found. Exposure of honeybees to glyphosate can alter the bee gut microbial community and increases a bee's susceptibility to infection by pathogens (Motta et al. 2018). Clearly, if we want to ensure healthy populations of honeybees, bumblebees, and wild pollinator insects, we must generally do something to promote biodiversity in our agricultural landscape.

Because bees are very mobile and collect pollen across the landscape, they have also been used as bioindicators of environmental pollution. Pollen from beehives contained up to 17 different toxic pesticides (Greenpeace 2014). In addition, 53 different other chemicals could be detected. In the samples from Germany, a bioaccumulative degradation product of DDT, which was banned decades ago, the neonic thiacloprid was frequently found in bee pollen.

Pesticide toxicity to honeybees is generally determined by the effects of sprays and residues directly applied to adult honeybees. However, the behaviour of social bees, their long-range foraging habits may lead to other types of exposures (Hooven et al. 2013). Bees might be exposed to one pesticide in one crop, but then fly to a new cropping system, then get exposed to a second pesticide, and so on. We have currently no idea about the role of additive, synergistic, chronic, or delayed effects from multiple sources and types of exposures play a role. Although fungicides are not thought to affect adult bees, certain fungicides, such as captan, iprodione, and chlorothalonil, have been shown to affect brood development, or affect the microorganisms that ferment bee bread in laboratory studies. Studies suggest that pesticide tank mixtures are more toxic to bees than the single substances, but more research is needed on this topic.

Neonics are found not only in the treated crops, but also in the field edge vegetation (Sánchez-Bayo 2014). This means that all insects that feed on these field edge flowers, such as butterflies or hoverflies, are affected. When seeds dressed with neonics are eaten by grain-eating birds, such as partridges, or rodents, a few grains are sufficient for a lethal dose. The probability that pesticide dressed seeds will be eaten by wild animals is quite high. This can easily be observed when the red- or blue-colored seeds are sometimes spilled while sowing with so-called seed drill machines. How often wild animals are poisoned by neonics by taking up such seeds is not documented.

The concentrations of neonics found in agricultural fields often do not have an acute lethal effect on insects, but it has been shown that behavior, learning ability, food collection, and orientation in space are impaired (Tosi and Nieh 2017). Since parasitic and predatory insects often have complex search and attack patterns, sublethal effects via neurotoxic insecticides can also affect this behavior and thus the efficiency of biological control performed by these species. These are also effects that are not considered in experiments for pesticide approval.

Meanwhile, it is clear from numerous studies that the risks of neonicotinoids have been understated in the past and the benefits have been overstated. The manufacturer is still convinced that their neonicotinoid products are safe for bees (Kerr 2017).

Generally, the appreciation and knowledge about the importance of insects is not too good among the public. Very few people have an idea of how many insects are romping around even in crop fields. In an unsprayed experimental wheat plot, we were able to collect up to 1400 arthropod specimens per square meter, including ants, bees, spiders, grasshoppers, leaf beetles, ground beetles, springtails, cicadas, flies, and lacewings (Zaller et al. 2014b). This shows that agricultural land must not necessarily be a green desert, but can in fact inhabit a high abundance and diversity of insects and spiders. Only very few of them are pests on agricultural crops. On the contrary, many of them decimate pests in the wheat fields and thus carry out biological pest control. Many of these animals are also valuable food sources for birds, which in turn kill many insect pests and ultimately ensure that agricultural crops remain healthy.

Occasional mosquito invasions should not obscure the fact that the quantity and diversity of insects and spiders in our cultural landscape has almost halved in the last 35 years (Dirzo et al. 2014). It is remarkable that the human population has doubled in the same period. Dave Goulson, one of the world's most prominent entomologists, warned that despite the media hype on insect declines ecologists and entomologists should be deeply concerned that they have done such a poor job of explaining the vital importance of insects to the general public (Goulson 2019). For many insects, we simply do not know what they do and we have not even given a name to the large majority of species, let alone studied what ecological roles they might perform.

Insects make up the bulk of known species and are intimately involved in all terrestrial and freshwater food webs. The use of pesticides is often justified with the production of food for an ever-increasing world population. However, the majority of the crop types grown by humans require pollination by insects; hence we could not feed the growing global human population without pollinators. Before we justified the importance of insects with the ecosystem services they provide, such as pollination service to crops, fruit trees, strawberry, tomato, rape, and sunflower plants at around 150 billion € per year (Gallai et al. 2009). These calculations do not include the effects that a decline in pollinators would have on agricultural plant production and thus also on animal production. The effects on wildflowers and all other ecosystem services that the natural flora and fauna provide for agriculture and society are also missing. A recently published insect atlas gives a great overview of the situation of insects in agriculture around the world (HBS and BUND 2020).

Poorly investigated are effects of pesticide adjuvants on bees. Organosilicon surfactants are used as spray adjuvants on many agricultural crops including wine grapes, tree nuts, and tree fruits. A study showed that these substances make bee larvae more susceptible to deadly viruses and can thus be responsible for the general death of honeybees (Fine et al. 2017). These results demonstrate again that adjuvants that are considered biologically inert during the pesticide registration procedures need to be investigated with more scrutiny.

The huge group of parasitic wasps and its susceptibility to pesticides are rarely considered either; perhaps it is the incredible diversity of this insect group that discourages people from studying them. As parasitoids, they lay their eggs on or in the bodies of other arthropods, sooner or later causing the death of these hosts. Some groups of parasitic wasps comprise as many as 500,000 species (Chalcid wasps), ichneumon wasps with 100,000 species, and the braconid wasps with up to 50,000 species (Gullan and Cranston 2014). Some of these parasitoids are only a few millimeters small, and thus, even the tiniest doses of neonicotinoids can have lethal effects of interfering with their mating and oviposition (Tappert et al. 2017). This has important consequences, since parasitic wasps play an eminent role as natural enemies of pest insects and as natural pest controllers in agricultural ecosystems.

The general decline in insect abundance will have consequences for birds and bats living of insects. Research from Scotland showed that agricultural change during almost three decades has influenced birds through changes in insect quantity and quality and may have contributed to the decline in farmland birds (Benton et al. 2002).

2.1.4 Birds and Bats with Nothing Left to Eat

Birds play a very important role in natural and agricultural ecosystems. They can serve as pollinators of crop plants, consume pestiferous insectts, carry and distribute seeds of wild plants over long distances. In recent decades, however, many bird species have experienced a dramatic decline in their populations. Of course, pesticides cannot solely be held responsible for this decline, but it is the system associated with pesticide-intensive agriculture with large monocultures and homogenous landscapes which makes a life for birds life difficult. Important drivers are also the decline in fallow land, plowing of meadows and conversion to arable land, intensification of grassland management with increasing mowing frequencies and the heavy use of fertilizers, changes in harvesting processes and times, and the destruction of natural and semi-natural landscape features, and on top of all these drivers climate change and other stressors.

Direct killing of birds via pesticides is forbidden in all countries. But nevertheless, birds are heavily affected by a variety of pesticides used either professionally or by private people. A classic example is that bird species find less food because of pest insects and nontarget insects have been killed by insecticides (Jahn et al. 2014). A comparison of long-term population trends of European insectivorous and seed-eating birds showed that both groups declined considerably throughout Europe within the last 25 years (Bowler et al. 2019). The decline in insectivores, that is insect feeding birds, was primarily associated with agricultural intensification and a loss of grassland habitats.

These detrimental effects are not only limited to the effects of insecticides. Also, herbicides remove wild herbs that provide insect food and provide protection for ground breeding or young birds. Moreover seed-eating birds suffer hunger when wild herbs are killed by herbicides. These interrelationships may seem trivial, but are often not considered because people have heard repeatedly that pesticides act very specifically—herbicides only kill plants and insecticide only insects. Other indirect pesticide effects on birds have been known since *Silent Spring* and are steadily uncovered.

Carbofuran, an insecticide of the group of carbamates, has a mode of action similar to the organophosphates and can kill birds at very low doses and has killed millions of birds each year in the USA. In its granular form, carbofuran was finally banned in in the USA in 1991 and 2009 also in its liquid form; it is also banned in Europe. Yet carbofuran is still finding its way into ecosystems and killing birds. For instance, Maryland wildlife officials announced in May 2019 that 7 bald eagles and one owl had been killed from suspected carbofuran poisoning (Holcombe 2019). Officials determined that the animals had ingested the pesticide—used legally and illegally today by farmers and breeders around the world, mainly as a form of pest control—after feeding from carbofuran-laced baits; the eagles were also found to have been eating a red fox carcass. Also in Austria, where carbofuran is banned for more than 10 years still dozens of protected birds of prey poisoned by this insecticide can be found every year (ORF 2019).

Carbofuran is not only lethal to birds but also dogs and other mammals can be severely poisoned by this insecticide. In Kenya, for example, the carbofuran has been reportedly used by herders as a deterrent to protect their flocks from lions (Wadhams 2010). Illegal placement of carbofuran poisoned bait to eliminate perceived pests, such as jackals (*Canis aureus*) or foxes (*Vulpes vulpes*), was responsible for high concentrations of carbofuran in the liver and kidney tissue of dead brown bear (*Ursus arctos*) in Croatia (Reljić et al. 2012). Experts have claimed that just one teaspoon of carbofuran is toxic enough to kill a full-grown bear.

Parathion is another insecticide with a huge toll on birds. Older studies mention that in a wheat field in the USA, for example, 1200 Canada geese were killed after the parathion was applied at a dose of only 0.08 g per square meter (White et al. 1982). Other studies report pesticide poisoning of 1000s of ducks and geese after cereal crop applications (Flickinger et al. 1980). This active ingredient was also very popular in Europe as a seed treatment in sugar beet cultivation, against insect pests in mushroom cultivation, onion cultivation, and commercial horticulture, but has been banned in the EU since 2007.

Another insecticide (diazinon), used on three golf courses in Canada, killed 700 Atlantic geese in a wintering population of only 2500 birds (Stone and Gradoni 1985). This active ingredient is still in use in the USA, together with the insecticides chlorpyrifos and malathion; diazinon is used in California alone on more than 400,000 ha (EPA 2016).

In the UK, the volume of seeds eaten by many bird species is large enough to pose a potential risk if the seeds are treated with one of the more toxic fungicides (Prosser and Hart 2005). Organophosphate insecticides, including disulfoton, fenthion, and parathion, are highly toxic to birds. These have frequently poisoned raptors foraging in fields (Mineau et al. 1999). Field studies have led to the conclusion that given the usual amounts of insecticide used direct mortality of exposed birds is both inevitable and quite frequent with the large number of insecticides currently registered (Mineau et al. 2005). In a small area of the Argentine Pampas, monocrotophos, an organophosphate, has killed 6000 Swainson's hawks (Goldstein et al. 1999). Insecticides generally had a negative effect on yellowhammer when spraying occurred during the breeding season. Spraying at this time may cause more damage than use throughout the year (Morris et al. 2005). Spraying insecticides within 20 days of hatching led to smaller brood size of yellowhammer, lower mean weight of skylark chicks, and lower survival of corn bunting chicks (Boatman et al. 2004). More frequent spraying of insecticides, herbicides, or fungicides was linked to a considerably smaller abundance of food invertebrates. This resulted in lower breeding success of corn buntings and may have contributed to their decline (Brickle et al. 2000).

Neonicotinoid insecticides impair not only bees, bumblebees, and soil organisms as mentioned before, but also birds. In the Netherlands, it was found that the more imidacloprid was present in surface water, the smaller the populations of 15 bird species were (Hallmann et al. 2014). Concentrations of unimaginably small amounts of 20 billionths of a gram of insecticide (neonic imidacloprid) per liter of water resulted in an annual decline of 3.5% of bird populations. A study collected feathers from house sparrows (*Passer domesticus*), an omnivorous bird from differently managed farms in Switzerland, and found at least one neonicotinoid in all 617 feathers sampled (Humann-Guilleminot et al. 2019). House sparrows living on conventional farms showed higher concentrations of neonicotinoids than individuals living on farms managed according to integrated or organic regulations.

Researchers also found that white-crowned sparrows (*Zonotrichia leucophrys*) eating seeds treated with neonicotinoids become anorexic, causing them to lose weight and delay their southward journeys (Eng et al. 2019). This is especially significant as these are migratory birds where a too late flight may mean that they miss the peak season for finding good food, a good mate, or a good nest site. The study might apply to other birds as well—and additionally help explaining the dramatic songbird decline of recent decades.

Herbicides, as mentioned, can decimate the nutritional basis of many birds. In sugar beet and winter rape, weed control with herbicides led to a decline of 17 primarily seed-eating field bird species (Gibbons et al. 2006). A study conducted in several European countries found that insecticides and fungicides in particular had a very negative effect on the diversity of soil-breeding field birds (Geiger et al. 2010). In Canada and France, a correlation was found between the use of herbicides and the decline of seed-eating birds (Gibbs et al. 2009).

The effects of fungicides on birds are subtler. Because of the possibility to use chemical fungicides, cereal plants can be sown more densely than in former times, so the farmland birds literally have less space to live in the dense cereal stands.

Overall farmland birds are in direct contact with pesticides during the whole season. A study considering 6500 ha of farmland in France showed that 71% of partridge (*Perdix perdix*) clusters are exposed to at least one pesticide (Bro et al. 2015). Contamination in birds involved 2–22 active ingredients with 63% fungicides, 25% herbicides, and 16% insecticides used on a variety of crops between April–June, when ground-nesting birds are breeding. These data show the huge pesticide exposure for birds in agricultural landscapes and also point to the fact that such scenarios are not adequately addressed in environmental risk assessments of pesticides.

A further aspect, which is little discussed, is the endangering of thousands of buzzards, owls, eagles, vultures, and other birds of prey when feeding on poisoned rodents in urban and rural areas. A study has shown that a single pesticide application against voles has killed 28 red kites and 16 buzzards (Coeurdassier et al. 2014). The true extent of pesticide poisoning in birds and other vertebrates is not known, as no systematic surveys are conducted.

As mentioned, particularly with such mobile organisms as birds, the pesticide problem cannot be discussed in isolation from other developments in agriculture. Those who pay attention to the developments of our landscapes will acknowledge the massive changes. Until a few decades at least in central Europe, meadows and pastures were a delight to the eye forming habitats for many animal and plant species. Today there are mainly monocultures of maize, cereals, apple orchards, or vineyards. Fallow land that is not used for agricultural purposes, hedges, solitary trees, small groups of trees, and all the other small structures have in many places disappeared from the landscape. For partridges and hares in Switzerland, as well as for gray hares and some other species in the German province of Brandenburg, it was estimated that stable population sizes can only be expected if non-crop habitats accounted for at least 10% of the agricultural landscape (Hoffmann et al. 2012).

Widespread population declines of birds over the past half century result in the cumulative loss of billions of breeding individuals across a wide range of species and habitats in North America (Rosenberg et al. 2019). Researchers show that declines are not restricted to rare and threatened species—those once considered common and widespread are also diminished. The most severely affected bird species are those that live in field and meadow landscapes. Their numbers are dwindling because prairie landscapes are being destroyed and instead agriculture is increasing with the use of pesticides. However, forest birds and species occurring in different habitats are also affected. Intensive agriculture is putting entire bird populations under pressure. More than 90% of the decline is accounted for by 12 bird species, including sparrows, finches, thrushes, and warblers.

Although these studies about pesticide effects on birds often do not provide clear evidence of a causal relationship, they are mostly concerned with correlative patterns and circumstantial evidence. Of course, a field experiment would provide more robust evidence, but exact testing on this issue is extremely costly since birds are very mobile and experiments would need to be very large. The question is whether the agrochemical industry should not be obliged to finance such extensive research conducted by independent scientists before a pesticide is authorized.

The impact of pesticides on food resources of terrestrial animal populations is generally not assessed in the context of the environmental risk assessment. It would be important that those assessments follow an ecosystem approach, because current procedures are obviously not sufficient to protect our biodiversity from the detrimental effects of pesticides (Brühl and Zaller 2019).

Another very mobile vertebrate group are bats (order Chiroptera), the only flying mammals. There are about 50 different species of bats in Europe, and around 1200 species worldwide with a biodiversity hotspot in the tropics. Bats often live in agricultural landscapes and feed on vast numbers of insects or are important pollinators for plants. Despite their living in agricultural landscapes, the impact of pesticides on bats is not covered at least in European environmental risk assessments. Currently, pesticide effects for birds are approximated to bats; however, a large variation in toxicological sensitivity and no relationship between sensitivity of bats and bird or mammal test species to pesticides could be found (EFSA et al. 2019b).

Populations of many rural bat species have declined and, in some cases, even collapsed since the increased use of insecticides in the 1960s and 1970s. Similar to birds, herbicides also affect bats by indirectly decimating insect prey. Pesticides destroy the bats' insects and many toxins accumulate in the bats' body fat. During winter, these fat deposits are mobilized and the pesticides distributed in the body. Bats can get almost 40 years old and their bodies still contain dangerous pesticides such as DDT and other chemicals such as PCBs (polychlorinated biphenyl used for instance as pesticide spray adjuvants), which have been banned for many years. Recent poisonings of bats in the USA with very high DDT concentrations (4081 ppm) in the brain of a dead bat also give reason to believe that this insecticide might still be in use (Buchweitz et al. 2018). The effects of pesticides include decimation of the nutritional basis, impaired the immune system, communication and learning ability. Of course, collisions with wind turbines, road kills, or the disappearance of breeding facilities (old trees, caves, open stable buildings) also contribute to the decline of bats.

Many probably consider the decline of bat species to be a more academic problem, but their service is also of great economic significance. It has been calculated for the USA that a single colony of 150 big brown bats (*Eptesicus fuscus*) kills about 1.3 million insect pests per year (Boyles et al. 2011). Other estimates show that a little brown bat (*Myotis lucifugus*) kills up to 8 g of insects per night. Projected at one million bats, 1320 tons of insects can be consumed in a single night. This corresponds to a contribution of about US$ 400 per hectare of insect control service provided by bats. Without bats, this amount would have to be spent on insecticides to kill the same number of insects. When this is extrapolated to the cotton-dominated agroecosystems in Texas and the harvest area in the USA in 2007, the value of the bats' performance for US agriculture would amount to an incredible US$ 23 billion per year. The estimates include the cost of reduced spending on chemical insecticides in cotton farming. Not included are the costs of possible pesticide resistance and necessary substitutes or the positive impact of bats on pests in neighboring forests. Even if these estimates are cautiously revised downwards, they clearly show that bats are also a very important economic component for agriculture.

A study on the association between bat activity and insecticide application in an apple orchard showed that pesticide residues present in arthropods can affect bats living in these apple orchards (Stahlschmidt and Brühl 2012). While no acute dietary risk was found for the recorded bat species, a potential reproductive risk for bat species that include foliage-dwelling arthropods in their diet was indicated. Another study in Germany assessed that 14 bat species were active in agricultural landscapes with almost 110,000 insects of suitable prey being active at the same area and time frame (Stahlschmidt et al. 2017). Less bat activity and lower numbers of nocturnal insects were collected above the vineyards compared to orchards, cereal and vegetable fields, and researchers suggest pesticide exposure of bats via ingestion of contaminated insect prey.

An evaluation of terrestrial mammals (22 species) and birds (27 species) in Germany shows that insecticides and herbicides have the most adverse impacts on both mammals and birds (Jahn et al. 2014). Insecticides negatively influenced the populations of more than half of the considered bird and mammal species because insect food resources were reduced by insecticides. Strictly herbivorous species like geese, the Linnet, or the brown hare were the only ones not directly affected by insecticide applications. Additional impacts might derive from insecticide contained in seed dressings; however, this was not considered in this study. Additionally, herbicides greatly impacted both bird and mammal species. The reduction of important weed species that function as host plants for insect populations decreases the food availability for many bird species such as the Gray partridge, corn buntings, or yellowhammer (Bradbury et al. 2008; Ewald et al. 2002). Besides, for mammals and for some bird species also a lack of cover and therefore shelter is important.

Poisoning by fungicides represents a risk for several species like wood mice (*Apodemus sylvaticus*) that feed on fungicide-treated seeds (Barber et al. 2003). Invertebrate populations may be also affected by fungicide applications which negatively influence the food availability for many bird and mammal species (Ewald and Aebischer 1999). Of course, rodenticides had the most negative impact on rodent species. Most active ingredients in rodenticides are anticoagulants, slowly acting toxins that block vitamin K regulating blood's ability to clot. The poisoned animals typically die as a result of internal bleeding a couple of days after ingesting a lethal dose of poison. Besides rodents, also other mammal species like hedgehogs (Dowding et al. 2010), brown hares, and wild boars have been contaminated with this pesticide by consuming the baits (de Snoo et al. 1999). Additionally, birds of prey like the red kite (*Milvus milvus*) and carnivorous small mammals like the Stoat (*Mustela erminea*) and least weasel (*Mustela nivalis*) are highly affected by the reduction of their food resource (Brakes and Smith 2005). Exposure with rodenticides stems from use in private gardens or professional use in storage facilities, woodlots, or Christmas tree plantations (Topping and Elmeros 2016).

Very little is known about the effects of molluscicides on birds or mammals, except for poisoning incidents with hedgehogs and wood mice (Joermann and Gemmeke 1994; Shore et al. 1997). Insectivorous species that feed on slugs, like shrews, may be at risk due to the reduction of food resources as well as secondary poisoning (Münch 2011).

2.2   Agroecosystems Lose Their Self-Regulation

When pesticides affect not only the target organisms, but more or less all living organisms in an ecosystem, then the function of the ecosystem is compromised. It is evident that pesticides are among the significant drivers of the worldwide loss in biodiversity (Dirzo et al. 2014; Sánchez-Bayo and Wyckhuys 2019). In Europe, the impact of pesticides on biodiversity should theoretically be considered when making the risk assessment for pesticides. However, this has not yet been put into practice. Still only the toxic effects of pesticides have been examined in a quite limited framework, but not the indirect effects on the environment. Despite overwhelming evidence, no serious political measures aiming to reduce the use of pesticides have yet been implemented at a European scale. However, some European countries, namely France, Sweden, Denmark, Germany, Luxembourg and Switzerland, have implemented some pesticide reduction programs.

We are currently facing the largest global meltdown of insect species and individuals (Samways 2019a). Insects are vital components of many ecosystems involved in countless ecological interactions, from herbivory and pollination to predation and parasitism (Samways 2019b). Which ecosystem functions are actually affected by pesticides? An important function of agroecosystems is to maintain the natural balance between pest organism and their antagonists. We have seen that insecticides and herbicides initially reduce the food supply for other organisms that perform biocontrol in ecosystems such as birds and bats. Fewer birds or bats also mean a reduced control of insects or small mammals, which tend to reproduce in large numbers and thus can easily become pestiferous. When insects and weeds are destroyed by pesticides, parasitoid insects or predatory spiders, which act as health police in the ecosystem, also suffer. Using pesticides to eradicate certain organism inevitably slashes a gap in the food web which has been evolutionary established over millions of years.

What is ultimately at stake are many additional ecosystem services such as pollination, drinking water purification, nutrient cycles, and soil fertility. These can only be provided by properly functioning ecosystems. Moreover, the resilience of ecosystems to climate and weather extremes is also periled because the more diverse an ecosystem is, the more resistant it is to climatic influences or other stressors.

The main disadvantage of many pesticides is clearly their lack of selectivity, although the agrochemical industry keeps reassuring that products are very specific. Ideally, they should only kill certain pests or pathogens and spare all other organisms, especially beneficial organisms. Pesticide products currently in use have not yet been achieved this, despite intensive research. Since beneficial, antagonistic insects often have more complex habitat requirements and longer generation times than pests, this development ultimately leads to an increase in pests and a decrease in the function of beneficial insects as pest controllers. Moreover, many insecticides are more toxic to natural enemies than to the pest organisms. It is estimated that antagonistic insect parasitoids are more than a 1000 times more sensitive than the pest insect itself (Pimentel 1991). A first use of pesticides therefore often requires further pesticide applications, and one inevitably ends up in a pesticide treadmill.

For arable crops, it has been calculated that parasitic insects, predatory organisms, and host plant resistance provide 80% of the nonchemical control of arthropods and plant pathogens (Pimentel et al. 1991). Additional pest control can be achieved by improved crop rotation, improved soil health, water and fertilizer management, optimal choice of seeding timing, seed density, catch crops, or push–pull approaches to move pest species to specific areas where they can be controlled, or to repel them from crop fields. Taken together, these so-called cultural pest control measures can effectively decimate the pest population by more than 50%, without negative impacts on crop yields or the quality of harvested products (Pimentel et al. 1993).

For forest pests in Canada, it has been documented that pest infestation increased as the treatment area increased in size. After the large-scale pesticide treatments were discontinued, the infestation decreased as natural enemies increased again (Moriarty 1988). We know that many ecosystems are adapted to a certain degree of damage and can resist pests flexibly. Chemical intervention deprives these systems of their natural regulatory potential. Either one accepts the permanent, little damage to crops or a permanent use of pesticides seems necessary.

Another function of ecosystems is the maintenance of nutrient cycles. Dead plant material and crop residues are decomposed by soil organisms and nutrients contained therein are made available again for crop growth. Generally, herbicides can influence decomposition and soil microorganisms in various ways; however, the long-term effects are unclear (Wardle et al. 1994).

In principle, pesticides are used to protect crops from pests and to increase yields. Sometimes, however, agricultural crops themselves are damaged by pesticides, for instance, when crop growth is reduced by pesticide application, when residues of herbicides in the soil damage the growth of subsequent crops, or when increased pesticide application leads to increased residue levels in the products making them no longer available for sale. These losses naturally lead to financial losses for farmers that are usually ignored when praising advantages of pesticides. In arable farming, plowing is often reduced for reasons of erosion and climate protection. In no-till farming weeds that were previously plowed are now sprayed dead with broadband herbicides. This herbicide spraying can however lead to an increase in fungal diseases in cereal crops (Hallmann et al. 2009). Because the parts of the plant affected by the fungus are no longer plowed after harvesting, the harmful pathogens survive longer and can infect the following crop.

A literature review testing whether glyphosate-based herbicides might predispose crops to pathogens and diseases showed effects of undermining crop health in a number of ways (Martinez et al. 2018). Among them was an impairment of the innate physiological defenses of glyphosate-sensitive crops by interruption of the shikimic acid pathway, the impairment of physiological disease defenses of crops, an increase in the population and/or virulence of some phytopathogenic microbial species in the crop rhizosphere, and via a reduced uptake and utilization of nutrients by crops. Especially the last aspect is very little studied.

2.2.1 Pesticide Drift and Pesticide Erosion: Universal Problems Affecting Everybody

Pesticide drift is the unintentional distribution of pesticides off-target during spray applications, as runoff from pesticide-treated plants and soil, or via soil erosion by wind or water. This can lead to damage in human health, environmental contamination, and also property damage (Matthews 2016). Some of the pesticide ingredients applied also evaporate directly from the surface of the fields. This can be experienced when cycling through areas where pesticide spraying is happening. The biting smell of these substances and the furry feeling on the tongue while cycling through a pesticide cloud is something that stays in the mind.

Pesticides sprayed on crop fields can easily be transported to neighboring non-crop fields and affect plant and insect diversity there (Schmitz et al. 2014). Moreover, pesticide residues have also been found in remote mountain regions far away from their place of application (Ríos et al. 2019) or in Antarctica (Corsolini et al. 2011) as a consequence of long-range atmospheric transport and deposition. Some substances used in the 1970s have accumulated in the ice of the Alpine glaciers through wind and snow. With global warming and the melting of the glaciers, these substances are released again and are found in supposedly untouched well water (Villa et al. 2003). Even in the fat of penguins in the Antarctic, traces of pesticides are measured that were applied thousands of kilometers away (Geisz et al. 2008).

The cold condensation occurring in high-altitude sites drives the accumulation of pollutants in mountain regions even up to the peaks of the Andes (Grimalt et al. 2004). Pesticides applied to intensive crop production in the Central Valley in California have been shown to volatilize under warm temperatures and be transported through the atmosphere to be deposited in the cooler, higher elevation regions of the Sierra Nevada Mountains (LeNoir et al. 1999). The highest residue concentrations were those of compounds with heavy summertime agricultural use. A significant drop in pesticide concentrations in both air and water samples was observed within a few 100-m elevations from the valley; however, levels remained relatively constant between 500–2000 m in altitude. Concentrations found in water may be harmful to amphipods but less severe for rainbow trout and stoneflies but not much is known about long-term effects of low concentrations.

Plants sprayed with pesticides are absorbed by bees and enter the human body through residues in the honey we put in our muesli or on breakfast bread. It seems irrelevant where the honey comes from; after all, pesticide residues have now been detected in honey samples from 80 different countries (Mitchell et al. 2017). Pesticides, which are sprayed with planes against mosquitoes, are absorbed by fish that eat mosquito larvae and ultimately end up in humans when this fish is eaten.

Drift occurs with all methods of pesticide application, but most strongly when applied by planes or helicopters where only 25–50% of the pesticides reach the harmful organism (Hall 1991). When sprayers are used on the ground, about 65–90% of the pesticides reach the target organisms. Although it is mandatory for farmers to have their sprayers regularly checked, nobody controls if they are applied at the appropriate weather conditions. Especially part-time farmers with a tight time budget apply pesticides when they have time and not necessarily at the best, wind still and dry weather conditions. Moreover, for this group of farmers, possible damage thresholds or the exact observation of the pest and disease infestation often falls victim to the lack of time.

Agricultural crops can also be affected by pesticide drift from neighboring fields, even if they are several kilometers away (Barnes et al. 1987). Damage to cultivated plants caused by pesticide drift is particularly noticeable in areas with a wide range of agricultural crops. Concrete data on this are known from cotton-growing regions, as cotton is one of the most pesticide-intensive agricultural crops in the world. In the southwest of Texas, for example, cotton worth of US$ 20 million was contaminated by drift from a neighboring wheat field that was treated with the herbicide 2,4-D from a helicopter (Hanner 1984).

In Europe too, the consequences are acute and in part even threatening economic existence of affected farmers. Especially for organic farmers neighboring a pesticide using conventional farmer, pesticide drift and contamination of the organic products can threaten their livelihoods. Pesticide drift from neighboring fields also causes problems in vineyards (Gerlock 2013) or producers of organic herbs and teas (Ackerman-Leist 2017). Broader buffer strips within the pesticide-treated agricultural field could considerably reduce spray drift to non-crop areas (de Jong et al. 2008).

For decades have pesticides under pressure from agrochemical industry been approved without considering potential long-distance transports. Meanwhile, at least in Germany, politics seems to be dealing with the issue (Agrarministerkonferenz 2015). The decisive factor was a court ruling by organic farmers whose harvest became unsaleable due to a contamination by pesticide drift of the herbicides pendimethalin and prosulfocarb (Klingenschmitt 2016). German federal states decided that this long-distance transport of certain pesticides should also be considered in the pesticide registration process. It remains to be seen whether and how this declaration of intent will really be reflected in the practice of approval.

An understudied aspect of pesticide drift is to what extent public or private land is affected. Nearly half of 71 studied playgrounds located in the surroundings of intensively managed apple and wine orchards were contaminated with at least one pesticide (Linhart et al. 2019). The majority of the pesticides were known to show some endocrine activity which is especially worrisome with children. Pesticide concentrations were positively associated with the proportion of apple orchards in the surroundings of the playgrounds, the amount of rainfall, and wind speed; the more distance there was to the next sprayed field the lower the pesticide residue contamination. Studies like this could be used to define puffer zones between pesticide treated fields and public land.

So even if all legal requirements for organic farming are complied with, pesticides that are not permitted in organic products may be detectable. These can get into the products in many different ways, for example through contaminated sites or shipments from conventional agriculture as well as through contamination from processing machines, storage facilities, transport containers, or packaging. The German association of organic foods and natural products defined an orientation value of 0.01 mg/kg of product with a maximum of two pesticide residues that is tolerated in these cases (BNN 2019). This value is 10 times lower than the default maximum residue level for conventional products (0.1 mg/kg). With these examples in the background, the question arises as to whether the coexistence of pesticide-intensive and organic farming is practically feasible at all. This is a politically very delicate matter, because as a consequence a certain way of farming would have to be prescribed for entire regions in order to avoid interference.

Pesticide residues present in the soil can also affect subsequent crop development. After the herbicide atrazine was applied in maize the subsequently sown soybeans were severely damaged because the herbicide was still active in the soil (Pimentel 2005). Neonicotinoid residues in the soil from seed dressing can be taken up from other plants and be found in their guttation liquid, i.e. small drops formed at the tip of leaves (Mörtl et al. 2017). These guttation drops are used by insects to take up water. These examples show that pesticides can have legacy effects into the following planting season; again a scarcely studied aspect.

We already saw that some pesticides are washed out and end up in groundwater, even if they are applied in the recommended dosages. Processes that can move pesticides are leaching, diffusion, volatilization, erosion and runoff, assimilation by microorganisms, and plant uptake (Pérez-Lucas et al. 2018). Pesticides are frequently leached through the soil by the effect of rain or irrigation water. Leaching is highest for weakly sorbing and/or persistent compounds, in climates with high precipitation and low temperatures, and soils with low organic matter and high sand content.

In the 1990s, pesticides residues of the insecticide aldicarb or the herbicides alachlor and atrazine were detected in more than 30,000 groundwater wells across the USA (USGS 1996). Pesticide contamination of groundwater is a national issue because groundwater is used for drinking water by about 50% of the US population. The proportions of sampled wells with pesticide detections ranged from 4% (nationwide, rural domestic wells) to 62% (corn-and-soybean areas of the northern midcontinent). Unfortunately, the situation did not improve much since then. In a later survey it was shown that at least one pesticide was detected in 97% of all streams in agricultural areas, 97% in urban areas, and 94% in streams from mixed-land-use watersheds (Gilliom and Hamilton 2006). Pesticides were less common in groundwater than in streams. They occurred most frequently in shallow groundwater beneath agricultural and urban areas, but also in groundwater beneath undeveloped areas. Although concentrations of pesticides in streams and groundwater were typically below water-quality benchmarks for human health, an overshoot of the water-quality benchmarks for aquatic life and/or fish-eating wildlife in more than 50% of the streams with substantial agricultural and urban areas was seen.

Such monitoring studies are particularly important not only for assessing the occurrence of pesticides in groundwater and streams in relation to human health and the environment but also for measuring the success of political actions regarding pesticide reduction. Especially in agricultural areas shallow groundwater is used for drinking water and groundwater contamination is difficult to reverse once it occurs. It is important to note that the US EPA drinking water standard is 3 µg/liter pesticide residue, while it is 0.1 µg/liter in Europe.

Pesticide residues in groundwater are also a great problem in Switzerland (BAFU 2019). In regions with intensive arable farming, 90% of the wells are contaminated. The situation is likely to be similar in other countries, especially in intensively farmed regions. Pesticide-contaminated groundwater can only be purified again with great effort. The pesticides that are most easily washed out are the water-soluble ones, with little adhesion to soil particles and those with long-term effects. If such pesticides reach sandy soils and the groundwater level is high, from heavy rainfall or irrigation, then there is a very high risk of pesticide leaching into the groundwater. This matter is also of global significance because groundwater is used at least by 50% of the global population for drinking water and accounts for 43% of all of the water used for irrigation (UNESCO 2015), and pesticides remain in this groundwater for a very long time.

In addition to the direct leaching of pesticides, they also enter water bodies through soil erosion. As seen earlier, pesticide residues can affect the growth and development of amphibians that live in these waters. Fish can also be killed directly by pesticides, or indirectly by chronic damage to particularly sensitive fish, or by the destruction of the food basis of fish in the form of insects, amphibians, or aquatic plants. Apart from fish, up to 42% fewer species were recorded in streams in heavily pesticide-polluted areas compared to regions untouched by pesticides (Beketov et al. 2013). Again, these findings call for a drastic reduction of our pesticide use and a complete revision of the current environmental risk assessment of pesticides as it obviously fails to protect biodiversity and human health. Runoff and soil erosion is also the main cause of off-site contamination with neonicotinoids (Niu et al. 2020).

So far, we have primarily mentioned pesticide effects in arable land, fruit growing, and viticulture. However, the use of herbicides can also have consequences for agricultural livestock production when pesticide-contaminated feed is offered. The example of the chronic botulism disease in cattle illustrates this (Krüger et al. 2013). Botulism has been used for over 200 years to describe a disease caused by poisoning that can affect both humans and other mammals. Recently, the disease has become increasingly common in connection with cattle, especially dairy cows. Between 1996 and 2010 alone, this disease was detected in over 1000 German cattle farms. Infected cattle die miserably or give birth to dead calves. In both cases, botulism is caused by toxins of the bacterium *Clostridium botulinum*. The poison, called BoNT (botulinum neurotoxin), is such a strong nerve poison that theoretically a quantity of 40 g would suffice to destroy the entire human population (Lorenzen 2013). For a long time, it remained unclear why more and more cattle are suffering from chronic botulism and why this affects mainly high-performance dairy cows. In the meantime, however, there is growing evidence that glyphosate is probably the main cause of the disease because glyphosate kills health-promoting bacteria in the stomach and intestines, causing considerable damage to the intestinal flora. Bacteria such as the botulism bacterium can no longer be sufficiently repelled by the cattle' immune system. Especially in high-performance dairy cows, about 78% of the feed in Europe consists of imported protein from genetically modified soybeans from South America that are cultivated with massive glyphosate inputs.

Pesticides used in agriculture and off-target pesticide drift expose workers and the public and can lead to chronic and acute illnesses. This aspect will be further elaborated in a following chapter.

Unlike agricultural uses for which pesticide use can be reasonably estimated, the amount of a pesticide applied in an urban setting is virtually impossible to estimate because almost no records of use are available. A long-term study from 27 urban stream sites found both upward (atrazine) and downward (simazine) trends for some herbicides between 1992 and 2008 (Ryberg et al. 2010). In the same time span, the insecticides chlorpyrifos, diazinon, and malathion decreased, but fipronil was more often found. These results mirrored the regulatory phaseout of residential uses for chlorpyrifos and diazinon and the introduction of fipronil in 1996 as a substitute for chlorpyrifos and diazinon. The downward trends in malathion is estimated to be caused by voluntary substitution of pyrethroids or fipronil for malathion. Unfortunately, glyphosate was not assessed in this survey.

2.2.2 Pests and Diseases Get Used to Pesticides and Become Resistant

When pesticides are no longer effective against the target organisms, this is referred to as resistance. Just as the mode of action of many pesticides is often not comprehensively known, the biological mechanisms of resistance formation are insufficiently understood. Basically, evolution is tricking us here with our attitude of pesticide applications. In every population of organisms there are few individuals who better tolerate pesticides than others. Since only individuals with the tolerance survive repeated applications of a pesticide, this pesticide treatment can be seen as a highly effective natural selection pressure for resistant individuals. Because of short reproduction periods of many pest organisms, the population then only consists of resistant specimens and the pesticide becomes ineffective. Therefore, resistance can be seen as a stable, heritable trait that results in a reduction in sensitivity to the applied pesticide.

Pesticides with a single mode of action are at a higher risk for resistance development than pesticides with multiple mode of actions. Most pesticides have a single mode of action because this often goes with less impact on non-target organisms. As early as the 1980s, a report by the UN Environment Programme identified pesticide resistance as one of the four most important environmental problems in the world; the situation has certainly not substantially improved in the meantime (UNEP 1980). Globally, around one thousand insect, mite, and rodent species and about 550 weed species and plant pathogens are resistant to pesticides (Miller and Spoolman 2011).

The standard approach of agrochemical industry to tackle increased resistance in pest populations is to use additional pesticide applications. However, these additional applications only further rotate the pesticide spiral, and other target organisms will sooner or later develop resistances. An impressive example of pesticide resistance is known from North America where extremely high pesticide resistance has developed over a longer period of time in populations of the cotton bollworm (*Helicoverpa armigera*), a moth species (Pimentel 2005). Ultimately, around 285,000 hectares of cotton had to be abandoned because the insecticides used were completely ineffective. The economic and social consequences for the affected regions in Texas and Mexico were devastating. According to calculations, in California alone about 10% of the cotton harvest was lost due to resistance symptoms (Pimentel 2005). If we assume that other important crops in the USA lose 10% of their yields due to pesticide resistance, this translates into an annual loss of about US$ 1.5 billion. For the USA, it is estimated that about 10% of pesticides are applied only to counter emerging resistance problems. The costs necessary in tropical countries to combat existing pesticide resistance are estimated to be even higher.

Resistance is also an issue with malaria control. In the case of malaria in India, malaria cases were reduced from 70 million to two million by 1985 through the use of insecticides. However, the mosquitoes that transmit malaria quickly developed resistance to the insecticides used, so that malaria cases are now back at around 60 million cases per year. Similar problems exist in other parts of Asia, Africa and South America (Pimentel 2005).

Although glyphosate was not considered to be a high-risk herbicide with regard to the development of resistances, at least 47 glyphosate-resistant weed species have been found so far (Heap 2020). Today, they infest millions of hectares and some of the resistant weed species are cross-resistant to other herbicides. Glyphosate-resistant weeds can withstand up to 19-fold the dose tolerated by ordinary sensitive weeds and exhibit a great diversity of molecular and genetic resistance mechanisms (Legleiter and Bradley 2008).

The increased incidence of glyphosate-resistant weeds has led to even more herbicide use. Agrochemical companies are developing genetically modified crops that are also resistant to other herbicides, e.g., 2,4-D. And this will certainly lead to even more herbicide consumption—a classic case of pesticide treadmills. About 30% of the net income of US farms is already being used for pesticides and seeds, with more resistances this share will most likely further increase (Benbrook 2016). The USA has a particularly serious problem with herbicide resistance, as large amounts of herbicides are used due to the systemic nature of the widespread use of genetically modified crops. It is estimated that about 24 million hectares of arable land are already covered with glyphosate-resistant superweeds (UCS 2013). In the southeast of the USA, 90% of cotton and soybean farmers are affected by these superweeds. The promotion of monocultures has also contributed to this development because the same crop is growing on the same field for years. By following basic principles of crop rotation and catch crops, herbicide use could be drastically reduced while at the same time profits for farmers also increase.

Pesticide resistance is also a huge issue in other regions, including Europe. This development is so grotesque that in some places weeds now have to be controlled mechanically again, or much worse, by the application of some old, very poisonous weed killers from old stocks.

The most toxic herbicide on the world market plays a special role in herbicide resistance: paraquat from the former Swiss now Chinese agrochemical company Syngenta. One teaspoon is enough to kill a human. The manufacturer is campaigning for the product's ability to be used in a way that protects the soil, because allegedly farmers no longer have to plow in order to prepare the seed bed. The product is also important for destroying weeds that have become resistant to other herbicides (Dyttrich 2015). It is perfidious, but agrochemistry earns money again from the problems it has caused itself because resistance creates a need for further pesticides! In Switzerland and the EU, paraquat is banned for a long time; however, it is still used in developing countries. A company spokesman's succinct analysis is that the "state of mind" in Western Europe seems to be somewhat different from that in the rest of the world.

House flies (*Musca domestica*) have been reported to be partly resistant to neonicotinoids (Kavi et al. 2014) and fipronil (Abbas et al. 2014). The development of insecticide resistance against neonicotinoids in the brown planthopper (*Nilaparvata lugens*) was first observed in Thailand in 2003 and has since been found in other Asian countries such as Vietnam, China, and Japan (Simon-Delso et al. 2015). This problem has exacerbated yield losses in rice production in eastern China.

Resistance to fungicides is an economically important issue as well. When resistance results from modification of a single major gene, pathogen subpopulations are either sensitive or highly resistant to the pesticide. Resistance in this case is seen as a complete loss of disease control that cannot be regained by using higher rates or more frequent fungicide applications. When fungicide resistance results from modification of several interacting genes, pathogen isolates exhibit a range in sensitivity to the fungicide depending on the number of gene changes. Fungal isolates that are resistant to one fungicide are often also resistant to other closely related fungicides, even when they have not been exposed to these other fungicides, because of a similar mode of action. Resistance management then often includes the use of fungicides with different mode of actions in combination with nonchemical control measures, such as disease-resistant cultivars or crop rotations (APS 2019).

2.2.3 The Pesticide Boomerang Is Always Coming Back

Pesticides can affect us and our environment although applied in great spatial and temporal distance. They come back to us in many ways just like a boomerang; however, they are often delayed by decades. When they are applied in the field, neighboring areas are often automatically "treated" through pesticide drift especially when pesticides are applied by planes or helicopters.

We have seen that wind and rain can distribute pesticides far beyond their application area. Some of the pesticides, especially the long-lived organochlorine compounds such as the insecticides DDT and lindane or the herbicide atrazine, can be detected in the remotest parts of the world. Often, they accumulate there for decades. In Austria, for example, the herbicide atrazine was banned in 1994, but together with its degradation products it is still one of the most frequent polluters of groundwater 20 years later.

"That too! Melting Alpine glaciers are toxic" was the headline of the biggest Swiss tabloid (Anon. 2018). The headline was triggered by reports on melting glaciers in the Alps releasing pesticides that had been trapped in the ice for decades (Bogdal et al. 2009). The researchers investigated sediment layers of Lake Oberaarsee, a reservoir located at 2300-m altitude. Sediment drill cores are archives of the past, in which the yearly deposition of air pollution can be analyzed. The researchers found many environmental toxins, including persistent organic substances, in the one-meter long drill core in the layers from the year 1960–1970. Pollutant levels declined after many environmental pollutants were banned in the early 1970s. In younger sediment layers, however, the proportion of toxins increased again. The reason given for this is that the lake is mainly fed by a melting glacier due to global warming and thereby releases large amounts of toxic substances that were deposited at the glacier years ago. This aspect has so far been largely ignored in the climate change debate.

Several organochloride pesticides (DDT, HCH, HCB) have also been found in ice cores in other Alpine glaciers for instance in Italy (Villa et al. 2003). Intensive agricultural crops in the Alps are often only a few kilometers away from glaciers, distances that can easily be overcome by pesticide drift. Evidence of pesticides in Tibets very remote grasslands is somewhat more surprising, and it is suggested that East Asian monsoon, the Indian monsoon, and western winds have brought DDT and other persistent pesticides from their locations of application (Wang et al. 2016).

In the sediments of a French lake surrounded by vineyards, researchers found pesticides from the entire twentieth century (Sabatier et al. 2014). Again, a sediment drill core covered the period from 1900 to 2011. The highest concentrations of DDT could be detected in layers from the 1990s. This is about 20 years after the ban of DDT, and it is assumed that soil erosion processes flushed DDT-contaminated soil layers into the lake.

The persistent organic pollutants evaporate and adhere to small dust particles in the air (Socorro et al. 2016). With half-lives ranging from several days to several years, the substances can be transported worldwide and transported to the remotest regions. In Adelaide penguins in Antarctica, for example, DDT contents have been detected for more than 30 years (Geisz et al. 2008). Here, too, it is assumed that meltwater from glaciers is the source of this new DDT contamination. According to calculations, about 1–4 kg of DDT per year are released into the seawater via glacier erosion, which actually does not seem very much. The tricky thing is, however, that DDT accumulates in the fatty tissue of penguins and remains in the body for decades.

Surprisingly high pesticide levels were even been found at the deepest point of the world's oceans (Jamieson et al. 2017). Pollutants that have been banned for almost 30 years such as polychlorinated biphenyl (PCB) used as coolant fluids and pesticide adjuvants or polybrominated diphenyl ethers (PBDE) used as flame retardants were found in amphipods from a depth of 10,000 m. The levels of pollutants in these small crustaceans are 50 times higher than those of crabs living in heavily polluted rivers in China. The pollutants were apparently carried via dead, sunken animals to the Mariana Trench in the western Pacific, the deepest point of the world's oceans. In any case, it was once again confirmed that humans leave their traces even in supposedly untouched areas.

Back to Europe. Rain is often regarded as the purest water, filtered from all impurities. That may have been the case once. Rainwater samples in Switzerland have shown that they were so heavily contaminated with pesticides (atrazine, alachlor) that it would even have been illegal to declare this rainwater as drinking water (MacKenzie and Pearce 1999). How do pesticides get into rainwater at all? It is assumed that the pesticides evaporate from the agricultural fields and have become part of the rain clouds. In the EU and Switzerland, a limit of one 100 nanograms applies to each pesticide per liter (0.1 µg/liter) of drinking water. In a rainwater sample, almost 4000 nanograms/liter of a widely used herbicide (2,4-dinitrophenol) were found. The highest pesticide concentrations were found in the first minutes of a violent storm, especially during rain events after a longer dry phase and when fields were sprayed with pesticides shortly before.

In another study from Switzerland, pesticides from agriculture were found both in rainwater and in wastewater coming from roofs (Bucheli et al. 1998). Pesticides are washed out of the atmosphere and seep into the groundwater when roof wastewater is discharged into gravelly soils. Since more and more roof wastewater is used to replenish groundwater reserves, these findings are of great significance. In addition to pesticides from rainwater, herbicides are also dissolved from the roofing materials, where moss growth is prevented by herbicide application. It is just another example of pesticide use, where it is not suspected.

The boomerang comes also via import of products that have been sprayed with pesticides that are banned in the importing country. The insecticide DDT was used worldwide in the 1950s and 1960s, in agriculture, households, and gardens, but also for malaria control in tropical countries. Although it has been banned in Europe for decades, it continued to be produced and abundantly used in the plantations of the countries of the global South. Since most of the plants grown there are intended for export, DDT eventually ends up in our food again.

Other pesticides than DDT are also regularly found in imported ornamental flowers, which are often produced under poor environmental conditions and deplorable social standards. In a bunch of roses from Kenya, 56 different pesticides were detected (Ökotest 2017). One in three of the tested bunches even contained pesticides that are suspected of causing cancer. Pragmatists may argue that we do not eat these roses and see this problem as negligible. However, when these contaminated roses end up on the compost heap and vegetable plants are later fertilized with this compost, then we might pollute our veggies with these pesticides. The greatest danger, however, certainly lurks for the people working in the large ornamental flower nurseries, often without sufficient protective equipment.

A boomerang loaded with the insecticide profenofos starts in Switzerland, touches down in Brazil, and is coming back to Switzerland again (Heidis Mist 2020). Profenofos is a potent insecticide used mainly in the cultivation of cotton, corn, sugar beet, soybeans, potatoes, and vegetables. It is highly toxic to aquatic organisms, birds, bees, and small mammals and can lead in humans at very high exposure to respiratory paralysis and death. Because of these adverse effects, profenofos has been banned in Switzerland since 2005. Nevertheless, the substance is still produced in a Swiss company and exported—for instance to Brazil (Tscherrig 2020). There (especially in the states São Paulo and Minas Gerais), it is among the most frequently detected pesticide in drinking water with concentrations that would be unacceptable for human consumption in Europe (Gaberell 2020). Switzerland is also affected because products produced abroad with profenofos can still be imported: in 2017, profenofos was detected in 41 food products, mainly in vegetables and various fruits and spices from Asia. This makes profenofos the most frequently detected prohibited pesticide in Swiss food (Tscherrig 2020).

Ultimately, pesticides will also appear in drinking water. Scientists analyzed data from more than 4000 measuring points on 223 chemicals from the catchment areas of 91 rivers, including the Danube and Rhine (Malaj et al. 2014). Results show that chemical pollution represents an ecological risk for about half of the water bodies. In around 15% of cases, acute toxic effects on aquatic organisms could even occur. The main causes of the pollution are agriculture and sewage treatment plants. The strongest pollution comes from pesticides. Other studies also found that the precautionary limit of 0.1 µg pesticide residue per liter water was exceeded. Okay, only 1% of a total of 134,080 tests showed excessive values; however, this still amounts to 1340 analyses exceeding this threshold. The most frequent exceedances still occur for the total herbicide atrazine and its degradation product (desethylatrazine), although its use has been banned since 1995 in most European countries.

In all these analyses, it is important to emphasize that only a small selection of pesticides is searched for. With analytical methods, that can be very complicated and complex; only the substance that are searched for can be detected. In other words, basically nobody can claim that our food, water, or environment is free of pesticide residues. There is no regulatory authority in the world that takes this task seriously and regularly monitors all food and the environment for residues of all pesticides authorized in the respective country.

So far, most of the examples in this book have dealt with pesticides or their active ingredients and their fate in the environment. However, sooner or later these active substances will be converted into degradation products, so-called metabolites. Much less is known what these metabolites do in our environment. For a long time, these metabolites were ignored, because according to the assertions of the manufacturers it was assumed that the pesticides would be rapidly degraded into harmless decomposition products. However, studies have shown that the degradation products of insecticides can be up to a 100 times more toxic to amphibians than the original products. Particularly in the case of the very widespread insecticide class of organophosphates to which pesticides as parathion, malathion, chlorpyrifos belong (Sparling and Fellers 2007).

2.3   Numerous Side Effects on Humans

How problematic pesticides can be for humans varies from substance to substance and is hotly discussed in individual cases. It is unequivocal that most pesticides have nontarget effects also on humans. It is also evident that some pesticide classes (e.g., organochlorines, organophosphates) are especially toxic to humans. No wonder, these substances were and are still used as chemical warfare agents. Testing effects of pesticides on humans is difficult and when data from people who were highly exposed to pesticides and are likely to affected by them are mixed with people who were less exposed you are diluting a potential effect (Robin 2014).

With the worldwide increase in synthetic pesticides since the 1940s health risks for humans also increased. A very comprehensive review by Mostafalou and Abdollahi (2017) assessed 448 studies and found a huge body of evidence showing possible roles of pesticide exposures and the increased incidence of human diseases such as cancers, Alzheimer's, Parkinson's, amyotrophic lateral sclerosis (ALS), asthma, bronchitis, infertility, birth defects, attention deficit hyperactivity disorder, autism, diabetes, and obesity. Most of the disorders are induced by insecticides and herbicides most notably organophosphorus, organochlorines, phenoxyacetic acids, and triazine compounds.

Ultimately, it is often not clear whether the active ingredients or rather an additive is responsible for the effect, especially in the case of long-term health effects. Moreover, in reality a broadly diversified combination of substances and influences have to be considered. Pesticide residues in the environment and in food now expose practically all social groups, and it is almost certain that we all have pesticide residues in our bodies, whether or not we work in agriculture.

Nevertheless, agricultural workers are at greater risk of pesticide exposure than non-agricultural workers. A study analyzed acute pesticide poisoning cases in agricultural workers between the ages of 15 and 64 years that occurred in California between 1998 and 2005 (Calvert et al. 2008). Of the 3271 cases included in the analysis, 71% (2334 cases) were employed as farmworkers, 12% as processing/packing plant workers, 3% were farmers, and others miscellaneous agricultural workers (19%). The majority of cases (87%) had low severity illness, while 12% were of medium severity and less than 1% of high severity. One case was fatal. Rates of illness among female agricultural workers were almost twofold higher compared to males.

The Californian Pesticide Illness Surveillance for instance lists during 2005–2009, 3162 occupational exposures and 1047 non-occupational exposures (Roberts and Reigart 2013). In a large prospective study including about 43,000 pesticide users in the USA, the Agricultural Health Study, it was estimated that about 4% of the cohort had at least one pesticide poisoning or an unusually high pesticide exposure episode in their lifetime (Payne et al. 2012). Pesticide drift including off-target movement of pesticide spray, volatiles, and contaminated dust resulted between 1998 and 2006 in 2945 reported illnesses in 11 US states (Lee et al. 2011). The annual incidence ranged from 1.39–5.32 per million persons over the 9-year period. Agricultural workers and residents in agricultural regions had the highest rate of pesticide poisoning from drift exposure; 14% of cases were children under the age of 15. Soil fumigations with various pesticides has been identified a major hazard, causing large drift incidents.

An indication how widely distributed pesticides are in our modern world are studies on glyphosate residues in urine samples. Glyphosate can be detected in the urine of people who are professionally involved with the substance as well as in people who are not consciously exposed to glyphosate. In a study of 182 urban dwellers from 18 European countries, 45% of people were found to have glyphosate in their urine (Krüger et al. 2015).

An industry-funded study also showed that American farmers who use glyphosate have it in their bodies (Acquavella et al. 2004). Urinary samples of 48 farmers, their spouses, and their 79 children (4–18 years of age) showed that 60% of farmers had detectable levels of glyphosate in their urine on the day of application. Farmers who did not use rubber gloves as protection had higher glyphosate urine concentrations than other farmers. Four percent of spouses and 12% of children had detectable glyphosate in their urine on the day of application. None of the systemic doses estimated in this study approached the US EPA reference dose for glyphosate of 2 mg/kg body weight and day.

By the way, I also had my urine tested for glyphosate residue and low but clearly detectable amounts were found. I am still puzzled where this might come from; the last application of glyphosate during an experiment is months away and of course I used safety equipment. So, I reckon that the intake came via the food and drinks I consume. Due to its ubiquitous use, it can be assumed that a large part of the population is continuously exposed to glyphosate. So far, there have been no studies on the health consequences of a long-term intake of glyphosate in small amounts—in other words, on the scenario that corresponds to the reality and everyday life of most people.

American farmers had up to 233 μg of glyphosate per liter in their urine and other family members of farmer households up to 29 μg per liter

(Whitford et al. 2012). In Germany, the values are much lower and range between 0.5 and 2 μg of glyphosate per liter for both farmers and city dwellers (Krüger et al. 2015). Agricultural contact with glyphosate resulted in significantly higher glyphosate concentrations in urine. Mixed food consumers had higher glyphosate residue levels than vegetarians and vegans; organic food consumers were highly significantly less contaminated than non-organic food consumers.

Decades after the ban on DDT in Bavaria, this substance can still be found in food, breast milk, human blood, house dust, in samples of indoor air, and building materials (Roscher and Juds 2016). Even newborns have a burden that is attributable to the fact that these substances can also pass through the placenta. Infants can thus absorb up to 0.8 μg/kg body weight per day through breast milk. DDT residues can also often be detected in the house dust of apartments in which the product was never used.

Generally, it is difficult to estimate the global extent of pesticide poisoning because reporting systems do not exist or do not function poorly in many places. In Germany, poisoning and health problems associated with chemical products are recorded in a monitoring system. According to the German Chemicals Act, every physician is obliged to report diseases or suspected cases of poisoning by certain substances to a federal risk assessment center. Pesticides also fall under this reporting obligation. Their medical reports are documented, evaluated, and analyzed in a standardized manner. From 1990–1999, there were a total of 7700 reports; 1303 concerned poisonings with pesticides, 462 cases of which concerned moderate to severe poisonings (Hahn et al. 2000). More than half of these poisonings occurred with insecticides.

It is also complicated that names for pesticides and poisoning phenomena are inconsistent, making it difficult to compare existing statistics. Despite these shortcomings in recording pesticide poisoning, the available data show that the consequences of dealing with pesticides represent a major global health problem that is particularly serious in developing countries.

2.3.1 Direct Poisoning

According to the UN World Health Organization (WHO), there were globally about 26 million cases of pesticide poisoning per year in the 1990s (WHO 1992); a more up-to-date estimate mentions about 41 million cases of poisoning per year (PAN Germany 2009). Of these pesticide poisonings, about three million cases are treated in hospitals, resulting in about 750,000 chronic diseases every year (Hart and Pimentel 2002). Although the USA has quite strict regulations to protect humans and the environment against pesticide poisonings, the US EPA reports over 300,000 nonfatal pesticide poisonings per year (Hansen and Donohoe 2003).

It is estimated that about 200,000 acute poisoning deaths by pesticides occur each year, 99% of which are in developing countries although only about 25% of the global pesticide quantity is used there (UNHRC 2017). The documented poisonings were caused by highly dangerous pesticides such as paraquat, lambda-cyhalothrin, chlorpyrifos, and additionally glyphosate (PAN Germany 2019).

Recently, a colleague of mine was emptying his office because he was leaving for his retirement and found a hand-written note he made after a message aired in August 1983 by the Austrian public radio. The message was put forward by a regional police department warning that whoever had stolen 20 heads of cabbage from a field near the city of Linz should not eat the cabbage because it was treated with E605 before and its consumption would be life-threatening. E605 was the brand of an insecticide containing the active ingredient parathion, an extremely hazardous pesticide not only to humans but also to wildlife. Admittedly, this case was almost 40 years ago, but it should illustrate the bizarre attitude of treating food with really dangerous chemicals. The use of parathion was finally banned in Europe in 2001; however, its production and export in other countries is still allowed.

In many developing countries, highly dangerous pesticides are also a common suicide method, especially in Asian countries. It is estimated that in the period 1995–2005 between 60%–90% of suicides in China, Malaysia, Sri Lanka, and Trinidad and Tobago were by pesticide ingestion (Bertolote et al. 2006). In those parts of the world about 300,000 deaths each year stem from intentional pesticide poisoning (Gunnell and Eddleston 2003). The discrepancy between this death number and the pesticide-related death numbers published by the UN mentioned above would demand further verification.

A growing number of suicides due to pesticide ingestion are also reported from many other Asian countries and countries from Central and South America (e.g., Brazil, El Salvador, Guatemala, Guyana, Nicaragua, and Paraguay). This may well mean that the global number of suicide deaths is considerably higher and clearly makes pesticide ingestion the most common method of suicide on a worldwide basis (Gunnell and Eddleston 2003). The most important prevention for this would be by more strictly controlling access to pesticides, which all too often are easily accessible and stored without any precautions in most rural households. Studies indicate that interventions to control access to pesticides are effective and work better when integrated into more comprehensive community education programs as well as into pesticide management programs. Especially dramatic is the situation in India in relation to genetically modified seeds and the financially hopeless situation of farmers (Shiva 2010).

Children's exposures to pesticides is especially precarious and should be limited as much as possible. In 2008, pesticides were the ninth most common substance reported to poison control centers in the USA, and approximately 45% of all reports of pesticide poisoning were for children (Roberts et al. 2012). Besides organophosphates and carbamate, numerous other pesticides that may cause acute toxicity, such as pyrethroid and neonicotinoid insecticides, herbicides, fungicides, and rodenticides, also have specific toxic effects. A growing body of epidemiological evidence suggests parental pesticide use and adverse birth outcomes including physical birth defects, low birth weight, and fetal death, although the data are less robust than for neurodevelopmental effects or cancer.

2.3.2 Chronic Poisoning: Cancer

Numerous studies demonstrate chronic effects of pesticides on the human nervous system, respiratory diseases, reproduction, and cancer. Farmers, other pesticide users, pesticide plane pilots, and workers in the agrochemical industry show increased rates of prostate cancer, melanoma and other skin cancers, and lip cancer. Approximately 40 chemicals classified by the UN International Agency for Research on Cancer (IARC) as likely carcinogens are still on the market today (PCP 2010), not only in developing countries with low environmental standards. It is hard to believe, but also in Europe and the USA, farmers are advised to apply pesticides that are "likely to be cancerogenic" according to the safety data sheets of the manufacturers.

Data from the USA for the late 1980s showed that 18% of all insecticides and 90% of all fungicides were carcinogenic (NRC 1987). This study may now appear somewhat outdated, but it should be remembered that many cancers only break out after decades of exposure to a carcinogenic substance. In addition, many pesticides from the 1980s are still in use in developing countries and return to us via imports of agricultural products. A study commissioned by US President Barack Obama found that cancer cases in American children under the age of 20 are on the rise. Leukemia, for example, is consistently higher in children who grow up on a farm or whose parents use pesticides in the garden (PCP 2010). Swedish researchers have blamed pesticides for the most common cancers in the Western world (Hardell and Eriksson 1999).

The most heavily discussed issue in this respect is the controversy over glyphosate, which the IARC has classified as "likely carcinogenic" to humans, while several national agencies came to a different conclusion. We already heard earlier that US EPA's evaluation was largely based on data from studies on technical glyphosate, whereas IARC's review placed more weight on the results of formulated glyphosate-based herbicides and the metabolite AMPA (Benbrook 2019). EPA's evaluation was focused on typical, general population dietary exposures assuming legal, food-crop uses, and did not consider nor address generally higher occupational exposures and risks. IARC's assessment encompassed data from typical dietary, occupational, and elevated exposure scenarios.

One of the most intensely debated issue is whether glyphosate can increase the risk of cancer, especially non-Hodgkin's lymphoid malignancies. An international study considered data of agricultural workers from the USA, France, and Norway (Leon et al. 2019) investigating 33 individual active chemical ingredients. Among 316,270 investigated farmers, 2430 non-Hodgkin's cases were analyzed. Results showed moderately increased risk for developing non-Hodgkin's and ever use of the insecticide terbufos; chronic lymphocytic leukemia/small lymphocytic lymphoma and the insecticide deltamethrin; and diffuse large B-cell lymphoma and the herbicide glyphosate. However, an unclear situation of exposure of the studied people to these pesticides hampered clear associations.

The most comprehensive survey with about 89,000 participating farmers and their spouses is the so-called Agricultural Health Study conducted in the US states Iowa and North Carolina (AHS 2019). Results of this study revealed that overall rates of cancer among participating farmers and their spouses remained lower than in the general population, especially cancers of the oral cavity, pancreas, and lung. The authors explained this by lower rates of smoking among farmers and related cancers such as lung cancer.

However, compared to the general population, farmers had higher rates of prostate cancer, lip cancer, thyroid cancer, testicular cancer, and peritoneal cancers, and multiple myeloma and acute myeloid leukemia (AHS 2019). More than 30% of women enrolled in the study reported using insecticides and about 8% reported using at least one organochlorine pesticide in their lifetime. The use of one or more organochlorine pesticide was associated with increased rates of glioma (malignant brain tumor), multiple myeloma, pancreatic cancer, and certain types of breast cancer, compared with those who did not use any organochlorine. In later years, organochlorines were largely replaced with organophosphate insecticides. Women who used the organophosphates chlorpyrifos and terbufos were observed to have higher rates of breast cancer.

Spouses who do not use pesticides can still be exposed from spray drift or handling pesticide-contaminated items on the farm. In one study, a higher rate of breast cancer in spouses whose husbands used the organophosphate fonofos was found (AHS 2019). Participating farmers who reported a high pesticide exposure event, such as getting a large amount of pesticides on their skin, were more likely to report a loss of sense of smell 20 years later. Loss of sense of smell is common with aging, especially in patients with Parkinson's and Alzheimer's diseases (AHS 2019).

An evaluation of glyphosate with cancer incidence within the Agricultural Health Study analysed 44,932 glyphosate users, including 5779 incident cancer cases (Andreotti et al. 2018). No association was apparent between glyphosate and any solid tumors or lymphoid malignancies overall. However, there was evidence of increased risk of acute myeloid leukemia among the highest exposed group. It has to be noted that in this study only licensed pesticide users were considered, users who had a special training how to apply pesticides.

Given the high media attention on glyphosate, it might surprise that no stronger associations between glyphosate use and cancer risk were found. Finding associations between pesticides and certain diseases is challenging. It includes recruiting the study population from applicants for a restricted-use pesticide, collecting detailed information on the frequency and duration of pesticides, as well as other farming-related and general population exposures. As mentioned, there are clear indications of potential carcinogenicity of glyphosate from toxicological and epidemiological studies (Guyton et al. 2015). However, epidemiological studies have inherent limitations as they generally detect increased cancer incidence and mortality cancer hazard decades after carcinogen exposure begins (Ward 2018).

Pesticides are most likely not the single factor for these diseases, but they appear to be clearly identified as a very important contributing factor. From the province of Punjab in India, a province in which agriculture is very pesticide-intensive in the course of the Green Revolution, over 33,000 cases of cancer have been reported in recent years alone; a link to the pesticides used there has been made (Shiva et al. 2013).

Strict compliance with the precautionary principle would mean that there is enough evidence to ban glyphosate. However, in March 2017 the European Chemicals Agency (ECHA) announced its verdict on glyphosate: a complete acquittal regarding cancer risk, possible genetic damage, and reproductive toxicity. However, soon afterward EU experts were accused of conflict of interest according to ECHA's own criteria because funding for their research on glyphosate was provided by the chemical industry (Johnston 2017). Also, it was criticized that the ECHA committee was using unpublished scientific evidence provided by the industry in formulating its opinions in addition to studies published in peer-reviewed journals. One would expect that these agencies, whose scientific opinions form the basis for regulatory action, should only consider scientific evidence that is publicly available so that any scientist can check the findings.

During this time a petition from the agrochemical industry was launched in the UK calling for people to stand up for UK agriculture and save glyphosate from being banned by false knowledge and ignorance. The petition has attracted several thousand signatures and states that whole areas of England will become overrun by weeds if glyphosate is banned (Johnston 2017). Such prophecies can also be heard in other countries where there is a risk of a glyphosate ban.

A review analyzed nearly three decades worth of epidemiologic research on the relationship between non-Hodgkin's lymphoma and occupational exposure to 80 active ingredients of pesticides. Results showed positive associations between the use of phenoxy herbicides, carbamate insecticides, organophosphorus insecticides, and the organochlorine insecticide lindane with non-Hodgkin's lymphoma (Schinasi and Leon 2014). It was criticized that data for this study came mainly from high-income countries while low- and middle-income countries where many pesticides are applied were missing.

Yet another study shows that a very low concentration of glyphosate (in the parts per trillion range and thus environmentally relevant for everyone) can trigger breast cancer when combined with another risk factor (Duforestel et al. 2019). In the study, scientists exposed noncancerous human breast cells to glyphosate in vitro over a course of 21 days. The cells were then placed in mice to assess tumor formation. Cells exposed to glyphosate alone did not induce tumor growth. However, cancerous tumors did develop after glyphosate was combined with a type of gene regulatory molecule that is present in all humans.

The difficulty in detecting a pesticide-induced cancer is the very long latency period, which in many cases can be between 30 and 50 years (Streissler 2016). This means that a cancer diagnosed today may be due to contact with a carcinogenic pesticide in the 1970s. Who can remember which pesticides he/she was in contact with so long ago?

Another analysis of the Agricultural Health Study and additional five case–control studies investigated whether there was an association between high cumulative exposures to glyphosate-based herbicides and increased risk of non-Hodgkin's lymphoma in humans (Zhang et al. 2019). Using the highest exposure groups when available in each study, authors report that the overall risk of non-Hodgkin's lymphoma in individuals exposed to glyphosate-based herbicides was increased by 41%. Overall, in accordance with findings from experimental animal and mechanistic studies, this new meta-analysis of human epidemiological studies suggests a compelling link between exposures to glyphosate-based herbicides and increased risk for non-Hodgkin's lymphoma.

The response of the German Federal Institute for Risk Assessment to these findings was quite appeasing and did not challenge the former assessment that glyphosate is not cancerogenous (BfR 2019). Like a mantra they state that glyphosate is not a carcinogen when used properly and as intended. They criticize that the study authors (Zhang et al. 2019) did not determine how much glyphosate the study participants were actually exposed to and that US farmers generally use more glyphosate than other farmers. Also, other pesticides may have been used at the same time as they were exposed to glyphosate. All in all, another position by an authority that ignores the precautionary principle.

A new review of carcinogenicity assessments of pesticide active ingredients shows 40% of them are not carried out in compliance with existing European guidelines, leading to possible continued exposure of farmers and consumers to cancer-causing pesticides (Clausing and Vicaire 2019). The study analyzed the carcinogenicity sections of the renewal assessments reports of ten pesticides: captan, chlorothalonil, chlorpropham, dimoxystrobin, diuron, folpet, forchlorfenuron, phosmet, pirimicarb, and thiacloprid. According to this report at least three of the pesticides (chlorothalonil, diuron, and forchlorfenuron) should have been classified as "presumed" human carcinogens, rather than just "suspected" human carcinogens. This is important as the carcinogenicity classification triggers the regulatory fate of a pesticide active ingredient. Pesticides classified as "suspected" human carcinogens can be marketed, while those classified as "presumed" human carcinogens must be withdrawn. Researchers further concluded that for three pesticides (folpet, pirimicarb, and thiacloprid) the classification should be upgraded, and for phosmet a severe data gap should have been identified by the European authorities and a flawed decisive carcinogenicity. For captan, chlorpropham, and dimoxystrobin, reports were not sufficiently informed to allow any conclusive external review. In 30% of the cases, significant details were missing from the dossiers, raising uncertainties about how European authorities concluded at all.

The Mississippi River basin produces about 80% of major US crops, about two-thirds of US pesticides are applied there. Historically, heavy pesticide application and agricultural irrigation were reported to result in high pesticide residues in surface water, fish, and wells of the Mississippi embayment.

Result of a study shows that elevated colorectal cancer risk in the Mississippi River floodplain is likely linked to historically high pesticide application (Sun 2018). Risk ratio of colorectal cancer incidence in 86 counties of the Mississippi River floodplain was about 29% higher than that of other counties in the 48 contiguous states. Risk ratio of colon cancer mortality in 63 counties of Mississippi embayment was 33% higher than that of other counties in the 48 states between 1999 and 2016. These risk ratios of colorectal cancer incidence and colon cancer mortality were still higher after smoking and diabetes factors were filtered off. Result here suggests that pesticide may be an independent risk factor directly associated with elevated cancer risks.

Even very small amounts of glyphosate can damage human embryonic and placenta cells as well as the DNA of humans and animals. A French study found that glyphosate kills human cells within 1 day, even when diluted 100,000 times (Gasnier et al. 2009). In human cells, a glyphosate-based herbicide (Roundup) can lead to complete cell death within 24 h.

Most health studies have focused on the safety of the active ingredient glyphosate, rather than the mixture of ingredients found in the commercial herbicides. But a study found that the adjuvants contained in glyphosate-based herbicides (again Roundup in this case) can amplify the toxic effect on human cells—even at concentrations much more diluted than those used by farmers or gardeners. Nearly 4000 inert ingredients are approved for use by the US EPA (Gammon 2009), about 1600 in Germany (BVL 2018). One specific inert ingredient, polyethoxylated tallow amine (POEA), was deadlier to human embryonic, placental, and umbilical cord cells than the herbicide itself. POEA is a surfactant, or detergent, lowering water surface tension. It is added to Roundup and other herbicides to help them penetrate plants' surfaces, making the weed killer more effective.

Monsanto scientists argue that the human cells in this study were exposed to unnaturally high levels of these adjuvants, and it is very unlike anything you had see in real-world exposure (Gammon 2009). However, the researchers tried multiple concentrations of the herbicide from recommended dosages down to concentrations 100,000 times more diluted than recommended. The worrying finding was that cell damage was seen also in the lowest concentrations.

Findings that inert ingredients have different effects on human health than the commercial formulation were not only found for glyphosate-based herbicides but also for others. Herbicide formulations containing atrazine and adjuvants caused DNA damage, which can lead to cancer, while atrazine alone did not (Zeljezic et al. 2006). Active ingredients of three herbicides (glyphosate, isoproturon, and fluroxypyr), three insecticides (pirimicarb, imidacloprid, and acetamiprid) and three fungicides (tebuconazole, epoxiconazole, and prochloraz) showed different effects on mitochondrial activities, membrane degradations, and caspases 3/7 activities on human cell lines than the respective commercial products (Mesnage et al. 2014). In this study, fungicides were the most toxic from concentrations 300–600 times lower than agricultural dilutions, followed by herbicides and then insecticides, with very similar profiles in all cell types. Despite its relatively benign reputation, Roundup was among the most toxic herbicides and insecticides tested. Most importantly, 8 formulations out of 9 were up to 1000-times more toxic than their active principles. Hence, researchers challenge the relevance of ecotoxicological studies considering only the toxicity of the active ingredients but ignoring the influence of adjuvants.

Studies with buccal epithelial cell lines also found genotoxic effects after short exposure to concentrations that correspond to a 450-fold dilution of spraying used in agriculture indicating that inhalation may cause DNA damage in exposed individuals (Koller et al. 2012).

At present, the health risk assessment of pesticides in the European Union and in the USA focuses almost exclusively on the stated active principle (Mesnage and Antoniou 2018). We have seen that adjuvants can also be toxic in their own right with numerous negative health effects having been reported in humans and on the environment. Despite the known toxicity of adjuvants, they are not subject to an acceptable daily intake and they are not included in the health risk assessment of dietary exposures to pesticide residues. Overall, it seems clear from a scientific standpoint that adjuvants should also be considered in regulatory precautionary measures and environmental risk assessments.

A lab study carried out with rats showed that glyphosate contamination of their feed led to the formation of a non-alcoholic fatty liver (Mesnage et al. 2017). The concentration of the herbicide Roundup used was 70,000 times lower than permitted residue levels in the EU. Non-alcoholic fatty liver now affects 25% of the population in the USA and Europe. This disease is usually caused by diabetes, obesity, high triglycerides, and high cholesterol. Nevertheless, more and more people fall ill with it, although they do not have any of the risk factors mentioned. Are pesticides to blame? This is not completely proven with the data available, but there are clear indications that this might be the case.

Any limit values for pesticide residues lead to the misconception that everything underneath is harmless to health. We already mentioned how limit values are created and learned that limit values are merely a legal framework, but certainly no limit to the health safety of a substance. It should also be remembered that we come into contact with thousands of potentially carcinogenic substances in our lives, but that a scientific evaluation of health risks of more than one substance does not exist. The precautionary principle would therefore be to keep any exposure to toxins as low as possible.

In addition to the pesticide problems that directly affect humans, there are thousands of pets (mainly dogs and cats) that are poisoned by pesticides applied in municipalities or private gardens. For example, there are many reports on dog poisoning with metaldehyde, commonly contained in slug pellets used in vegetable farming and private gardens (Dolder 2003). Poisoning can be detected because the vomit or breath of pets has a strong smell similar to formaldehyde or acetylene.

2.3.3   Disorders of the Nervous System

Again, Mostafalou and Abdollahi (2017) provide a great overview of studies considering the effects of pesticides on neurotoxicity. Studies show that pesticides can influence the development of the nervous system and the brain development of fetuses or newborns below concentrations that had been classified by authorities (Crumpton et al. 2000).

Many pesticides are similar to substances important in brain biochemistry. Thus, concerns have been raised that the developing brain may be particularly vulnerable to adverse effects of neurotoxic pesticides. A systematic evaluation of published evidence on neurotoxicity of pesticides in current use shows associations with neurodevelopmental deficits, but mainly deals with mixed exposures to pesticides (Bjørling-Poulsen et al. 2008). Laboratory experimental studies suggest that many pesticides currently used in Europe—including organophosphates, carbamates, pyrethroids, ethylenebisdithiocarbamates, and chlorophenoxy herbicides—can cause neurodevelopmental toxicity. The occurrence of residues in food and other types of human exposures should be prevented for these neurotoxic pesticides.

However, there are hardly any official, publicly accessible statistics on chronic poisoning diseases caused by the use of pesticides or residues in food. The increasing number of cancers, hormonal effects, and neurological disorders (such as Parkinson's, Alzheimer's, or amyotrophic lateral sclerosis—ALS, a disease that is characterized by stiff muscles, and gradually worsening weakness due to muscles decreasing in size) has been shown to be associated with the use of some pesticides in agriculture.

Studies have found an association between the herbicides paraquat and maneb and Parkinson's disease. A study revealed that low-level exposure to the pesticides disrupts cells in a way that mimics the effects of mutations known to cause Parkinson's disease (Stykel et al. 2018). Adding the effects of the chemicals to a predisposition for Parkinson's disease drastically increases the risk of disease onset. People exposed to these pesticides are at a 250% higher risk of developing Parkinson's disease than the rest of the population. Frighteningly enough, paraquat is still used in more than 100 countries worldwide, sold by the Switzerland-based agrochemical company Syngenta. The herbicide is used in banana, coffee, palm oil, cotton, rubber, fruit, and ananas plantations. Additionally, it is used in maize and rice fields.

In the US state of Washington, a study examined the relationship between estimated residential exposure of more than 4600 people who had died from Parkinson's disease and pesticide use in their surroundings (Caballero et al. 2018). Therefore, mortality records for 2011–2015 were geocoded using residential addresses and classified as having exposure to agricultural land use within a 1000 m radius. Results showed that individuals exposed to land use associated with glyphosate herbicides had 33% higher odds of premature mortality than those that were not exposed. Exposure to cropland associated with all pesticide application or paraquat application was not statistically significantly associated with premature death from Parkinson's (but the trend was in the hypothesized direction). No significant associations were observed between exposure to the herbicides atrazine or diazinon and premature death from Parkinson's. This research cannot demonstrate a causal link between pesticide use and Parkinson's, but it is definitely worrying and should be investigated in more detail. Often studies suffer from disclosed pesticide application data, despite the obligation of farmers to keep records on where, which, and what amount of pesticides have been applied.

A recent study suggests that glyphosate may pave the way for diseases such as Alzheimer's, diabetes, as well as depression, heart attacks, and infertility. Glyphosate is even referred to as the wrecking ball for the human body because of its wide use and health benefits (Pompa 2016).

Another risky insecticide is chlorpyrifos, which is commonly used in cotton, corn, almonds, and fruit trees, including oranges, bananas, and apples. An industry-funded study found no evidence of selective developmental neurotoxicity following exposure to chlorpyrifos (Maurissen et al. 2000).

However, based on independent epidemiological studies, the evidence points to adverse health effects of chlorpyrifos exposure on the developing nervous system, associated with lowered intelligence at school age, at current levels of exposure (Grandjean and Landrigan 2014). Researchers obtained the industry-funded data based on the freedom of information legislation, reanalyzed the data, and found treatment-related changes in a brain dimension measure for chlorpyrifos at all dose levels tested, which had not been reported in the original test summary (Mie et al. 2018). Researchers observed a difference between raw data and conclusions in the test reports that indicates a bias that was not detected by the regulatory agencies. In an interview researchers dispute the adequacy of the current pesticide approval system, in this case especially to protect pregnant women and the development of the child's brain (Achinger and Wreschniok 2018). The Spanish authority INIA, which is responsible for the approval procedure of this pesticide in Europe, refused a public statement to this case because of the confidentiality of industry data.

In some cases, these scientific warnings lead to political actions. The current authorization of chlorpyrifos in the European Union expired on January 31, 2020. The European Commission announced it will put forward a proposal to European governments to not renew the market approval of this pesticide (Vicaire 2019). It was acknowledged that chlorpyrifos is responsible for an average IQ reduction by 2.5 points of every child living in Europe. Research also points at chlorpyrifos being an endocrine disruptor of the normal functioning of thyroid hormone. Pesticide critical NGOs made an immense pressure with collecting more than 200,000 signatures under a petition to ban chlorpyrifos.

Until its ban chlorpyrifos remained one of the most commonly used pesticides in the EU and other parts of the world and consequently is one of the most frequently found residues in food samples (EFSA 2018a). Chlorpyrifos is most prominently present in citrus fruit, bananas, apples, peaches, and pears, all of them are popular among children too. Southern European countries are among the countries of origin with highest frequency of detection of chlorpyrifos residues. However, also in other countries residues of chlorpyrifos are frequently found on fruits (Vicaire and Lyssimachou 2019): 60% of fruit samples from the USA contained chlorpyrifos residues, 53% of samples from Malta, 51% from China, 50% from Cyprus, 46% from Turkey, 44% from Egypt, and 40% from Serbia. These data show that banning chlorpyrifos in Europe would be even better news for public health when the ban would also include all fruit imports with chlorpyrifos residues.

There is also increasing evidence that pesticides can lead to sensory disorders such as lack of sensitivity and cognitive effects such as memory loss, language problems, and learning difficulties (Hart and Pimentel 2002). Strangely enough, there is a disease specifically named after a pesticide group organophosphate-induced delayed polyneuropathy (OPIDP). This disease is well documented and causes irreversible neurological damage affecting multiple nerves. A drastic example of influencing the drawing abilities of 4- and 5-year-old children is described in a study from Mexico (Guillette et al. 1998). The children in the experiment had the same diet and came from a similar genetic and cultural environment. One group lived in an agricultural village in the valley and was exposed to pesticide drift; the other group lived on the mountains far away from agriculture and pesticides. When asked to draw people on a sheet of paper, it was striking that the pesticide-contaminated children were much less advanced in their brain development and scribbled figures corresponding to those of much younger kids. In addition to these neurological effects, pesticides can affect the respiratory system and reproductive organs.

A case of pesticide poisoning among some 40 Canadian and US diplomats and their families while stationed in Cuba was recently reported (Chartrand et al. 2019). These Canadian and Americans living in Havana showed unexplained illnesses starting in 2016. Symptoms included headaches, dizziness, and difficulty concentrating but also hearing strange sounds; the symptoms were called Havana symptoms. Soon it was also believed that the symptoms were the result of attacks by mysterious acoustic weapons using sonic or microwave technology. However, a scientific study revealed that most likely insecticides used to kill insects that carry infectious diseases were responsible. Like many tropical countries, Cuba launched aggressive campaigns to stop the spread of the Zika virus and frequently sprayed embassies with pyrethroids and organophosphates in very high dosages (Friedman et al. 2019).

2.3.4 Endocrine Disruption and Problems with Fertility

As seen for chlorpyrifos some pesticides are not only toxic in the classical meaning but may act as endocrine disruptors; that is, they interfere with our hormone system. One of the most influential books in this regard is authored by Theo Colborn, Dianne Dumanoski, and John Peterson Myers entitled *Our Stolen Future: Are We Threatening Our Fertility, Intelligence, and Survival?* (Colborn et al. 1996). This scientific detective story, as it is stated in the subtitle, was also an eye-opener for me as a student.

Generally, endocrine disruptors (EDCs) are substances that alter the function of the endocrine system and consequently cause adverse health effects in an intact organism or its offspring and can therefore affect populations and communities of organisms. Hormones regulate very important body functions such as physical and mental development, metabolism, growth, and psychological sensitivities. EDCs influence or disturb the sensitive hormonal system by imitating natural hormones or blocking their effectiveness.

Meanwhile, more than 1000 chemicals have been identified as EDCs or at least potential EDCs by the United National Environment Programme (UNEP 2017). At least 43 frequently used pesticides are suspected of being hormonally effective. Among them are many that can be regularly found in our drinking and groundwater. In fact, 73% of surface water samples, 53% of groundwater samples, and 50% of drinking water samples in France contained pesticides with hormonal activity (HEAL 2017).

EDCs have also been found in beverage can coatings and cosmetics. Many animal and toxicological studies suggest that multiple pesticides may have thyroid-disrupting properties. Both persistent organochlorine pesticides and nonpersistent pesticides such as organophosphates, carbamates, and pyrethroids may interfere with thyroid function (WHO 2012). The persistent chemicals DDT (and the metabolite DDE), hexachlorobenzene (HCB), and nonylphenol a surface-active substance used in pesticide aerosols, are among the most studied with regard to thyroid-disrupting effects.

Although some of these chemicals have long been banned in many countries, they are still present in the environment due to their long environmental half-lives. Exposure occurs at home, in the office, on the farm, in the air we breathe, the food we eat, and the water we drink. Biomonitoring, i.e., the measurement of chemicals in body fluids and tissues, shows nearly all humans have a chemical body burden based on detectable levels in blood, urine, placenta and umbilical cord blood, and body tissues (Gore et al. 2014). In addition to the known EDCs, there are countless suspected EDCs or chemicals that have never been tested.

EDCs are especially critical in children whose hormone system is in development. Therefore, findings such as that 90% of EDC pesticides found on public playgrounds due to drift from neighboring apple orchards or vineyards appear very worrisome (Linhart et al. 2019). While in the animal kingdom feminizations of certain populations are observed, in humans there is a decrease in sperm activity in men, breast cancer, diabetes, and abnormal genital development. Many pesticides for instance act like estrogens and increase breast cancer rates in women. Breast cancer rates in the USA rose from 1 in 20 in 1960 to one in eight women in 1995 (Colborn et al. 1996).

Disorders caused by pesticides in reproduction include testicular malfunctions or sterility. A study of men in the Faroe Islands, an Island in the midst of the European Nordic ocean between Iceland, Norway, and Scotland, showed that the amount of pesticides (e.g., PCB or DDT) in fish and whale meat correlated with that in the male body (Perry et al. 2016). The more the men ingested during their life, the more frequently their sperm had abnormal chromosome numbers.

Moreover, slight sexual deviations seem to be increasing in recent years, such as hypospadias. This is the diagnosis when the penis is not fully developed and the urethra does not end at the tip of the penis. A surgeon in France, in an documentary of the public television cooperation ARTE from 2010, says that they now operate four to ten times a week with this diagnosis (ARTE 2012). The interviewed surgeon suspects that pesticides and adjuvants in plastics adversely affect the development of the penis in boys. Additionally, hormone-containing drug residues, for example residues from the birth control pill, also enter the food chain via drinking water and can thus impair the sexual development of the fetus even in the womb of mothers.

Not only the development of the external genitals can be affected by pesticides. We know very well that substances such as DDT interfere with the interaction of sex hormones and the brain, explains a neurobiologist in the abovementioned documentation. Recent studies from the USA show that there are connections between exposure to harmful substances during pregnancy and children's playing behavior. A meta-analysis of studies assessing the association of pesticide exposure with hypospadias found elevated (but only marginally significant) risks associated with maternal occupational exposure and paternal occupational exposure (Rocheleau et al. 2009).

Exposure to a substance is always the first step toward a health risk. In my opinion, pesticides—no matter how small amounts—do not belong in the human body. If current pesticide use cannot guarantee that their substances gets in our bodies this can be seen as an interference with our personal rights. In the USA up to 232 chemicals, including many pesticides, were found in the umbilical cord of newborns (PCP 2010)—what an inheritance! This demonstrates that many pesticides also break through the placental barrier, a kind of filter that separates the mother's bloodstream from that of the fetus. Glyphosate has also been shown to do this. About 15% of the glyphosate was transferred from the mother's side of the bloodstream to the child's side (Mose et al. 2008). Experiments with lab rats carried out during pregnancy and lactation showed that male offspring were damaged by a glyphosate herbicide (Dallegrave et al. 2007). Additionally, sperm production was influenced and testosterone levels dropped, so that it must be assumed that also the ability to reproduce can be impaired by glyphosate. The study authors suggest a negative influence on the fertility of people who actively or passively come into contact with the glyphosate-containing herbicides.

Of utmost concern are also so-called multigenerational effects. Experiments with lab rats showed that a short-term contact with a fungicide (vinclozolin) during pregnancy led to increased cyst growth in the ovaries in the third generation (Skinner et al. 2013). So exposure of the grandma can have effects on grandchildren who never were in contact with this substance. Vinclozolin is a dicarboximide fungicide that is commonly used to control *Botrytis* and *Sclerotinia* diseases in vineyards, raspberries, lettuce, kiwi, snap beans, and onions or on golf courses.

When pregnant rats were exposed to an insect repellent (DEET) contained in most sprays and lotions against mosquitoes, it was found that even grandchildren and great-grandchildren had defects in their sexual organs that were attributable to this insecticide exposure (Manikkam et al. 2012). It is worrying that these insect repellents are often used very carelessly in the form of insect sprays, lotions, or also as mosquito repellents in the nursery or by pregnant women. Of course, one can say for oneself what applies to rats does not have to apply to humans. But rats are also vertebrates, and they are the established standard organisms to test for adverse effects of pesticides. It would be quite fatal if we have to admit later, when there was sufficient data, that the rat studies indeed provided important information.

Other birth problems are reported from Paraguay. It was shown that women living within a radius of 1 km of pesticide-sprayed soy fields were twice as likely to give birth to a child with malformations (Benítez-Leite et al. 2009). In Argentina (the province of Córdoba) where a great deal of glyphosate is used, the largest spectrum and highest number of malformations of the country are observed (MSPPC 2010).

Studies from Denmark, Brazil, and Spain show that gardening workers and their children suffer more from fertility disorders and genital deformities (PAN Germany 2013). Children of gardeners who have been exposed to high levels of pesticides often have damage to the testicles and their functions. In Denmark, cases of newborns with absence of one or both testes from the scrotum (cryptorchidism) and malformation of the urethra (hypospadias) have been repeatedly documented. These changes in sex development are regarded as feminization symptoms and indicate disturbances of the hormonal system. These findings clearly show that the legal regulations on pesticide testing are not sufficient for preventive health protection.

Reduced fertility is defined by the WHO as the absence of pregnancy after more than 12 months of unprotected sexual intercourse. According to the

WHO, 15% of all couples in the so-called Western world are now affected. Surveys showed that around 28% of gardeners became pregnant after 1 month, compared with 31% of non-gardeners. A much greater difference in expected pregnancy was found when comparing gardeners who did not wear gloves or protective clothing while spraying pesticides and those who always wore gloves (Abell et al. 2000). There was also a difference between women who sprayed pesticides themselves and those who only handled plants that had previously been sprayed with pesticides.

The negative effects of pesticides on male fertility have been mentioned. The effect can lie in damage to the germ cells (sperm stages) or the nutrient and supporting cells located between the germ cells, which are indispensable for the maturation of the sperm (Bretveld et al. 2006). In total, dozens of studies on the reduction of male fertility after occupational exposure of agricultural workers are now available (Bingham and Monforton 2010). These studies come from regions with intensive industrial and conventional agriculture or from countries in which pesticide production facilities are located. In Costa Rica, about 1500 male workers in banana plantations were diagnosed with infertility as a result of DBCP exposure as early as the 1990s, corresponding to about 25% of plantation workers (Thrupp 1991). DBCP (1,2-dibromo-3-chloropropane) was used until the mid-1980s in many agricultural crops to destroy nematodes, small worm-like animals, living in the soil. DBCP is now banned worldwide because of its sterilizing effect on men. Nevertheless, the people affected still suffer from the consequences up to our days.

In industrialized countries, more and more couples are involuntarily childless. In the 1950s, the proportion of childless couples was still 5–8%; in 2000s, it is 15–20% according to an environmental health physician (ORF 2010). Of course, there are also social reasons for this. But it is also clear that the number of sperms in the ejaculate, for example, has decreased. The situation gets even more complex as the standard values for sperm counts were decreased, rather than searching for the underlying causes. Whereas in 1951 120 million sperm/ml of semen were considered normal, in 1989 the WHO established a normal value of only 20 million sperm/ml. The cocktail of pesticides, wood preservatives, heavy metals, and other chemicals that is constantly present in the modern world is considered to be an important cause of reduced fertility.

Negative health effects of pesticides are naturally much more severe in children than in adults. First, children have higher metabolic rates and their ability to absorb, detoxify, and excrete pesticides is different from adults. Secondly, children consume more food per body weight than adults and therefore also consume more pesticides than adults (the latter is also considered in the uncertainty factors applied for defining residue levels). This is important because children's brains are more than five times larger than those of adults in relation to body weight. In California, for example, children who have worked in agriculture have 40% lower blood cholesterol levels, a strong indication of poisoning with organophosphates and carbamates (Repetto and Baliga 1996).

Taken collectively, besides pesticides we are confronted with thousands of chemical substances every moment of our lives. Through every breath, every meal, or every drink, through clothes and cosmetics, at work, at home, and on journeys, the chemicals enter our bodies. Because of false kneeling before industry interests and hiding application and exposure data from the public, our knowledge on a global basis is still weak and no overall scientific assessment about its severity as for other control variables of our planetary boundaries can be made (Steffen et al. 2015). Moreover, this overall chemical burden is not considered when assessing possible health risks of pesticides. Some call this chronic poisoning of the planet by chemicals the event with probably the greatest impact on humanity (Cribb 2016). In any case, it is definitely a historically unique phenomenon that is mostly ignored in the debates about man-made global change.

2.4   Critical Scientists Get Into Trouble

Anyone who takes a critical look at the use of pesticides risks of being placed in the activist corner by the agrochemical lobby and like-minded people. Even colleagues would appear at agrochemical lobby events and claim that warning from pesticides is pure panic and that these warners are afraid of the wrong things (Obermüller 2015). Of course, the reference to the meanwhile refuted statement of Paracelsus, according to which the dose makes the poison, must not be missing at such events. It is even claimed that this is the only scientific theory that has survived 500 years and is still undisputed today. The accusations are that critical scientists or environmental organizations do not take the facts so seriously as the pesticide manufacturers or simply perceive the facts very selectively. The journalist, book author, and documentary filmmaker Marie-Monique Robin describes in her uncompromising investigation many of the methods applied by big agrochemical corporations to achieve these goals (Robin 2009, 2010, 2014).

From personal experience and stories told by fellow scientists who take a critical look at the effects of pesticides, it is known that they are very quickly denounced in public and accused of not properly mastering their subject. There is no hesitation in intervening also via the management of research institutions. People would complain, for instance that my research is doing harm to agriculture and that I am bashing farmers and literally recommend "to turn me off." Fortunately, in Austria we have the constitutional freedom of speech, teaching, and research. This situation can be hard to handle especially for young scientists without permanent contracts. With this pressure, I personally also feel the scissors in my head, self-censoring and thinking twice what consequences a scientifically watertight statement might have for the public. A very famous pesticide-critical colleague from France told me that the management of his university withdrew his lectures for the first semester students because it was feared that he would too much influence the young students with his critical ideas. Independent public funding agencies are contacted by industry representatives recommending to stop funding because of the low quality of the proposed research. Another colleague was openly accused of destroying the life's work of some pesticide developer when talking about a scientific paper that found adverse effects on nontarget organisms.

Meanwhile, I also learned that at least one agrochemical company spies on me and closely watches my social media postings regarding pesticides. This was also documented for hundreds of pesticide-critical journalists, politicians, and NGO activists across Europe (Klawitter 2019).

Internationally popular became the case of Tyrone Hayes, professor at UC Berkeley. Hayes was hired by the agrochemical industry to study the effects of the company's herbicide atrazine on amphibians and assumed he would not have been hired if the company thought there was anything to find (UCS 2017). But when his studies discovered that atrazine exposure is linked with changing genetically male frogs into functional females, he was harassed by the manufacturer of this herbicide in an effort to both discredit his science and tarnish his reputation as a researcher. We heard before that atrazine has been banned in the European Union over concerns that it is a groundwater contaminant with potential effects for wildlife and human health. Yet it is still the second most widely used herbicide in the USA, with millions of tons applied per year. Atrazine often washes into water supplies, particularly in the Midwest, and in 2012, the manufacturer reached a US$ 105-million-settlement with community water systems over the presence of atrazine in drinking water.

The pesticide manufacturer accused Hayes of not having analyzed the data correctly. Then a legal battle developed. A report published by the US Union of Concerned Scientists (UCS 2017) states that unsealed documents of this legal case showed that the manufacturer had a plan including "publication of third-party critique of his science," "systematic rebuttals of all TH appearances," "making emails public," "contact Berkeley," and even "investigate wife." An email from a communications consultant recommended getting hold of Hayes' calendar of speaking engagements so the manufacturer could provide audiences with "irrefutable evidence of his polluted message." A later proposal aired the idea of purchasing Hayes' name as a search word on the Internet, presumably to tarnish his reputation.

Definitely, this case is extreme but symptomatic to what scientists who represent critical views on our pesticide use could expect. When giving public talks I, and some of my critical colleagues as well, frequently notice groups of people asking pointed questions which is of course all right, but often they just want to make mockery of my statements or conclude that my research is outdated and pure alarmism.

Other serious cases are reported from the GMO industry, for which the safety or harmlessness of their products in comparison to other products is key. Otherwise the authorities would have to restrict their distribution. Árpád Pusztai was until his involuntary retirement an internationally renowned biochemist and nutritionist working on plant lectins. In 1995, he conducted feeding studies with rats testing genetically modified potatoes. In 1998, Pusztai said in an BBC interview that his group had observed damage to the intestines and immune systems of rats that were fed the GM potatoes. He actually stated that "If I had the choice I would certainly not eat it," and "I find it's very unfair to use our fellow citizens as guinea pigs. We have to find guinea pigs in the laboratory." (Robin 2009). Well, a few days after this interview was aired massive interventions by industry and high political authorities from the USA and UK resulted in the dissolution of Pusztai's research team, the confiscation of the data, and a gag order for press contacts.

Another example of the GMO research is Gilles-Éric Séralini, a French molecular biologist who published a study showing development of tumors in rats in a 2-year feeding trial with rats of GM maize and glyphosate-based Roundup (Séralini et al. 2012). This peer reviewed study was retracted by the original journal and then after protests by numerous scientists about this unprecedented case was republished in another journal (Séralini et al. 2014).

These are only a couple of prominent cases of industry and/or political interventions against scientists. Interestingly, there are even consulting firms that publish studies commissioned by the chemical industry for so-called product defense. Scientists who uncover pesticide effects on health and the environment run the risk of being damaged in their reputation by these companies. This became known as the manufacturer of atrazine that launched a campaign to discredit scientists who had criticized this herbicide (UNHRC 2017).

Of course, acquiring funding for research projects on pesticides is increasingly difficult if critical studies have been published. While we got support for pesticide projects from a ministry in Austria, after publishing my book, a fellow with good contacts to the ministry told me that I should not be so naïve expecting further project funding any more. To be fair I have to admit that I did not try to submit a grant proposal there since then either.

A few weeks after one of our earthworm studies appeared, I was also pilloried on the Internet (PSIRAM 2015). The keywords for this article are agriculture, pseudoscience, science, and bad science. The motto of this website is "Realism as Opportunity," it is considered the largest Internet pillory in the German-speaking world. Meanwhile, more than 2000 articles have one goal: to defame people and websites. The makers work under the protection of anonymity, and the servers steadily change locations. To be mentioned there is quite a doubtful achievement. In the meantime, numerous cases have been documented where researchers who critically dealt with pesticides or genetic engineering were even put on the sidelines by false claims or discredited with anonymous emails (Robin 2014).

As so many, I am a great fan of Wikipedia, but one has to be aware that agrochemical lobbyists are also manipulating entries there (Oppong 2014). The Wikipedia database is based on swarm intelligence. Anyone can create, edit, or supplement an article. However, there is a corrective behind that checks entries. The theoretical purpose of this structure is to ensure that the database remains neutral and scientifically correct. Wikipedia is often seen as a totally democratic online encyclopedia, but in reality, it is dominated by very strong hierarchies. While there are millions of active users of Wikipedia, only a relatively small circle of Wikipedians, the so-called administrators, participate in internal discussions, votes, and approvals of entries.

In previous votes of German-language Wikipedia pages, only between 50 and 350 votes were cast. Nevertheless, as far as the quality of content is concerned, Wikipedia still scores surprisingly well compared to other encyclopedias. However, it should be noted here that the quality of an article also depends on its subject area. With increasing popularity, Wikipedia was also increasingly mentioned in the press as a source for further reading, but also directly as a source for historical facts and figures. Even respected media link online to Wikipedia; about 95% of all high school students in Germany get their knowledge from Wikipedia.

We also wanted to include the results of our earthworm studies on the negative effects of glyphosate in Wikipedia. However, this was the subject of intense discussions filling several pages. Finally, the main reason for rejecting to cite our study was that, according to an administrator, Wikipedia readers should not be confronted with scientific studies, i.e., so-called primary sources. Rather, secondary sources, such as newspaper articles, would be preferred. I know this is definitely not the case for all Wikipedia articles; some really read like popular-science articles in the best sense. After this experience I am much more cautious when referring to Wikipedia.

Public denunciation does not even stop at scientists as historically heroic as Rachel Carson. The Internet is full of slander from alleged scientific institutions that portray Carson as a mass murderer, worse than Hitler. Why? According to these sources, millions of Africans died of malaria as a result of the ban on DDT imposed by *Silent Spring*. The same critics describe DDT as the most important chemical ever developed to prevent diseases, and it was only banned by a hysteria inspired by Carson. Behind such activities are often the same institutions and industry-related lobbying groups that call themselves institutes and opposed tobacco regulations, the ban on ozone-depleting chemicals, restrictions on fossil fuel use, or connections between human activity and climate change (Oreskes and Conway 2010). The goal is to scatter false information and making people doubt on scientific findings. People would ask: Perhaps there is not so much to the apocalyptic talk? Perhaps only a few self-obsessed, mediocre scientists want to get attention and secure funding for their research projects with these catastrophe reports? Unfortunately, many editorial offices of independent newspapers and independent specialist magazines now also lack investigative journalists who could uncover these ties between lobbyists and media with the appropriate meticulousness, flair, expertise, and tenacity. What remains is a certain skepticism toward science in general. And that is the goal of these lobby activities. This also supports policy makers in their hesitant actions.

Campaigns by the pesticide industry and farmers' representatives in Germany are likely to have similar motivations when pesticide-critical NGOs are accused of unsettling consumers and defaming conventional farmers mainly for the purpose of collecting donations (Beste 2017). I too was blamed several times—also by official representatives of agriculture—to only spreading panic among people using fake news in order to better sell my book.

With the increasing privatization of research, another important aspect is emerging. There are public subsidies for private companies, which make large profits by implementing the research results and pass any losses or environmental costs on to society. I also experienced during panel meetings of funding bodies that millions of € in public research funds were spent on the further development of pesticide-intensive cultivation practices.

Another critical aspect is the increased establishment of company-financed endowed professorships at public universities. For universities, this naturally means a reduction in research expenditure. For research on pesticide effects, however, this should be viewed critically, because the burden of proof has always been reversed in the field of pesticide research. It is not the manufacturers has a conflict of interest to carry out a transparent, comprehensible risk analysis of its product; critical research could come instead from independent researchers it they declare that there are no ties to industry. The question is whether this really works when clauses in research contracts between the funder from industry require that every report must be double-checked before it can be published.

Even one of the most renowned universities in Switzerland takes generous grants from pesticide and seed companies who want to promote sustainable agriculture (Haller 2011). A circumstance that also becomes an ethical component because the donor continues to produce and exports pesticides that are banned in Switzerland for its health-related concerns, thereby putting the health of many users in developing countries at risk.

The fact that agrochemistry tries to intervene out of business interests can perhaps be understood. If, however, representatives of farmers intervene, the matter becomes tricky. Farmer representatives reportedly wanted to pay Austrian beekeepers a compensation for pesticide damage to bees if, in return, bee mortality is not publicly discussed (ORF 2013). Finally, beekeepers have not complied with this. A further point of criticism is research into the causes of bee mortality with the financial participation of the agrochemical industry that actually produces bee toxic pesticides. Additionally, as a critical scientist one would get many amoral offers from the industry for lucrative research projects.

In his book, *Purchased research—Science in the service of corporations*, the author Christian Kreiß concludes that a major problem is that industrial and thus financial interests determine more and more the contents, subjects, and agendas of our research (Kreiß 2015). Therefore, research is being steered on a large scale into interest-driven channels and not as open as it should be.

References

Abbas N, Khan HA, Shad S (2014) Cross-resistance, genetics, and realized heritability of resistance to fipronil in the house fly, *Musca domestica* (Diptera: Muscidae): a potential vector for disease transmission. Parasitol Res 113:1343–1352

Abell A, Juul S, Bonde JPE (2000) Time to pregnancy among female greenhouse workers. Scand J Work Environ Health 26:131–136

Achinger E, Wreschniok L (2018) Pestizidzulassungen in der EU. Verbraucherschutz Fehlanzeige. Bayerischer Rundfunk. Report München, www.br.de/fernsehen/das-erste/sendungen/report-muenchen/videos-und-manuskripte/pestizide-verbraucherschutz-116.html. Accessed 02 Nov 2019

Ackerman-Leist P (2017) A precautionary tale. How one small town banned pesticides, preserved its food heritage, and inspired a movement. Chelsea Green Publishing, White River Junction, VT

Acquavella JF et al (2004) Glyphosate biomonitoring for farmers and their families: results from the Farm Family Exposure Study. Environ Health Perspect 112:321–326

AGES (2015) AGES Stellungnahme zur Regenwurm-Studie der BOKU. Glyphosat und das Risiko für Regenwürmer. In: Ernährungssicherheit AfGu (ed). https://www.ages.at/fileadmin/AGES2015/Themen/Pflanzenschutzmittel_Dateien/BOKU_Regenwurmstudie_Stellungnahme.pdf. Accessed 10 Oct 2019

Agrarministerkonferenz (2015) Ergebnisprotokoll der Agrarministerkonferenz am 2. Oktober 2015 in Fulda, 69 p

AHS (2019) Agricultural Health Study. Study Update 2019. National Cancer Institute, National Institute of Environmental Health Sciences, U.S. Environmental Protection Agency, National Institute for Occupational Safety and Health. https://aghealth.nih.gov/news/AHSUpdate2019.pdf. Accessed 02 Nov 2019

Andreotti G et al (2018) Glyphosate use and cancer incidence in the agricultural health study. J Natl Cancer Inst 110:509–516

Anon (2018) Auch das noch! Schmelzende Alpen-Gletscher sind giftig. In: Blick. https://www.blick.ch/news/schweiz/auch-das-noch-schmelzende-alpen-gletscher-sind-giftig-id33662.html. Accessed 20 Oct 2019

APS (2019) What are fungicides? In: American Phytopathological Society (APS) (ed) https://www.apsnet.org/edcenter/disimpactmngmnt/topc/Pages/Fungicides.aspx. Accessed 07 Dec 2019

ARTE (2012) Tabu Intersexualitaet – Menschen zwischen den Geschlechtern. https://www.youtube.com/watch?v=rNg8NhVwb5s. Accessed 02 Nov 2019

BAFU (2019) Pflanzenschutzmittel im Grundwasser. https://www.bafu.admin.ch/bafu/de/home/themen/wasser/fachinformationen/zustand-der-gewaesser/zustand-des-grundwassers/grundwasser-qualitaet/pflanzenschutzmittel-im-grundwasser.html. Accessed 30 Dec 2019

Baier F et al (2016a) Non-target effects of a glyphosate-based herbicide on Common toad larvae (*Bufo bufo*, Amphibia) and associated algae are altered by temperature. PeerJ 4:e2641. https://doi.org/10.7717/peerj.2641

Baier F, Jedinger M, Gruber E, Zaller JG (2016b) Temperature-dependence of glyphosate-based herbicide's effects on egg and tadpole growth of Common Toads. Front Environ Sci 4:51. https://doi.org/10.3389/fenvs.2016.00051

Bandow C, Karau N, Römbke J (2014) Interactive effects of pyrimethanil, soil moisture and temperature on *Folsomia candida* and *Sinella curviseta* (Collembola). Appl Soil Ecol 81:22–29

Barber I, Tarrant KA, Thompson HM (2003) Exposure of small mammals, in particular the wood mouse *Apodemus sylvaticus*, to pesticide seed treatments. Environ Toxicol Chem 22:1134–1139

Barnes CJ, Lavy TL, Mattice JD (1987) Exposure of non-applicator personnel and adjacent areas to aerially applied propanil. Bull Environ Contam Toxicol 39:126–133

Beketov MA, Kefford BJ, Schäfer RB, Liess M (2013) Pesticides reduce regional biodiversity of stream invertebrates. Proc Natl Acad Sci U S A 110:11039–11043

Belda I, Zarraonaindia I, Perisin M, Palacios A, Acedo A (2017) From vineyard soil to wine fermentation: microbiome approximations to explain the "terroir" concept. Front Microbiol 8:821

Benbrook CM (2016) Trends in glyphosate herbicide use in the United States and globally. Environ Sci Eur 28:3

Benbrook CM (2019) How did the US EPA and IARC reach diametrically opposed conclusions on the genotoxicity of glyphosate-based herbicides? Environ Sci Eur 31:2

Benítez-Leite S, Macchi M, Acosta M (2009) Malformaciones congénitas asociadas a agrotóxicos. Rev Chil Pediatr 80:377–378

Benton TG, Bryant DM, Cole L, Crick HQP (2002) Linking agricultural practice to insect and bird populations: a historical study over three decades. J Appl Ecol 39:673–687

Bertolote JM, Fleischmann A, Eddleston M, Gunnell D (2006) Deaths from pesticide poisoning: are we lacking a global response. Br J Psychiatry 189:201–203

Beste A (2017) Vergiftet. Pestizide in Boden und Wasser – das Beispiel Glyphosat. In: AgrarBündnis eV (ed) Der kritische Agrarbericht 2017. Schwerpunkt Wasser. ABL Bauernblatt Verlags-GmbH, Konstanz, pp 204–208

BfR (2019) Neue Meta-Analyse zu glyphosathaltigen Pflanzenschutzmitteln ändert die Bewertung des Wirkstoffs nicht. Stellungnahme Nr 008/2019 des BfR vom 3. April 2019. https://www.bfr.bund.de/cm/343/neue-meta-analyse-zu-glyphosath-altigen-pflanzenschutzmitteln-aendert-die-bewertung-des-wirkstoffs-nichtpdf. Accessed 18 Dec 2019

Bingham E, Monforton C (2010) The pesticide DBCP and male infertility. In: Agency EE (ed) Lessons from health hazards. The pesticide DBCP and male infertility, vol No 1/2013, pp 203–214

Bjørling-Poulsen M, Andersen HR, Grandjean P (2008) Potential developmental neurotoxicity of pesticides used in Europe. Environ Health 7:50–50

BNN (2019) BNN-Orientierungswert für Pestizide – Eine Leitlinie zur Beurteilung von Pestizidnachweisen in Bio-Produkten. https://n-bnn.de/sites/default/dateien/190409_BNN-Orientierungswert_DE.pdf. Accessed 24 Jan 2020

Boatman ND et al (2004) Evidence for the indirect effects of pesticides on farmland birds. Ibis 146:131–143

Bogdal C, Schmid P, Zennegg M, Anselmetti F, Scheringer M, Hungerbühler K (2009) Blast from the past: melting glaciers as a relevant source for persistent organic pollutants. Environ Sci Technol 43:8173–8177

Bonmatin J-M et al (2015) Environmental fate and exposure; neonicotinoids and fipronil. Environ Sci Pollut Res 22:35–67

Bowler DE, Heldbjerg H, Fox AD, de Jong M, Böhning-Gaese K (2019) Long-term declines of European insectivorous bird populations and potential causes. Conserv Biol 33:1120–1130

Boyles JG, Cryan PM, McCracken GF, Kunz TH (2011) Economic importance of bats in agriculture. Science 332:41–42

BR (2016) Glyphosat im Weinberg: Regenwürmer auf Rückzug. BR Fernsehen, Unser Land, Christoph Schuster, www.youtube.com/watch?v=dIhan0Qw0vM. Accessed 10 Oct 2019

Bradbury RB, Bailey CM, Wright D, Evans AD (2008) Wintering Cirl Buntings *Emberiza cirlus* in southwest England select cereal stubbles that follow a low-input herbicide regime: Capsule Birds selected stubbles preceded by crops with reduced pesticide inputs over those grown conventionally. Bird Study 55:23–31

Brakes CR, Smith RH (2005) Exposure of non-target small mammals to rodenticides: short-term effects, recovery and implications for secondary poisoning. J Appl Ecol 42:118–128

Bretveld R, Zielhuis GA, Roeleveld N (2006) Time to pregnancy among female greenhouse workers. Scand J Work Environ Health 32:359–367

Brickle NW, Harper DGC, Aebischer NJ, Cockayne SH (2000) Effects of agricultural intensification on the breeding success of corn buntings Miliaria calandra. J Appl Ecol 37:742–755

Bro E, Millot F, Decors A, Devillers J (2015) Quantification of potential exposure of gray partridge (Perdix perdix) to pesticide active substances in farmlands. Sci Tot Environ 521–522:315–325

Brühl CA, Zaller JG (2019) Biodiversity decline as a consequence of an inadequate environmental risk assessment of pesticides. Front Environ Sci 7:177

Brühl CA, Schmidt T, Pieper S, Alscher A (2013) Terrestrial pesticide exposure of amphibians: an underestimated cause of global decline? Sci Rep 3:1135

Bucheli TD, Müller SR, Heberle S, Schwarzenbach RP (1998) Occurrence and behavior of pesticides in rainwater, roof runoff, and artificial stormwater infiltration. Environ Sci Technol 32:3457–3464

Buchweitz JP, Carson K, Rebolloso S, Lehner A (2018) DDT poisoning of big brown bats, Eptesicus fuscus, in Hamilton, Montana. Chemosphere 201:1–5

Butler D (2018a) EU pesticide review could lead to ban. Nature 555:150–151

Butler D (2018b) Scientists hail European ban on bee-harming pesticides. Nature, 27 April 2018. https://doi.org/10.1038/d41586-018-04987-4

BVL (2018) Beistoffe in zugelassenen Pflanzenschutzmitteln. In: Bundesamt für Verbraucherschutz und Lebensmittelsicherheit. https://www.bvl.bund.de/SharedDocs/Downloads/04_Pflanzenschutzmittel/zul_info_liste_beistoffe.html. Accessed 29 Sept 2019

BVL (2019a) Notfallzulassung nach Artikel 53 der Verordnung (EG) Nr. 1107/2009 für das Pflanzenschutzmittel: Exirel. https://www.bvl.bund.de/SharedDocs/Downloads/04_Pflanzenschutzmittel/01_notfallzulassungen/Exirel_

Kirschessigfliege_Weinbau_2019.pdf?__blob=publicationFile&v=2. Accessed 25 Dec 2019

BVL (2019b) Verzeichnis zugelassener Pflanzenschutzmittel, apps2.bvl.bund.de/psm/jsp/index.jsp. In: Bundesamt für Verbraucherschutz und Lebensmittelsicherheit. Accessed 30 Sept 2019

Caballero M, Amiri S, Denney JT, Monsivais P, Hystad P, Amram O (2018) Estimated residential exposure to agricultural chemicals and premature mortality by Parkinson's disease in Washington state. Int J Environ Res 15(12):2885. https://doi.org/10.3390/ijerph15122885

Calvert GM et al (2008) Acute pesticide poisoning among agricultural workers in the United States, 1998–2005. Am J Ind Med 51:883–898

Capowiez Y, Rault M, Costagliola G, Mazzia C (2005) Lethal and sublethal effects of imidacloprid on two earthworm species (*Aporrectodea nocturna* and *Allolobophora icterica*). Biol Fertil Soils 41:135–143

Carriger J, Rand G (2008) Aquatic risk assessment of pesticides in surface waters in and adjacent to the Everglades and Biscayne National Parks: I. Hazard assessment and problem formulation. Ecotoxicology 17:660–679

Carson R (1962) Silent Spring. Houghton Mifflin, New York, NY

Chartrand L, Movilla M, Ellenwood L (2019) Havana syndrome: exposure to neurotoxin may have been cause, study suggests. CBC News, www.cbc.ca/news/canada/havana-syndrome-neurotoxin-enquête-1.5288609. Accessed 12 Jan 2020

Clausing P, Vicaire Y (2019) Carcinogenicity assessment was flawed for 4 out of 10 pesticides, new report shows, www.env-health.org/carcinogenicity-assessment-was-flawed-for-4-out-of-10-pesticides-new-report-shows/. Accessed 02 Nov 2019

Coeurdassier M et al (2014) Unintentional wildlife poisoning and proposals for sustainable management of rodents. Conserv Biol 28:315–321

Colborn T, Myers JP, Dumanoski D (1996) Our stolen future: are we threatening our fertility, intelligence, and survival? A scientific detective story. Dutton, New York, NY

Corsolini S, Borghesi N, Ademollo N, Focardi S (2011) Chlorinated biphenyls and pesticides in migrating and resident seabirds from East and West Antarctica. Environ Int 37:1329–1335

Cribb J (2016) Surviving the 21st century: humanity's ten great challenges and how we can overcome them. Springer, Zurich

Crumpton T, Seidler F, Slotkin T (2000) Developmental neurotoxicity of chlorpyrifos in vivo and in vitro: effects on nuclear transcription factors involved in cell replication and differentiation. Brain Res 857:87–98

Dallegrave E, Mantese FD, Oliveira RT, Andrade AJM, Dalsenter PR, Langeloh A (2007) Pre- and postnatal toxicity of the commercial glyphosate formulation in Wistar rats. Arch Toxicol 81:665–673

de Jong FMW, de Snoo GR, van de Zande JC (2008) Estimated nationwide effects of pesticide spray drift on terrestrial habitats in the Netherlands. J Environ Manag 86:721–730 de Snoo GR, Scheidegger NM, de Jong FMW (1999) Vertebrate wildlife incidents with pesticides: a European survey. Pestic Sci 55:47–54

DiBartolomeis M, Kegley S, Mineau P, Radford R, Klein K (2019) An assessment of acute insecticide toxicity loading (AITL) of chemical pesticides used on agricultural land in the United States. PLoS One 14:e0220029

Dinehart SK, Smith LM, McMurry ST, Anderson TA, Smith PN, Haukos DA (2009) Toxicity of a glufosinate- and several glyphosate-based herbicides to juvenile amphibians from the Southern High Plains, USA. Sci Tot Environ 407:1065–1071

Dirzo R, Young HS, Galetti M, Ceballos G, Isaac NJB, Collen B (2014) Defaunation in the Anthropocene. Science 345:401–406

Dolder LK (2003) Metaldehyde toxicosis. Vet Med 98:213–215

Dowding CV, Shore RF, Worgan A, Baker PJ, Harris S (2010) Accumulation of anticoagulant rodenticides in a non-target insectivore, the European hedgehog (*Erinaceus europaeus*). Environ Pollut 158:161–166

Duforestel M et al (2019) Glyphosate primes mammary cells for tumorigenesis by reprogramming the epigenome in a TET3-dependent manner. Front Genet 10:885

Dyttrich B (2015) Giftiger Bodenschutz à la Syngenta. https://www.woz.ch/1521/agrochemie/giftiger-bodenschutz-a-la-syngenta. Accessed 20 Oct 2019

EASAC (2015) Ecosystem services, agriculture and neonicotinoids. EASAC policy report 26, 70pp. easac.eu/publications/details/ecosystem-services-agriculture-and-neonicotinoids/. Accessed 23 Oct 2017

EC (2019) EU pesticides database. https://ec.europa.eu/food/plant/pesticides/eu-pesticides-database/public/?event=homepage&language=EN. Accessed 23 Nov 2019

EEA (2018) European waters assessment of status and pressures 2018. EEA Report No. 7/2018, https://www.eea.europa.eu/publications/state-of-water. Accessed 12 Jan 2020

EFSA (2018a) The 2016 European Union report on pesticide residues in food. EFSA J 16:139p

EFSA (2018b) Evaluation of the data on clothianidin, imidacloprid and thiamethoxam for the updated risk assessment to bees for seed treatments and granules in the EU. EFSA Supporting Publications 15:1378E

EFSA (2018c) Peer review of the pesticide risk assessment for bees for the active substance clothianidin considering the uses as seed treatments and granules. EFSA J 16:e05177

EFSA (2018d) Peer review of the pesticide risk assessment for bees for the active substance imidacloprid considering the uses as seed treatments and granules. EFSA J 16:e05178

EFSA (2018e) Peer review of the pesticide risk assessment for bees for the active substance thiamethoxam considering the uses as seed treatments and granules. EFSA J 16:e05179

EFSA et al (2019a) Peer review of the pesticide risk assessment for the active substance sulfoxaflor in light of confirmatory data submitted. EFSA J 17:e05633

EFSA et al (2019b) Scientific statement on the coverage of bats by the current pesticide risk assessment for birds and mammals. EFSA J 17:e05758

Eng ML, Stutchbury BJM, Morrissey CA (2019) A neonicotinoid insecticide reduces fueling and delays migration in songbirds. Science 365:1177–1180

EPA (2016) Response to comments of the draft biological evaluations for chlorpyrifos, diazinon, and malathion. https://www3.epa.gov/pesticides/nas/final/response-to-comments.pdf. Accessed 25 Dec 2019

EPRS (2018) Directive 2009/128/EC on the sustainable use of pesticides. In: Service EPR (ed) http://www.europarl.europa.eu/RegData/etudes/STUD/2018/627113/EPRS_STU(2018)627113_EN.pdf. Accessed 13 Oct 2019

Ewald JA, Aebischer NJ (1999) Pesticide use, avian food resources and bird densities in Sussex. Joint Nature Conservation Committee Report No 296, 103 p

Ewald JA, Aebischer NJ, Brickle NW, Moreby SJ, Potts GR, Wakeham-Dowson A (2002) Spatial variation in densities of farmland birds in relation to pesticide use and avian food resources. Avian Landscape Ecology IALE (UK):305–312

Fine JD, Cox-Foster DL, Mullin CA (2017) An inert pesticide adjuvant synergizes viral pathogenicity and mortality in honey bee larvae. Sci Rep 7:40499

Fischer J, Müller T, Spatz A-K, Greggers U, Grünewald B, Menzel R (2014) Neonicotinoids interfere with specific components of navigation in honeybees. PLoS One 9:e91364

Flickinger EL, King KA, Stout WF, Mohn MM (1980) Wildlife hazards from furadan 3G applications to rice in Texas. J Wildl Manag 44:190–197

Friedman A, Calkin C, Bowen C (2019) Havana syndrome: neuroanatomical and neurofunctional assessment in acquired brain injury due to unknown etiology. Brain Repair Center, de.scribd.com/document/426438895/Etude-du-Centre-de-traitement-des-lesions-cerebrales-de-l-Universite-de-Dalhousie, 41pp. Accessed 12 Jan 2020

Gaberell L (2020) In der Schweiz verbotenes Pestizid verschmutzt brasilianisches Trinkwasser. In: Eye P (ed) https://www.publiceye.ch/de/themen/pestizide/profenofos?fbclid=IwAR2j8-LGJ-3Wmc_HqzqfC50YvCthjHmU9cR3U9BQd1Dw4O_b9PfWLHoZbfc. Accessed 26 Jan 2020

Gallai N, Salles J-M, Settele J, Vaissière BE (2009) Economic valuation of the vulnerability of world agriculture confronted with pollinator decline. Ecol Econ 68:810–821

Gammon C (2009) Weed-whacking herbicide proves deadly to human cells. Scientific American, vol June 23, 2009. https://www.scientificamerican.com/article/weed-whacking-herbicide-p/. Accessed 02 Jan 2020

Gasnier C, Dumont C, Benachour N, Clair E, Chagnon M-C, Séralini G-E (2009) Glyphosate-based herbicides are toxic and endocrine disruptors in human cell lines. Toxicology 262:184–191

Gaupp-Berghausen M, Hofer M, Rewald B, Zaller JG (2015) Glyphosate-based herbicides reduce the activity and reproduction of earthworms and lead to increased soil nutrient concentrations. Sci Rep 5:12886

Geiger F et al (2010) Persistent negative effects of pesticides on biodiversity and biological control potential on European farmland. Basic Appl Ecol 11:97–105

Geisz HN, Dickhut RM, Cochran MA, Fraser WR, Ducklow HW (2008) Melting glaciers: a probable source of DDT to the Antarctic marine ecosystem. Environ Sci Technol 42:3958–3962

Gerlock G (2013) Herbicide drift threatens Midwest vineyards. Harvest Public Media.   https://www.kcur.org/post/herbicide-drift-threatens-midwest-vineyards#stream/0. Accessed 20 Oct 2019

Gibbons DW et al (2006) Weed seed resources for birds in fields with contrasting conventional and genetically modified herbicide-tolerant crops. Proc Roy Soc B 273:1921–1928

Gibbs KE, Mackey RL, Currie DJ (2009) Human land use, agriculture, pesticides and losses of imperiled species. Divers Distrib 15:242–253

Gilliom RJ, Hamilton PA (2006) Pesticides in the nation's streams and ground water, 1992-2001 – a summary. Fact Sheet, 6pp, pubs.er.usgs.gov/publication/fs20063028. Accessed 17 Oct 2017

Goldstein MI et al (1999) Monocrotophos-induced mass mortality of Swainson's Hawks in Argentina, 1995–96. Ecotoxicology 8:201–214

Gore AC, Crews D, Doan LL, Merrill ML, Patisaul H, Zota A (2014) Introduction to endocrine disrupting chemicals (EDCs). A guide fro public interest organizations and policy-makers. In: Society E (ed) Introduction to endocrine disrupting chemicals (EDCs), A guide for public interest organizations and policy-makers. https://www.endocrine.org/-/media/endosociety/files/advocacy-and-outreach/important-documents/introduction-to-endocrine-disrupting-chemicals.pdf. Accessed 02 Nov 2019

Goulson D (2013) An overview of the environmental risks posed by neonicotinoid insecticides. J Appl Ecol 50:977–987

Goulson D (2019) The insect apocalypse, and why it matters. Curr Biol 29:R967–R971

Grandjean P, Landrigan PJ (2014) Neurobehavioural effects of developmental toxicity. Lancet Neurol 13:330–338

Greenpeace (2014) Gift im Bienen-Gepäck. Analyse von Pestizidrückständen in Bienenbrot und Pollenhöschen von Honigbienen (Apis mellifera) aus 12 europäischen Ländern, www.greenpeace.de/sites/www.greenpeace.de/files/publications/20140415-pollenreport-gift-im-bienengepaeck.pdf. Accessed 19 Oct 2019

Grimalt JO, Borghini F, Sanchez-Hernandez JC, Barra R, Torres García CJ, Focardi S (2004) Temperature dependence of the distribution of organochlorine compounds in the mosses of the Andean mountains. Environ Sci Technol 38:5386–5392

Guillette EA, Meza MM, Aquilar MG, Soto AD, Garcia IE (1998) An anthropological approach to the evaluation of preschool children exposed to pesticides in Mexico. Environ Health Perspect 106:347–353

Gullan PJ, Cranston PS (eds) (2014) The insects: an outline of entomology, 5th edn. Wiley-Blackwell, London

Gunnell D, Eddleston M (2003) Suicide by intentional ingestion of pesticides: a continuing tragedy in developing countries. Int J Epidemiol 32:902–909

Guyton KZ et al (2015) Carcinogenicity of tetrachlorvinphos, parathion, malathion, diazinon, and glyphosate. Lancet Oncol 16:490–491

Hahn H, Michalak K, Preussner K, Engler A, Heinemayer G, Gundert-Remy U (2000) Ärztliche Mitteilungen bei Vergiftungen nach §16e Chemikaliengesetz 1999. In: Bericht der "Zentralen Erfassungsstelle für Vergiftungen, gefährliche Stoffe und Zubereitungen, Umweltmedizin" im Bundesinstitut für gesundheitlichen Verbraucherschutz und Veterinärmedizin für das Jahr 1999. Bundesinstitut für gesundheitlichen Verbraucherschutz und Veterinärmedizin, mobil.bfr.bund.de/cm/350/aerztliche_mitteilungen_bei_vergiftungen_1999.pdf, 62pp. Accessed 02 Nov 2019

Hall FR (1991) Pesticide application technology and integrated pest management (IPM). In: Pimentel D (ed) Handbook of pest management in agriculture, vol II. CRC Press, Boca Raton, FL, pp 135–170

Haller D (2011) Maya Graf geht gegen Millionen-Spenden der Syngenta vor. Basellandschaftliche Zeitung. http://www.aargauerzeitung.ch/basel/baselbiet/maya-graf-geht-gegen-millionen-spenden-der-syngenta-vor-114988856. Accessed 02 Nov 2019

Hallmann J, Quadt-Hallmann A, Tiedemann AV (2009) Phytomedizin. Grundwissen Bachelor, 2. Auflage edn. Verlag Eugen Ulmer, Stuttgart, Deutschland

Hallmann CA, Foppen RPB, van Turnhout CAM, de Kroon H, Jongejans E (2014) Declines in insectivorous birds are associated with high neonicotinoid concentrations. Nature 511:341–343

Hanner D (1984) Herbicide drift prompts state inquiry. Dallas Morning News

Hansen E, Donohoe M (2003) Health issues of migrant and seasonal farm workers. J Health Care Poor Underserved 14:153–164

Hardell L, Eriksson M (1999) A case–control study of non-Hodgkin lymphoma and exposure to pesticides. Cancer 85:1353–1360

Hart K, Pimentel D (2002) Public health and costs of pesticides. In: Pimentel D (ed) Encyclopedia of pest management. Marcel Dekker, New York, NY, pp 677–679

Hayasaka D, Korenaga T, Suzuki K, Saito F, Sánchez-Bayo F, Goka K (2012) Cumulative ecological impacts of two successive annual treatments of imidacloprid and fipronil on aquatic communities of paddy mesocosms. Ecotox Environ Safe 80:355–362

Hayes TB, Falso P, Gallipeau S, Stice M (2010a) The cause of global amphibian declines: a developmental endocrinologist's perspective. J Exp Biol 213:921–933

Hayes TB et al (2010b) Atrazine induces complete feminization and chemical castration in male African clawed frogs (*Xenopus laevis*). Proc Natl Acad Sci USA 107:4612–4617

HBS and BUND (2020) Insektenatlas. Daten und Fakten über Nütz- und Schädlinge in der Landwirtschaft, Heinrich-Böll-Stiftung & BUND, www.boell.de/sites/default/files/2020-01/WEB_insektenatlas_2020.pdf?dimension1=ds_insektenatlas. Accessed 24 Jan 2020

HEAL (2017) Endocrine disrupting pesticides in tap and surface water. http://www.env-health.org/resources/press-releases/article/endocrine-disrupting-pesticides-in

Heap I (2020) The international survey of herbicide resistant weeds. www.weedscience.org. Accessed 05 Jan 2020

Hébert M-P, Fugère V, Gonzalez A (2019) The overlooked impact of rising glyphosate use on phosphorus loading in agricultural watersheds. Front Ecol Environ 17:48–56

Hegde G, Krishnamurthy SV, Berger G (2019) Common frogs response to agrochemicals contamination in coffee plantations, Western Ghats, India. Chem Ecol 35:397–407

Heidis Mist (2020) Des Schweizers weisse Weste: Export von giftigen Pestiziden. https://heidismist.wordpress.com/2020/01/26/des-schweizers-weisse-weste-export-von-giftigen-pestiziden/. Accessed 26 Jan 2020

Helander M, Saloniemi I, Omacini M, Druille M, Salminen JP, Saikkonen K (2018) Glyphosate decreases mycorrhizal colonization and affects plant-soil feedback. Sci Tot Environ 642:285–291

Hesselbach H, Scheiner R (2018) Effects of the novel pesticide flupyradifurone (Sivanto) on honeybee taste and cognition. Sci Rep 8:4954

Heute (2018) Glyphosat bedroht die Qualität unserer Weine, heute, www.heute.at/s/glyphosat-bedroht-die-qualitat-der-weine-53218999. Accessed 24 Dec 2019

Hinck JE, Blazer VS, Schmitt CJ, Papoulias DM, Tillitt DE (2009) Widespread occurrence of intersex in black basses (*Micropterus* spp.) from U.S. rivers, 1995–2004. Aquat Toxicol 95:60–70

Hoffmann J et al. (2012) Bewertung und Verbesserung der Biodiversität leistungsfähiger Nutzungssysteme in Ackerbaugebieten unter Nutzung von Indikatorvogelarten, Julius Kühn-Institut, Federal Research Centre for Cultivated Plants, Braunschweig, www.openagrar.de/receive/openagrar_mods_00009677. Accessed 17 Oct 2017

Holcombe M (2019) Poisonings have killed an owl and 7 bald eagles. Now, there is an investigation into the mysterious deaths. CNN. https://edition.cnn.com/2019/05/03/us/maryland-eagle-owl-deaths-poison-trnd/index.html. Accessed 25 Dec 2019

Hooven L, Sagili R, Johansen E (2013) How to reduce bee poisoning from pesticides. In: A Pacific Northwest Extension Publication, vol. PNW 591. http://ipm.ucanr.edu/PDF/PMG/pnw591.pdf. Accessed 04 Jan 2020

Humann-Guilleminot S, Clément S, Desprat J, Binkowski ŁJ, Glauser G, Helfenstein F (2019) A large-scale survey of house sparrows feathers reveals ubiquitous presence of neonicotinoids in farmlands. Sci Tot Environ 660:1091–1097

IGP (2015) IGP zu NGO-Sturmlauf: Wurm-Studie von Global 2000 ohne Aussagekraft. https://www.igpflanzenschutz.at/igp-zu-ngo-sturmlauf-wurm-studie-von-global-2000-ohne-aussagekraft/. Accessed 10 Oct 2019

IPBES (2017) The assessment report of the Intergovernmental Science-Policy Platform on Biodiversity and Ecosystem Services on pollinators, pollination and food production. In: Potts SG, Imperatriz-Fonseca VL, Ngo HT (eds) Secretariat of the Intergovernmental Science-Policy Platform on Biodiversity and Ecosystem Services, Bonn. doi:https://doi.org/10.5281/zenodo.3402856. Accessed 24 Dec 2019, p 552

ISO (2008) Soil quality – Avoidance test for determining the quality of soils and effects of chemicals on behaviour – Part 1: Test with earthworms (*Eisenia fetida* and *Eisenia andrei*). In: ISO 17512-1

Jahn T, Hötker H, Oppermann R, Bleil R, Vele L (2014) Protection of biodiversity of free living birds and mammals in respect of the effects of pesticides, vol 30/2014. NABU. https://www.umweltbundesamt.de/publikationen/protection-of-biodiversity-of-free-living-birds. Accessed 20 Oct 2019

Jamieson AJ, Malkocs T, Piertney SB, Fujii T, Zhang Z (2017) Bioaccumulation of persistent organic pollutants in the deepest ocean fauna. Nat Ecol Evol 1:0051

Jensen OP (2019) Pesticide impacts through aquatic food webs. Science 366:566–567

Joermann G, Gemmeke H (1994) Meldungen über Pflanzenschutzmittelvergiftungen von Wirbeltieren. Nachrichtenblatt Deutscher Pflanzenschutzdienst, Biologische Bundesanstalt für Land- und Forstwirtschaft, pp 295–297

Johnston I (2017) EU experts accused of conflict of interest over herbicide linked to cancer. Independent, London

Kavi LAK, Kaufman PE, Scott JG (2014) Genetics and mechanisms of imidacloprid resistance in house flies. Pestic Biochem Physiol 109:64–69

Kerr JT (2017) A cocktail of poisons. Science 356:1331–1332

Klawitter N (2019) Skandal bei Bayer-Tochter. Monsanto spionierte Kritiker in ganz Europa aus. Spiegel Online, 13 May 2019. https://www.spiegel.de/wirtschaft/unternehmen/bayer-konzern-unter-druck-monsanto-hat-kritiker-in-ganz-europa-ausgespaeht-a-1267229.html. Accessed 10 Oct 2019

Klingenschmitt E (2016) Gifte belasten auch Bio-Äcker. SWR2 Impuls. http://www.swr.de/swr2/wissen/impuls-pestizide-protest/-/id=661224/did=17849754/nid=661224/eshk2f/index.html. Accessed 04 Oct 2019

Kniss A (2015) Dead plants are probably bad for earthworms, vol 2017. http://weed-controlfreaks.com/2015/09/dead-plants-are-probably-bad-for-earthworms/. Accessed 10 Oct 2019

Kniss A (2019) Conflict of interest statement. https://plantoutofplace.com/conflict-of-interest-statement/. Accessed 10 Oct 2019

Köhler HR, Triebskorn R (2013) Wildlife ecotoxicology of pesticides: can we track effects to the population level and beyond? Science 341:759–765

Koller VJ, Fürhacker M, Nersesyan A, Mišík M, Eisenbauer M, Knasmueller S (2012) Cytotoxic and DNA-damaging properties of glyphosate and Roundup in human-derived buccal epithelial cells. Arch Toxicol 86:805–813

Kolok AS (2016) Modern poisons. A brief introduction to contemporary toxicology. Island Press, Washington, DC

Kreiß C (2015) Gekaufte Forschung. Wissenschaft im Dienste der Industrie - Irrweg Drittmittelforschung. Europa Verlag, Berlin, Deutschland

Krüger M, Shehata AA, Schrodl W, Rodloff A (2013) Glyphosate suppresses the antagonistic effect of *Enterococcus* spp. on *Clostridium botulinum*. Anaerobe 20:74–78

Krüger M, Lindner A, Heimrath J (2015) Nachweis von Glyphosat im Urin freiwilliger, selbstzahlender Studien- teilnehmer – "Urinale 2015". http://www.urinale.org/wp-content/uploads/2016/03/PK-Text-Handout.pdf. Accessed 02 Nov 2019

Lau ET, Karraker NE, Leung KM (2015) Temperature-dependent acute toxicity of methomyl pesticide on larvae of three Asian amphibian species. Environ Toxicol Chem 34:2322–2327

Lee S-J et al (2011) Acute pesticide illnesses associated with off-target pesticide drift from agricultural applications: 11 States, 1998-2006. Environ Health Perspect 119:1162–1169

Legleiter TR, Bradley KW (2008) Glyphosate and multiple herbicide resistance in common waterhemp (*Amaranthus rudis*) populations from Missouri. Weed Sci 56:582–587

LeNoir JS, McConnell LL, Fellers GM, Cahill TM, Seiber JN (1999) Summertime transport of current-use pesticides from California's Central Valley to the Sierra Nevada Mountain Range, USA. Environ Toxicol Chem 18:2715–2722

Leon ME et al (2019) Pesticide use and risk of non-Hodgkin lymphoid malignancies in agricultural cohorts from France, Norway and the USA: a pooled analysis from the AGRICOH consortium. Int J Epidemiol 48:1519–1535

Linhart C et al (2019) Pesticide contamination and associated risk factors at public playgrounds near intensively managed apple and wine orchards. Environ Sci Eur 31:28

Lorenzen S (2013) Nervengift für Rinder. Chronischer Botulismus und der Einsatz von Glyphosat – ein Lehrbeispiel für politisches Versagen, Der kritische Agrarbericht 2013. www.kritischer-agrarbericht.de/fileadmin/Daten-KAB/KAB-2013/Lorenzen.pdf, pp 226–230. Accessed 20 Oct 2019

MacKenzie D, Pearce F (1999) It's raining pesticides. New Scientist. https://www.newscientist.com/article/mg16221803-100-its-raining-pesticides/

Main AR, Headley JV, Peru KM, Michel NL, Cessna AJ, Morrissey CA (2014) Widespread use and frequent detection of neonicotinoid insecticides in wetlands of Canada's Prairie Pothole Region. PLoS One 9:e92821

Malaj E et al (2014) Organic chemicals jeopardize the health of freshwater ecosystems on the continental scale. Proc Natl Acad Sci USA 111:9549–9554

Mandl K, Cantelmo C, Gruber E, Faber F, Friedrich B, Zaller JG (2018) Effects of Glyphosate-, Glufosinate- and Flazasulfuron-Based Herbicides on Soil Microorganisms in a Vineyard. Bull Environ Contam Toxicol 101:562–569

Manikkam M, Tracey R, Guerrero-Bosagna C, Skinner MK (2012) Pesticide and insect repellent mixture (permethrin and DEET) induces epigenetic transgenerational inheritance of disease and sperm epimutations. Reprod Toxicol 34:708–719

Mart M (2015) Pesticides, a Love Story. America's Enduring Embrace of Dangerous Chemicals. University Press of Kansas, Lawrence, Kansas, USA

Martinez DA, Loening UE, Graham MC (2018) Impacts of glyphosate-based herbicides on disease resistance and health of crops: a review. Environ Sci Eur 30:2

Matthews G (2016) Pesticides: health, safety and the environment. Wiley Blackwell, Chichester

Maurissen JPJ, Hoberman AM, Garman RH, Hanley TR Jr (2000) Lack of selective developmental neurotoxicity in rat pups from dams treated by gavage with chlorpyrifos. Toxicol Sci 57:250–263

Mesnage R, Antoniou MN (2018) Ignoring adjuvant toxicity falsifies the safety profile of commercial pesticides. Front Public Health 5:361

Mesnage R, Defarge N, de Vendômois JS, Séralini G-E (2014) Major pesticides are more toxic to human cells than their declared active principles. BioMed Res Int. Article ID 179691:8 p

Mesnage R, Renney G, Séralini G-E, Ward M, Antoniou MN (2017) Multiomics reveal non-alcoholic fatty liver disease in rats following chronic exposure to an ultra-low dose of Roundup herbicide. Sci Rep 7:39328

Mie A, Rudén C, Grandjean P (2018) Safety of Safety Evaluation of Pesticides: developmental neurotoxicity of chlorpyrifos and chlorpyrifos-methyl. Environ Health 17:77

Miller GT, Spoolman SE (2011) Living in the environment, 17th edn. Cengage, New York

Mineau P et al (1999) Poisoning of raptors with organophosphorus and carbamate pesticides with emphasis on Canada, US and UK. J Raptor Res 33:1–37

Mineau P, Downes CM, Kirk DA, Bayne E, Csizy M (2005) Patterns of bird species abundance in relation to granular insecticide use in the Canadian prairies. Ecoscience 12(267–278):212

Mitchell EAD, Mulhauser B, Mulot M, Mutabazi A, Glauser G, Aebi A (2017) A worldwide survey of neonicotinoids in honey. Science 358:109–111

Moriarty F (1988) Ecotoxicology. The study of pollutants in ecosystems. Academic, London

Morris AJ, Wilson JD, Whittingham MJ, Bradbury RB (2005) Indirect effects of pesticides on breeding yellowhammer (*Emberiza citrinella*). Agric Ecosyst Environ 106:1–16

Mörtl M, Darvas B, Vehovszky A, Gyori J, Szekacs A (2017) Occurrence of neonicotinoids in guttation liquid of maize – soil mobility and cross-contamination. Int J Environ Analyt 97:868–884

Mose T, Kjaerstad M, Mathiesen L, Nielsen J, Edelfors S, Knudsen L (2008) Placental passage of benzoic acid, caffeine, and glyphosate in an ex vivo human perfusion system. J Toxicol Environ Health A 71:984–991

Mostafalou S, Abdollahi M (2017) Pesticides: an update of human exposure and toxicity. Arch Toxicol 91:549–599

Motta EVS, Raymann K, Moran NA (2018) Glyphosate perturbs the gut microbiota of honey bees. Proc Natl Acad Sci USA 115:10305–10310

MSPPC (2010) Informe Comisión de Contaminates del Agua. In: Ministerio de Salud Pública de la Pcia del Chaco A (ed), p 14

Münch C (2011) Schädigung einer Population des Wiedehopfes (*Upupa epops*) im nördlichen Ortenaukreis durch Mesurol-Schneckenkorn. Naturschutz am südlichen Oberrhein 6:50–52

Muth F, Leonard AS (2019) A neonicotinoid pesticide impairs foraging, but not learning, in free-flying bumblebees. Sci Rep 9:4764

Niu YH, Li X, Wang HX, Liu YJ, Shi ZH, Wang L (2020) Soil erosion-related transport of neonicotinoids in new citrus orchards. Agric Ecosyst Environ 290:106776

NRC (1987) Regulating pesticides in food. The Delaney paradox. In: Innovation. NRCUCoSaRIUPUPaA (ed) https://www.ncbi.nlm.nih.gov/books/NBK218045/. Accessed 02 Nov 2019, Washington, DC

Obermüller E (2015) Fürchten wir uns vor den falschen Dingen? science.orf.at, sciencev2.orf.at/stories/1764654/index.html. Accessed 02 Nov 2019

OECD (2004) Guidelines for testing of chemicals. 222: earthworm reproduction test (*Eisenia fetida/Eisenia andrei*). Organisation for Economic Co-operation and Development, Paris

Ökotest (2017) Alles andere als rosig. Rosensträuße. Öko-Test 5/2017:127–133

Oppong M (2014) Verdeckte PR in Wikipedia. Das Weltwissen im Visier von Unternehmen. In: OBS-Arbeitsheft 76. Eine Studie der Otto Brenner Stiftung, Frankfurt/Main

Oreskes N, Conway EM (2010) Merchants of doubt. How a handful of scientists obscured the truth on issues from tobacco smoke to global warming. Bloomsbury Publishing, London

ORF (2010) Unfruchtbarkeit durch Schadstoffbelastung? science.orf.at, sciencev1.orf.at/science/news/44726. Accessed 02 Nov 2017

ORF (2013) Imker schlugen "Schweigegeld" aus, orf.at/v2/stories/2182011/2181992/. Accessed 02 Nov 2019

ORF (2018) Glyphosat und Co. bedrohen Weinqualität, science.orf.at, science.orf.at/stories/2919820/. Accessed 24 Dec 2019

ORF (2019) Seeadler mit Carbofuran vergiftet, noe.orf.at, noe.orf.at/v2/news/stories/2972663/. Accessed 25 Dec 2019

PA (2015) Anfrage der Abgeordneten Wolfgang Pirklhuber, Freundinnen und Freunde an den Bundesminister für Land- und Forstwirtschaft, Umwelt und Wasserwirtschaft betreffend Wissenschaftliche Erkenntnisse über auf Glyphosat basierende Herbizide. https://www.parlament.gv.at/PAKT/VHG/XXV/J/J_06793/fnameorig_473461html. Accessed 10 Oct 2019

PAN Germany (2009) Pestizide und Gesundheitsgefahren. Daten und Fakten, Pesticide action network, www.pan-germany.org/download/Vergift_DE-110612_F.pdf, 16pp. Accessed 02 Nov 2019

PAN Germany (2013) Endokrine Wirkung von Pestiziden auf Landarbeiter, insbesondere auf Beschäftigte in Gewächshauskulturen und Gärtnereien, Pesticide action network, www.pan-germany.org/download/pan_studie_endokrine_pestizide_1303.pdf. Accessed 02 Nov 2017

PAN Germany (2019) Giftige Exporte. Die Ausfuhr hochgefährlicher Pestizide aus Deutschland in die Welt. Pestzide action network, Hamburg, 20pp. Accessed 23 Dec 2019

Payne K, Andreotti G, Bell E, Blair A, Coble J, Alavanja M (2012) Determinants of high pesticide exposure events in the agricultural health cohort study from enrollment (1993-1997) through phase II (1999-2003). J Agric Safety Health 18:167–179

PCP (2010) Reducing environmental cancer risk. What we can do now? President's Cancer Panel, deainfo.nci.nih.gov/advisory/pcp/annualReports/pcp08-09rpt/PCP_Report_08-09_508.pdf, 278pp. Accessed 17 Oct 2017

Pelosi C, Joimel S, Makowski D (2013) Searching for a more sensitive earthworm species to be used in pesticide homologation tests – a meta-analysis. Chemosphere 90:895–900

Pérez-Lucas G, Vela N, Aatik AE, Navarro S (2018) Environmental risk of groundwater pollution by pesticide leaching through the soil profile. In: Larramendy M, Soloneski S (eds) Pesticides – use and misuse and their impact in the environment. IntechOpen. https://www.intechopen.com/books/pesticides-use-and-misuse-and-their-impact-in-the-environment/environmental-risk-of-groundwater-pollution-by-pesticide-leaching-through-the-soil-profile. Accessed 20 Oct 2019

Perry MJ et al (2016) Sperm aneuploidy in faroese men with lifetime exposure to dichlorodiphenyldichloroethylene (p,p'-DDE) and polychlorinated biphenyl (PCB) pollutants. Environ Health Perspect 124:951–956

Pimentel D (1991) CRC handbook of pest management in agriculture. CRC Press, Boca Raton, FL

Pimentel D (2005) Environmental and economic costs of the application of pesticides primarily in the United States. Environ Dev Sustain 7:229–252

Pimentel D et al (1991) Environmental and economic impacts of reducing US agricultural pesticide use. In: Pimentel D (ed) Handbook on pest management in agriculture. CRC Press, Boca Raton, FL, pp 679–718

Pimentel D et al (1993) Assessment of environmental and economic impacts of pesticide use. In: Pimentel D, Lehman H (eds) The pesticide question: environment, economics and ethics. Chapman & Hall, New York, NY, pp 47–84

Pisa LW et al (2015) Effects of neonicotinoids and fipronil on non-target invertebrates. Environ Sci Pollut Res 22:68–102

Pompa D (2016) The dangers of glyphosate: an interview with Dr. Stephanie Seneff, drpompa.com/additional-resources/health-tips/the-dangers-of-glyphosate-an-interview-with-dr-stephanie-seneff. Accessed 13 Oct 2017

Prosser D, Hart A (2005) Assessing potential exposure of birds to pesticide-treated seeds. Ecotoxicology 14:679–691

PSIRAM (2015) Glyphosat, die BOKU und der Regenwurm. http://blog.psiram.com/2015/09/glyphosat-die-boku-und-der-regenwurm/. Accessed 02 Nov 2019

Reljić S et al (2012) A case of a brown bear poisoning with carbofuran in Croatia. Ursus 23(86–90):85

Relyea RA (2003a) How prey respond to combined predators: a review and an empirical test. Ecology 84:1827–1839

Relyea RA (2003b) Predator cues and pesticides: a double dose of danger for amphibians. Ecol Appl 13:1515–1521

Relyea RA (2005) The lethal impact of roundup on aquatic and terrestrial amphibians. Ecol Appl 15:1118–1124

Relyea RA (2009) A cocktail of contaminants: how mixtures of pesticides at low concentrations affect aquatic communities. Oecologia 159:363–376

Repetto R, Baliga SS (1996) Pesticides and the immune system: the public health risks. World Resources Institute, Washington, DC

Ríos JM et al (2019) Occurrence of organochlorine compounds in fish from freshwater environments of the central Andes, Argentina. Sci Tot Environ 693:133389

Roberts JR, Reigart JR (2013) Recognition and management of pesticide poisonings, 6th edn. United States Environmental Protection Agency. http://www2.epa.gov/pesticide-worker-safety. Accessed 21 Dec 2019

Roberts JR, Karr CJ, Council On Environmental Health (2012) Pesticide exposure in children. Pediatrics 130:e1765–e1788

Robin M-M (2009) Mit Gift und Genen: Wie der Biotech-Konzern Monsanto unsere Welt verändert, 2nd edn, Deutsche Verlags Anstalt, München

Robin M-M (2010) The World According to Monsanto: Pollution, Corruption, and the Control of the World's Food Supply. The New Press, New York, NY

Robin M-M (2014) Our daily poison. From pesticides to packaging, how chemicals have contaminated the food chain and are making us sick. The New Press, New York, NY

Rocheleau CM, Romitti PA, Dennis LK (2009) Pesticides and hypospadias: a meta-analysis. J Pediatr Urol 5:17–24

Rohr JR, Sesterhenn TM, Stieha C (2011) Will climate change reduce the effects of a pesticide on amphibians?: partitioning effects on exposure and susceptibility to contaminants. Glob Chang Biol 17:657–666

Roscher E, Juds V (2016) Belastung verschiedener Medien mit DDT. http://www. vis.bayern.de/ernaehrung/lebensmittelsicherheit/unerwuenschte_stoffe/ddt_ vorkommen.htm. Accessed 02 Nov 2019

Rosenberg KV et al (2019) Decline of the North American avifauna. Science 366:120–124

Rundlöf M et al (2015) Seed coating with a neonicotinoid insecticide negatively affects wild bees. Nature 521:77–80

Ryberg KR, Vecchia AV, Martin JD, Gilliom RJ (2010) Trends in pesticide concentrations in urban streams in the United States, 1992-2008, USGS Scientific Investigations Report, pubs.er.usgs.gov/publication/sir20105139. Accessed 17 Oct 2017

Sabatier P et al (2014) Long-term relationships among pesticide applications, mobility, and soil erosion in a vineyard watershed. Proc Natl Acad Sci U S A 111:15647–15652

Samways MJ (2019a) Addressing global insect meltdown. Ecol Citizen 3:23–26

Samways MJ (2019b) Insect conservation. A global synthesis. CABI Publishing, Wallingford, UK

Sánchez-Bayo F (2014) The trouble with neonicotinoids. Science 346:806–807

Sánchez-Bayo F, Wyckhuys KAG (2019) Worldwide decline of the entomofauna: a review of its drivers. Biol Conserv 232:8–27

Sattelberger R (2001) Einsatz von Pflanzenschutzmitteln und Biozid-Produkten im Nicht-Land- und Forstwirtschaftlichen Bereich. Monographien Band 146. Umweltbundesamt Wien, 106pp

Schinasi L, Leon ME (2014) Non-Hodgkin lymphoma and occupational exposure to agricultural pesticide chemical groups and active ingredients: a systematic review and meta-analysis. Int J Environ Res 11:4449–4527

Schmitz J, Hahn M, Brühl CA (2014) Agrochemicals in field margins – an experimental field study to assess the impacts of pesticides and fertilizers on a natural plant community. Agric Ecosyst Environ 193:60–69

Séralini G-E et al (2012) RETRACTED: Long term toxicity of a Roundup herbicide and a Roundup-tolerant genetically modified maize. Food Chem Toxicol 50:4221–4231

Séralini G-E et al (2014) Republished study: long-term toxicity of a Roundup herbicide and a Roundup-tolerant genetically modified maize. Environ Sci Eur 26:14

Shiva V (2010) India: Monsanto in India – a story of corruption, biopiracy, seed monopoly and farmers suicides. In: Shiva V, Barker D, Lockhart C (eds) The GMO emperor has no clothes: a global citizens report on the state of GMOs. Navdanya, New Delhi, pp 143–185

Shiva V, Shiva M, Shiva V (2013) Poison in our foods: links between pesticides and diseases. Natraj Publishers, New Delhi

Shore RF, Feber RE, Firbank LG, Fishwick SK, Macdonald DW, Norum U (1997) The impacts of molluscicide pellets on spring and autumn populations of wood mice *Apodemus sylvaticus*. Agric Ecosyst Environ 64:211–217

Silva V, Mol HGJ, Zomer P, Tienstra M, Ritsema CJ, Geissen V (2019) Pesticide residues in European agricultural soils – a hidden reality unfolded. Sci Tot Environ 653:1532–1545

Simon-Delso N et al (2015) Systemic insecticides (neonicotinoids and fipronil): trends, uses, mode of action and metabolites. Environ Sci Pollut Res 22:5–34

Siviter H, Brown MJF, Leadbeater E (2018a) Sulfoxaflor exposure reduces bumblebee reproductive success. Nature 561:109–112

Siviter H, Koricheva J, Brown MJF, Leadbeater E (2018b) Quantifying the impact of pesticides on learning and memory in bees. J Appl Ecol 55:2812–2821

Skinner MK, Haque CG-BM, Nilsson E, Bhandari R, McCarrey JR (2013) Environmentally induced transgenerational epigenetic reprogramming of primordial germ cells and the subsequent germ line. PLoS One 8:e66318

Smith SE, Read DJ (2008) Mycorrhizal symbiosis, 3rd edn. Academic, London

Socorro J, Durand A, Temime-Roussel B, Gligorovski S, Wortham H, Quivet E (2016) The persistence of pesticides in atmospheric particulate phase: an emerging air quality issue. Sci Rep 6:33456

Sparling D, Feller G (2009) Toxicity of two insecticides to California, USA, anurans and its relevance to declining amphibian populations. Environ Toxicol Chem 28:1696–1703

Sparling DW, Fellers G (2007) Comparative toxicity of chlorpyrifos, diazinon, malathion and their oxon derivatives to larval Rana boylii. Environ Pollut 147:535–539

Stahlschmidt P, Brühl CA (2012) Bats at risk? Bat activity and insecticide residue analysis of food items in an apple orchard. Environ Toxicol Chem 31:1556–1563

Stahlschmidt P, Hahn M, Brühl CA (2017) Nocturnal risks-high bat activity in the agricultural landscape indicates potential pesticide exposure. Front Environ Sci 5:62

Steffen W et al (2015) Planetary boundaries: guiding human development on a changing planet. Science 347:736–746

Stone WB, Gradoni PB (1985) Wildlife mortality related to the use of the pesticide diazinon. Northeast Environ Sci 4:30–38

Stoner KA (2016) Current Pesticide Risk Assessment Protocols Do Not Adequately Address Differences between Honey Bees (*Apis mellifera*) and Bumble Bees (*Bombus* spp.). Front. Environ Sci 4:79

Stoner KA, Eitzer BD (2012) Movement of soil-applied imidacloprid and thiamethoxam into nectar and pollen of squash (*Cucurbita pepo*). PLoS One 7:e39114

Straub L et al (2016) Neonicotinoid insecticides can serve as inadvertent insect contraceptives. Proc R Soc Lond B Biol Sci 283:20160506

Streissler C (2016) Krebserzeugende Arbeitsstoffe: Besserer Schutz. Wirtschaft & Umwelt 4/2016:22–25

Stykel MG et al (2018) Nitration of microtubules blocks axonal mitochondrial transport in a human pluripotent stem cell model of Parkinson's disease. FASEB J 32:5350–5364

Sun H (2018) Pesticide in the Mississippi River floodplain and its possible linkage to colon cancer risk in the US. Toxicol Environ Chem 100:794–814

Tappert L, Pokorny T, Hofferberth J, Ruther J (2017) Sublethal doses of imidacloprid disrupt sexual communication and host finding in a parasitoid wasp. Sci Rep 7:42756

Thrupp LA (1991) Sterilization of workers form pesticide exposure: the causes and consequences of DBCP-induced damage in Costa Rica and beyond. Int J Health Serv 21:731–757

Topping CJ, Elmeros M (2016) Modeling exposure of mammalian predators to anticoagulant rodenticides. Front Environ Sci 4:80. https://doi.org/10.3389/fenvs.2016.00080

Tosi S, Nieh JC (2017) A common neonicotinoid pesticide, thiamethoxam, alters honey bee activity, motor functions, and movement to light. Sci Rep 7:15132

Tscherrig T (2020) Schweiz exportiert gefährliche Pestizide. INFOsperber. https://www.infosperber.ch/Artikel/Umwelt/Pestizide-Schweiz-exportiert-gefahrliches-Nervengift. Accessed 26 Jan 2020

Tsvetkov N et al (2017) Chronic exposure to neonicotinoids reduces honey bee health near corn crops. Science 356:1395–1397

Tu C et al (2011) Effects of fungicides and insecticides on feeding behavior and community dynamics of earthworms: Implications for casting control in turfgrass systems. Appl Soil Ecol 47:31–36

UBA Berlin (2019) Pflanzenschutzmittelverwendung in der Landwirtschaft, www.umweltbundesamt.de/daten/land-forstwirtschaft/landwirtschaft/pflanzen-schutzmittelverwendung-in-der#textpart-1. Accessed 04 Oct 2019

UCS (2013) "Superweeds" resulting from Monsanto's products overrun U.S. Farm Landscape. Union of Concerned Scientists, www.ucsusa.org/news/press_release/superweeds-overrun-farmlands-0384.html#.WKcKCxiX_MU. Accessed 17 Sept 2019

UCS (2017) Syngenta harassed the scientist who exposed risks of its herbicide atrazine, Union of Concerned Scientists, www.ucsusa.org/resources/syngenta-harassed-scientist-who-exposed-risks-its-herbicide-atrazine. Accessed 13 October 2019

UNEP (1980) The state of the environment: selected topics – 1980. United National Environment Program, Nairobi

UNEP (2017) Overview report I: Worldwide initiatives to identify endocrine disrupting chemicals (EDCs) and potential EDCs. In. United Nations Environment Programme. The International Panel on Chemical Pollution (IPCP). https://wedocs.unep.org/bitstream/handle/20.500.11822/25633/EDC_report1.pdf?sequence=1&isAllowed=y. Accessed 02 Jan 2020

UNESCO (2015) The United Nations World Water Development Report 2015. http://www.unesco.org/new/fileadmin/MULTIMEDIA/HQ/SC/images/WWDR2015Facts_Figures_ENG_web.pdf. Accessed 24 Jan 2020

UNHRC (2017) Report of the Special Rapporteur on the right to food, vol. A/HRC/34/48. United Nations Human Rights Council. https://documents-dds-ny.

un.org/doc/UNDOC/GEN/G17/017/85/PDF/G1701785.pdf?OpenElement, 24 Jan 2017, Accessed 29 Sept 2019

USGS (1996) Pesticides in ground water. current understanding of distribution and major influences. US Geological Survey Fact Sheet FS-244-95:4pp

Van Dijk TC, Van Staalduinen MA, Van der Sluijs JP (2013) Macro-invertebrate decline in surface water polluted with imidacloprid. PLoS One 8:e62374

van Hoesel W et al (2017) Single and combined effects of pesticide seed dressings and herbicides on earthworms, soil microorganisms, and litter decomposition. Front Plant Sci 8:215

Vicaire Y (2019) There is only one way to protect children from the toxic pesticide chlorpyrifos: an EU-wide ban, www.env-health.org/there-is-only-one-way-to-protect-children-from-the-toxic-pesticide-chlorpyrifos-an-eu-wide-ban/. Accessed 02 Nov 2019

Vicaire Y, Lyssimachou A (2019) Chlorpyrifos residues in fruits. The case for a Europe-wide ban to protect consumers, www.env-health.org/wp-content/uploads/2019/06/June-2019-PAN-HEAL-Briefing-chlorpyrifos_web.pdf. Accessed 02 Jan 2020

Villa S, Vighi M, Maggi V, Finizio A, Bolzacchini E (2003) Historical trends of organochlorine pesticides in an alpine glacier. J Atmos Chem 46:295–311

Vonesh J, Kraus J (2009) Pesticide alters habitat selection and aquatic community composition. Oecologia 160:379–385

Wadhams N (2010) Lions, Hyena killed with poisoned meat. National Geographic. https://www.nationalgeographic.com/news/2010/4/100412-lions-poisoned-furadan-kenya/. Accessed 27 Dec 2019

Wang X, Gong P, Wang C, Ren J, Yao T (2016) A review of current knowledge and future prospects regarding persistent organic pollutants over the Tibetan Plateau. Sci Tot Environ 573:139–154

Ward EM (2018) Glyphosate use and cancer incidence in the agricultural health study: an epidemiologic perspective. J Natl Cancer Inst 110:446–447

Wardle DA, Nicholson KS, Rahman A (1994) Influence of herbicide applications on the decomposition, microbial biomass, and microbial activity of pasture shoot and root litter. NZ J Agric Res 37:29–39

White DH, Mitchell CA, Wynn LD, Flickinger EL, Kolbe EJ (1982) Organophosphate insecticide poisoning of Canada geese in the Texas Panhandle. J Field Ornithol 53:22–27

Whitehorn PR, O'Connor S, Wäckers FL, Goulson D (2012) Neonicotinoid pesticide reduces bumble bee colony growth and queen production. Science 336:351–352

Whitford F et al. (2012) Farm family exposure to pesticides. A discussion with farm families. Purdue extension PPP-72, Purdue University, www.extension.purdue.edu/extmedia/PPP/PPP-72.pdf. Accessed 02 Nov 2019

WHO (1992) Our planet, our health. Report of WHO Commission on Health and Environment. https://apps.who.int/iris/handle/10665/37933. World Health Organization, Geneva, 51pp. Accessed 02 Nov 2019

WHO (2012) Possible developmental early effects of endocrine disrupters on child health. https://apps.who.int/iris/bitstream/handle/10665/75342/9789241503761_eng. pdf;jsessionid=1A5F3DBA6BD8AB9D6CC4D8D87B3761F4?sequence=1. Accessed 03 Jan 2020

Willer H, Lernoud J (eds) (2019) The world of organic agriculture. Statistics and emerging trends 2019. Research Institute of Organic Agriculture (FiBL). IFOAM – Organics International, Frick, Switzerland and Bonn, Germany

Woodcock BA et al (2017) Country-specific effects of neonicotinoid pesticides on honey bees and wild bees. Science 356:1393–1395

Zaller JG, Heigl F, Ruess L, Grabmaier A (2014a) Glyphosate herbicide affects belowground interactions between earthworms and symbiotic mycorrhizal fungi in a model ecosystem. Sci Rep 4:5634

Zaller JG et al (2014b) Future rainfall variations reduce abundances of aboveground arthropods in model agroecosystems with different soil types. Front Environ Sci 2:44

Zaller JG et al (2016) Pesticide seed dressings can affect the activity of various soil organisms and reduce decomposition of plant material. BMC Ecol 16:37. https://doi.org/10.1186/s12898-12016-10092-x

Zaller JG et al (2018) Herbicides in vineyards reduce grapevine root mycorrhization and alter soil microorganisms and the nutrient composition in grapevine roots, leaves, xylem sap and grape juice. Environ Sci Pollut Res 25:23215–23226

Zeljezic D, Garaj-Vrhovac V, Perkovic P (2006) Evaluation of DNA damage induced by atrazine and atrazine-based herbicide in human lymphocytes in vitro using a comet and DNA diffusion assay. Toxicol In Vitro 20:923–935

Zhang L, Rana I, Shaffer RM, Taioli E, Sheppard L (2019) Exposure to glyphosate-based herbicides and risk for non-Hodgkin lymphoma: a meta-analysis and supporting evidence. Mutat Res/Rev Mutat Res 781:186–206

3

Where Are the Solutions to the Pesticide Problem?

The good thing first: there are plenty of ways to avoid pesticides—for agriculture, consumers, and municipalities. Compared to tackling human-made climate change, the pesticide issue seems rather straightforward and much easier to solve—would not there be the multinational agrochemical industry and their allies with an enormous economic and political influence.

If we want to avoid exposure to pesticides, we should primarily consume food produced with as few pesticides as possible. In the example of apples, we have seen that the mere washing of fruits is not always sufficient to get rid of pesticide residues, especially when systemic pesticides have been used. A study of apples coated with the fungicide thiabendazole and the insecticide phosmet, two pesticides frequently used in intensive apple production, was conducted to test the best way to get rid of pesticide residues (Yang et al. 2017). Rinsing the fruit in a baking soda solution for 12 min was most effective for removing thiabendazole, they found, while a 15-min baking soda rinse was most effective for getting rid of phosmet. Some of the pesticide passed beyond the apple's surface, with thiabendazole going four times deeper than phosmet. Generally, the US EPA requires apple producers to soak the fruit for 2 min in chlorine mixed with water (Suslow 2000). However, this procedure is intended to remove bacteria and other organic matter, not to wash off pesticides.

In principle, organic agriculture works without synthetic pesticides. Regular inspections by independent bodies along the entire production chain from the field to the supermarket shelf monitor compliance with the rules. Often one hears people joking that organic farmers differ from their conventional colleagues only in the fact that they apply pesticides at night, while the conventional farmer dares to do so during the day. Then it is referred to some

© Springer Nature Switzerland AG 2020
J. G. Zaller, *Daily Poison*, https://doi.org/10.1007/978-3-030-50530-1_3

scandals in organic farming, where pesticide contamination of organic product was found or conventional products were mislabeled as organic. However, these scandals were almost always the result of a contamination caused during transport by wholesalers, by pesticide drift, or by business tricks of criminal organizations. And these scandals were uncovered through regular controls within the organic farming monitoring system. However, despite the high-quality and pesticide-free guarantee, organic products still have a niche existence in our grocery stores and supermarkets.

Because organic products are more expensive than conventionally produced products, they often have an elitist appearance. This feeling is often reinforced by advertising, which is understandably not well received by consumers with low income. In fact, the price differences between organic and conventional products only exist because conventional agriculture is subsidized much more with taxpayer money than organic agriculture. More details on this are provided in the next chapter. If one only compares the prices of conventionally produced branded or premium products with organic products, the price difference is negligible. An excursion with students to a wholesaler was very enlightening for me in this context. We saw there that the potatoes in a packaging line were once packaged for sale in a discounter and the same potatoes were filled in a fancier package as a higher priced premium product for a supermarket. Of course, this is fraud on consumers, moreover as farmers do not get different prices whether the product is sold as a discounter or premium product.

The scientific literature is full of studies comparing the nutritional quality of organic and conventional products. Often there are no differences in the contents of nutrient, vitamins, or other parameters. The organic farming sector has also been involved trying desperately to prove that organic products are better than conventional ones. However, why should there be differences since the conventionally produced fruits were not produced under deficient conditions and sometimes even the same varieties are cultivated? The main difference between organic and conventional products becomes obvious when the farming systems as a whole are compared, including their effects on the environment, biodiversity, human health, and ideally also regarding socioeconomic aspects.

In 2009, the World Bank and the United Nations published the World Agricultural Report where more than 800 experts from all continents and disciplines assessed the most important ecological, economic, social, and cultural aspects of agriculture over the past 50 years and describe a reorientation for the next 50 years (McIntyre et al. 2009). People living in industrialized societies easily forget that agriculture is still the world's largest economic sector. A third of all working people are employed in agriculture, often as small family farmers or subsistence farmers who only produce food for their families and a couple of other families.

Especially in the exuberance of the *Green Revolution* these small structures were regarded as "phase-out models." The report on the state of world agriculture clearly questions the ever-increasing intensification of agriculture as a remedy for food security (IAASTD 2009). In the report, especially small-scale and diversity-based structures are seen as the agriculture of the future. Agriculture always operates in the area between the environment, the economy, and society, and its services go far beyond mere food production. Therefore, experts speak of the so-called multifunctional agriculture. However, over the past 50 years, the focus has been mainly on increasing efficiency and production. Important contributions of agriculture such as its service for a healthy environment, the creation of interesting jobs, and the preservation of an attractive landscape have received only little attention. Since these ecosystem services provided by agriculture are not compensated for on any market, they should be promoted through subsidies or, even better, reflected in fair, unsubsidized product prices.

The consequences of the worldwide use of pesticides for humans and the environment have also prompted a warning of the United Nations Human Rights Council (UNHRC 2017). The report harshly criticizes the use of pesticides and the business practices of agrochemical companies and calls for a global agreement to regulate and gradually move away from pesticides in agriculture toward agro-ecological cultivation practices. As mentioned earlier, excessive pesticide use not only results in hundreds of thousands of acute pesticide poisonings worldwide, but also in contaminated soils and water resources, a decline in biodiversity, and the destruction of natural enemies of pests and diseases. The report also deplores the persistent refusal of the pesticide and agricultural industry and representatives of agriculture to recognize the extent of the damage caused by pesticides.

Pesticide residue monitoring in food across Europe gives hope that less pesticides land on our plates. Overall, 96% of the 88,247 samples analyzed fell within the legal limits (84,627, samples). In 54% of the samples, no quantifiable residues were reported, while 42% of the samples contained quantified residues at or below the maximum residue levels (EFSA 2019a). Interesting in this context are the results of Eurobarometer surveys that are commissioned at regular intervals by the European Commission with the intention to provide policy makers with a picture of opinions and attitudes of EU citizens. The responses are representative of more than 500 million European consumers: According to the survey, 1% represents the views of five million consumers! In the spring of 2019, 27,655 people were questioned regarding possible food risks (EFSA 2019b). Here are a few results that are relevant to our topic: Food safety is important for 50% of the Europeans when buying food. People are most likely to be concerned about antibiotic, hormone, or steroid residues in meat (44%), followed by pesticide residues in food (39%), environmental pollutants in fish, meat, or dairy (37%), and additives like colors, preservatives, or flavorings used in food or drinks (36%). For information on food-related risks, Europeans most likely trust scientists (82%) and consumer organizations (79%), followed by farmers (69%), national authorities (60%), EU institutions (58%), NGOs (56%), and journalists (50%). Such surveys are important and are hopefully considered in agricultural policies rather than pursuing the current way of agriculture.

The merits of the *Green Revolution* in terms of yield increases over the last 50–60 years can be acknowledged; however, this comes at very high environmental and health costs. In addition to greenhouse gas emissions, agriculture influences nutrient cycles (nitrogen, phosphorus, water), the release of pesticides, and, of course, the decline in biodiversity in the landscape through the transformation of natural ecosystems for agriculture (Conway 1997). The consequences of man-made climate change rightly occupy and get a lot of attention in public debate, but the impact of agriculture should receive at least as much attention (Vitousek et al. 1997). Fortunately, it is increasingly realized that agriculture and climate change are closely intertwined.

We have already indicated that a renewal of agriculture is not only necessary for reasons of health and environmental protection, but would also be appropriate for purely economic reasons. Current industrialized agriculture is not really sustainable and resources are still being used in a very wasteful way. Agricultural machines consume an incredible amount of fossil fuels. Valuable plant nutrients are used inefficiently, washed out into the groundwater, and have to be removed using complex technology in the treatment of drinking water. Also, with pesticides, only a tiny fraction actually reaches the harmful organism; the rest contaminates the environment, reduces biodiversity, and causes a breakdown of the ecological interactions between pests and beneficial insects. Although these problems exist worldwide, measures must nevertheless be tackled regionally. The encouraging thing is that improvements at a regional scale are immediately visible—much more directly than in climate change activities.

In the following chapter, we will explore the opportunities for an agriculture without or at least with a dramatically reduced pesticide input.

3.1 Agriculture Without Pesticides, Is That Even Possible?

Just to make it clear right at the beginning of this chapter, of course is agriculture possible without synthetic pesticides! The German forester and bestselling author Peter Wohlleben once stated: Forests exist since more than 300 million years, humans for 300,000 years, and foresters for 300 years. How could our forest survive for so long without the care by foresters? Adapting this statement for our topic could be: Agriculture exists since more than 10,000 years, synthetic pesticides for 100 years, and glyphosate for 50 years. How did agriculture ever work without these substances? If we need to feed more people in the coming decades and fully rely on pesticide-intensive agriculture, then the use of pesticides will certainly also increase (Pimentel et al. 1991). But with all we have heard so far about pesticides, we must ask ourselves how responsible it would be to continue along this path into the future.

Perhaps, readers living in countries with pesticide-intensive agriculture cannot imagine that agriculture without pesticide use is possible. However, it is important to note that only about one-third of the agricultural products are produced by using pesticides (Zhang 2018); the majority is produced without. The agrochemical industry, many agronomists, and agricultural lobbyists are promoting the myth that pesticides are an essential part of modern agriculture and that their benefits will definitely outweigh any effects on the environment or human health. We already mentioned that even the UN Human Rights Council stated in refreshingly clear language that the agrochemical industry's assertions that pesticides are necessary to safeguard our food supply are not only false but also negligently dangerous (UNHRC 2017).

Many different agricultural systems exist worldwide that work well without synthetic pesticides, for instance, organic, ecological, or biodynamic agriculture (Altieri 1995; Freyer 2016), regenerative agriculture that focuses on sequestering carbon in the soil (Gosnell et al. 2019), or permaculture that simulates natural ecosystems (Mollison 1993). Apart from banning pesticides, synthetic mineral fertilizers are replaced by organic inputs such as compost in these forms of agriculture. Intrinsically connected with these forms of agriculture are the focus on closed nutrient and energy cycles within the farm system in addition to environmental protection and high ethical standards regarding animal husbandry and people working on and with the farm. The roots of some of these agricultural methods go back to the beginning of the twentieth century, also as a reaction to the increasing urbanization and industrialization at that time.

When farmers, for whatever reason, do not completely want to avoid pesticides, they should work according to the principles of integrated pest management (IPM); at least that is the plan in the European Union since 2014 (Lefebvre et al. 2015). The aim of IPM is not to use pesticides as a preventive measure, but only when certain damage thresholds for pathogens or pests have been exceeded. Currently, IPM is successfully applied mainly in greenhouses and horticulture. However, IPM is usually more complicated than conventional methods. In principle, the concept of integrated pest control is also applied in organic farming.

3.1.1 Organic and Related Ways of Agriculture

Worldwide, 69.8 million hectares of organic agricultural land including areas in conversion from conventional farming were recorded in 2017 (Willer and Lernoud 2019). The regions with the largest areas of organic agricultural land are Oceania (35.9 million hectares, which is half the world's organic agricultural land) and Europe (14.6 million hectares, 21%). Latin America has eight million hectares (11%) followed by Asia (6.1 million hectares, 9%), North America (3.2 million hectares, 5%), and Africa (2.1 million hectares, 3%). The countries with the most organic agricultural land are Australia (35.6 million hectares), Argentina (3.4 million hectares), and China (three million hectares). Almost a quarter of the world's organic agricultural land (16.8 million hectares) and more than 87% (2.4 million) of the farms were in developing countries and emerging markets.

Generally, organic agriculture has a very long and successful history. Nevertheless, there is a persistent belief or well-maintained myth that it is too inefficient to feed humanity or that the product quality is not sufficient for modern consumer needs. Admittedly, organic farming often yields less than conventional farming. But it is usually more profitable for the farmer because of higher product prices, it is friendlier to the environment, and it produces higher quality food that is free of pesticide residues (Reganold and Wachter 2016). In addition, organic farming provides more ecosystem services and social benefits. A United Nations report also states that small-scale organic farming is the only way to sustainably feed the world (UNCTAD 2013).

All agricultural systems that operate without pesticides have one thing in common: they adopt agro-ecological principles and consider the entire system, including the ecological, economic, and social dimensions. This promotes agricultural practices that consider local biodiversity and the environment. An important component there is to promote natural pest control through the promotion of ecological interactions between pests and diseases and their antagonists, including a sustainable soil fertility and long-term soil health.

Agriculture without synthetic pesticides is scientifically proven by numerous examples. In a long-term field trial, maize and soybeans were grown with and without pesticides. After 22 years, there was no difference in yield between the two cultivation methods (Pimentel et al. 2005). Since pesticide expenditures are saved while yields remain the same, the no-pesticide version was also economically feasible. In another 21-year study conducted in Switzerland, pesticide-free yields were 20% lower than those from sprayed fields. However, under organic farming, the use of fertilizers was also reduced by 34%, that of energy by 53%, and that of pesticides by 97%—a few biological pesticides were used (Mäder et al. 2002). At the end, the soils in the organically cultivated fields were healthier with a higher fertility and biodiversity. In addition, it has now been shown that organic farming systems are also better in buffering climatic extremes such as drought periods or heavy rainfalls.

The yield-gap between pesticide-intensive and organic farming is not only the result of different inputs used in the field, but certainly also the result of decades of underfunding of research and development in organic farming. It is estimated that globally some 43 billion € are spent annually on food and agricultural research. Funding on research, technology, and development, which specifically goes into organic farming, is estimated to be less than 1% of this amount (Niggli 2014). As a result, innovations in organic farming still come primarily from farmers themselves rather than from scientists or consultants. Germany is making a laudable exception here and is making extra funds available for broad research in organic farming from the field to the fork.

In the public debate, it is usually presented as if organic farming can peacefully exist aside from pesticide-intensive conventional agriculture. The truth is that conventional agriculture, especially due to widespread drift of pesticides, can actually endanger the coexistence of organically farming colleagues in their neighborhood. The pragmatic—but politically impossible—solution to this dilemma would be to designate pesticide-free regions. In the European Union, the principle of mutually unaffected coexistence of conventional and organic farming is currently in place. However, from 2021 on this concept will be replaced by a concept where organic farmers are demanded to actively defend their cultures and products against spray drift from their conventional neighbors in order to prevent pesticide contaminations (Schmidt 2018). Then, even very low traces of pesticide contaminations in organic products will trigger a stop to their being marketed as organic, whenever it is observed that not enough preventive measures against pesticide contamination by organic operators have been taken.

The following measures can reduce the use of pesticides independent from organic farming:

- Selecting disease-resistant crop and livestock varieties and designing diverse and wide crop rotations.
- Increasing biodiversity in the field (through mixed crops, intercropping, agroforestry, multispecies livestocks) and in the landscape (including non-crop habitats, hedges, solitary trees, woodlots, streams, ponds, or piles of fieldstones).
- Increasing soil organic matter in order to improve soil fertility and soil health by input of compost and other organic fertilisers, or including cover crops in the crop rotation.
- Easing the optical (cosmetic) requirements/standards for market fruits and vegetables.

It should be clear that these activities cannot be imposed on agriculture alone. Policy makers and society should make this a priority. In the end, everyone would benefit, except the agrochemical companies.

When giving talks to the general public, I often see on one side that many people have a rather naïve impression of organic farming, as if the organic farmer only needs to watch his/her fruits growing without any intervention. On the other side, there is an ongoing debate about the use of pesticides in organic agriculture. Many proponents of pesticides state that organic agriculture is just another way of producing food, but that there is not really a great difference between these two forms of agriculture—at the end, they say, high-quality products are produced by both. Let us try to shed some light into these opinions.

No synthetical pesticides are allowed in organic farming. Pesticides allowing in organic farming are mainly produced from naturally occurring substances or organisms who generally degrade more readily and are less harmfull for non-target organisms including humans than synthetic pesticides. Occassionally, I am asked from consumers how much glyphosate is actually allowed in organic farming. When I answer that no herbicides are allowed at all in organic farming, people begin to realize that a complete ban of glyphosate and other herbicides would actually not be such a disaster for agriculture, as organic farmers successfully demonstrate that they can work without.

Generally, the pest and disease management approach used in organic farming is very different from that used in conventional agriculture. In organic farming, it is only when all other methods of dealing with pests and diseases have been exhausted and the farmer is faced with a potential loss of crops that one of the approved pesticides can be used. The application of theses substances is a controlled process and is never the first line of action. Moreover, organic farming considers the whole agroecosystem and actively avoid the use of pesticides through the use of crop rotation, mechanical weed control, maintaining a high biodiversity, and other management techniques. Table 3.1 shows a comparison between active ingredients approved for conventional versus organic farming in the European Union. Accordingly, in 2016, 389 active ingredients were approved as pesticides in agriculture, of those 35 of which (equals 9% of total) were approved for use in organic agriculture (PAN UK 2017a).

Table 3.1 Number of pesticides approved for use in EU conventional vs. organic agriculture and their risk assessments

Parameter	Number of pesticide active ingredients approved in European Union			
	Conventional agriculture		Organic agriculture	
Number of approved active ingredients	389		35	8.9% of total for conventional
No identified toxicity of total numbers	49	12.5% of total	24	68.5% of total for organic
Risk classifications				
Acutely toxic class 1 + 2 + 3 + 4	5 + 17 + 26 + 78 = 126	32.4%	0 + 0 + 2 + 2 = 4	11.4%
Carcinogenicity category 2	28	7.2%	0	0%
Germ cell mutagenicity cat. 2	2	0.5%	0	0%
Reproductive toxicity cat. 1B + 2	5 + 23 = 28	7.2%	0	0%
Candidate for substitution 1				
Low ADI/ARfD/ AOEL	20	5.1%	0	0%
Two persistent bioaccumulative toxicant criteria fulfilled 1	57	14.7%	1	2.9%
Endocrine disruption	5	1.3%	0	0%

Data according to PAN UK (2017a). *ADI*, acceptable daily intake refers to the amount that can be consumed over a life time without risk to health; *ARfD*, acute reference dose refers to the amount that can be ingested over a short period of time (one meal or 1 day) without health risks; *AOEL*, acceptable operator exposure level

Of the 35 pesticides approved for organic agriculture, 11% (4 pesticides) are of low toxicological concern to humans or the environment. This is the result of their intrinsic nature or the method in which they are used. Of the 389 pesticides approved for conventional agriculture, 26% are in the same low toxicity class, but additionally 22 pesticides (5.6%) are in the highest toxicity classes. It is important to note that none of the pesticides approved for organic farming are considered carcinogenic, mutagenic, or reproductive toxic, while 58 pesticides (14.9%) approved for conventional farming are listed in these toxicity categories.

While 57 conventional pesticides are bioaccumulative, only one (copper as a fungicide) of those approved for organic agriculture is. Indeed, copper used to control fungal diseases is an inglorious example where a heavy metal with long persistence is still allowed also in organic farming. However, it is important to note that more than 90% of the copper used is applied by conventional farmers at least in countries like Austria or Switzerland. Pyrethrin, an insecticide derived from *Chrysanthemum* plants, is used for hundreds of years in agriculture and also approved for use in modern organic farming. Although it biodegrades well, it is seen critically after reports on various harmful effects on humans and pets (Wagner 2000). Intensive research is going on aiming to phase out both copper and pyrethrin in organic agriculture. This is a good example demonstrating that organic farming and other forms of sustainable agriculture are asked to steadily evaluate established methods.

Pesticides approved for organic agriculture are primarily derived from natural substances such as spearmint oil, citronella, and quartz sand. Other substances such as iron, potassium, beeswax, whey, baking soda, or vegetable oils are all part of the human diet and have no toxicological issues. Pyrethroids and pheromones are only allowed to be used in insect traps and are not openly applied to soil or crops. Pheromones are used to confuse the insects and disrupt mate localization, thus preventing mating and blocking the reproductive cycle. Pheromones are increasingly used also from conventional farmers, especially in viticulture. However, studies also showed that minimal neonicotinoid quantities on crops in conventional field can alter the chemical communication within insects and reduce the efficacy of this pesticide-free management methods (Navarro-Roldán et al. 2019).

All pesticides approved for organic farming go through a regulatory approval process to ensure they are not harmful to the environment and human health. These products also require prior approval from the organic farming certifier (the associations that award the organic farming labels). This is different from conventional farming where the farmer is generally free to use all approved pesticides according to the appropriate regulations. Although some farmers also have contracts with supermarkets and must comply with stricter pesticide residue levels.

Organic farming relies largely on preventative measures to control pests and diseases, so pesticide use is significantly lower than in conventional farming. It is estimated that if all agriculture would be organic, the amount of pesticides would decrease by 98% (PAN UK 2017a).

The situation for the USDA National Organic Standard is similar to that in Europe allowing only natural or nonsynthetic pesticides for organic farming. However, under very special circumstances the use of synthetic pesticides under highly controlled applications is allowed. The list of exceptions to the "no synthetics" rule contains substance such as chlorine dioxide, boric acid, sulfur, or copper (ECFR 2019). Organic farmers in the USA have restricted access to 25 synthetic active pest control products, while over 900 are registered for use in conventional farming (GlobalOrganics 2018). These allowed pesticide products have such low toxicity levels that the EPA considers them "exempt from tolerance."

We have seen that pesticide residues are occasionally also found on organic products. When found, the residue levels on organic products are consistently lower than those on conventional products due to the preventive measures organic growers are required to take to protect their crops. Contamination can come from nearby conventional farms that use pesticides in the form of inadvertent spray drift, carryover of residues in the soil, or contaminated irrigation water. Organic farmers are required to set up buffer zones to make sure that prohibited substances do not contaminate organic crops. Some organic farmers use high hedgerows on their field borders, in extreme cases even several meter high water curtains are installed (Schiebel 2017) to form a barrier between their farm and the pesticide spraying neighbors.

A meta-analysis based on 342 peer-reviewed publications covering nutritional quality and safety parameters in organic and conventional plant-based foods has shown that organic products generally contain lower pesticide residues than conventional products (Barański et al. 2014). Consequently, an Australian study revealed that eating 80% organic food for a week can reduce organophosphate residues in human urine by nearly 90% (Oates et al. 2014).

Although it seems a promising and definitely a sustainable way of doing agriculture, one needs to be realistic and see that organic farming is performed on only 1.4% of the total agricultural land worldwide (Willer and Lernoud 2019).

Of course, the share of organic land varies from country to country; Liechtenstein and Samoa are the world leaders with both around 38% organic share of their total agricultural land; it follows Austria with 24% in the third place (Willer and Lernoud 2019). I still remember the spirit of optimism among environmentalists when I worked at the University of Bonn in the early 2000. At that time, the German Minister of Agriculture of the Green party set the ambitious goal of bringing organic farming to 20% by 2010. Currently in 2020 Germany has still only 8.2% organic share.

Organic farming is not only good for the environment but also regarding the income of farmers (BMEL 2019). Calculations for Germany showed that organic farms earned on average 40,004 € in the 2017/18 marketing year (profits plus labor costs per man-work unit) which is 22% more than the conventional reference farms. Germany is the second largest organic market in the world after the USA, and in 2015, German households purchased organic food for 8.6 billion € (Willer and Lernoud 2016). Demand is still significantly greater than the supply. Again, it would be a political task to create attractive framework conditions for organic farming to thrive. Making synthetic fertilizers and pesticides more expensive would be one step and would get products from organic farming more affordable for more people.

Consumers for whom organic is too expensive or perhaps too ideologically overloaded are increasingly turning to regional foods. It is understandable that a conventionally produced apple from the region is preferred to an organically produced apple from thousands of kilometers away. However, one must be aware that neither the regional labels on the products nor the claims on the shelves can be relied upon. Also, regionally produced food is not automatically of better quality. Regional tomatoes are often also produced soilless on rock wool with a fertilizer solution just like tomatoes from other very intensively managed areas in other parts of the world. Well, at least the greenhouse gas emissions during long transports are eliminated.

Admittedly, in recent years, rethinking regarding pesticide use has developed. At least in Europe, after the pioneer village of Mals in South Tyrol's Venosta region, thousands of communities in Europe have committed themselves to ban pesticides. Mals became famous in environmental circles, because for the first time in Europe, the citizens of a municipality voted in a referendum against the use of pesticides within their municipality borders by an overwhelming 75% majority. This was explosive, as Mals is located in a region which is threatened by the expansion of pesticide-intensive apple production. The Venosta region is among the most intensive apple producing area in Europe. The discussion in Mals was ignited by the fact that organic farms, animals, and people were increasingly affected by pesticide drift. A hay analysis, for example, revealed relevant residues of nine pesticide active ingredients. Meanwhile, interesting books on this case were published in German (Schiebel 2017) and English (Ackerman-Leist 2017) and also a German movie was very successful (directed by Alexander Schiebel: *Das Wunder von Mals*). This public attention of the pesticide critics immediately led to enormous hostilities on the part of pesticide advocates in the village and the regional government. It went so far that the author was recently even sued by the regional Council for Agriculture for "slander and spreading false information" (Mumelter 2017). An apple farmer who explained in the movie how he had been successfully producing apples without pesticides in the region for over 30 years found his apple trees sprayed and killed with glyphosate in an act of sabotage. The apple farmer has suffered a loss of several thousand € as a result, and also lost his certification as an organic grower.

Despite all the euphoria about these positive developments, it is important not to allow oneself to be lulled in. For instance, shortly after the publication of the World Agricultural Report on the state of agriculture, the World Bank was given additional billions to invest in the long-term fight against hunger. The lion's share of the money ran into conventional and classical large-scale projects and subsidies for agrochemicals (UNHRC 2017).

But the ban on pesticides is not only based on the power of the agrochemical lobby. There are many financial ties which makes it difficult to get out of these pesticide application patterns. Many retirement insurance companies, for example, also rely on the returns of agrochemical and oil companies (VBZ NRW 2016). This then links the income from the pensions to environmental health and climate. One wonders whether this economic intertwining is not also a reason why changes are so slow and difficult to achieve.

The success of organic farming is a clear testimony to the fact that agriculture also functions without synthetic herbicides. This proves nothing less than that the most widely used pesticides in the world, such as glyphosate-based herbicides, are actually not necessary! Therefore, it is puzzling what panic farmers and the agrochemical industry have regarding a ban on glyphosate announcing that this would jeopardize our food security. If it were, then a business model for agrochemical groups would collapse, but that would probably be more of a business risk. Interestingly, not only laymen or agrochemical lobbyists but also many scientists react in a similar way and believe that a ban on glyphosate would possibly lead to the use of even worse herbicides. However, I lack the logic behind this fear. Do they seriously believe that supertoxic pesticides withdrawn from the market will suddenly be taken out of the moth box and allowed to be used again? If environmental legislation works, this can be ruled out.

3.1.2   Agro-ecological Landscapes

We heard in previous chapters that many pesticides are used wastefully. In other words, there is consideration for potential pesticide reduction. The amount of pesticides needed to protect crops depends on the individual crop variety, the variety of different crops grown in an area, crop rotation, the location of the crop field with its soil type, the weather, and the surrounding landscape structure in which the farm is located. When crops are cultivated in unfavorable locations, they are more often infested by pests or diseases than in more favorable locations. We can experience this ourselves when gardening or with indoor plants. Plants in shady corners in the garden or flat are more often visited by pest species (aphids, slugs) and diseases; hot and dry spots show mite infestations.

Winegrowers tell me that certain grape varieties, so-called fungi-resistant varieties, are perfectly resistant to common plant fungal diseases and rarely have to be sprayed with fungicides at all. But they are not cultivated more widely because the wine market is geared to a standardized taste of consumers. Perhaps a change in consumer attitudes could be brought about just by communicating that fewer pesticides are required for certain varieties.

Moreover, in the last few decades our agricultural landscapes have been cleared in order to make them machine-friendly; wayside vegetation, fallows, and solitary field trees have been removed. Remaining non-crop ecosystems are affected by pesticide drift from the nearby fields (Schmitz et al. 2014). The result is a loss of ecosystem services such as non-crop biological pest control by natural enemies and a degradation of soil health affecting nutrient and water cycles.

When breeding new varieties, the main long-term focus was on high yields, while breeding for disease resistance was neglected, as it was assumed that pests and diseases could be treated with pesticides anyway. But these high-yielding varieties have also been shown to be more susceptible to pests and diseases because they are also more palatable to pests due to high fertilization levels. Since most seed producers are now owned by large agrochemical companies, there is naturally little interest in breeding disease-resistant varieties, as this would undermine the pesticide business.

In order for agriculture to function better again without pesticides, a reestablishment of more diversity in agriculture and the landscape would be necessary, away from large monocultures toward small-scale and diverse units with sufficient non-crop elements (Badenhausser et al. 2020). It is obvious that pests or diseases spread more easily when huge areas of the same crop are available. The efficacy of using natural enemies to control pests under field conditions largely depends on their mobility and, more specifically, on their capacity to quickly locate pest infestation. Although natural enemies such as insect parasitoids (e.g., Braconidae) that control pest insects by parasitizing can be very mobile flying about 6 km in 1.5 h (Yu et al. 2009), small-scale structures will definitely increase their efficiency. We have also seen in our own studies that a diverse landscape offers more habitats for beneficial insects, but also that soil fertility can affect pest occurrence (Zaller et al. 2008). High proportions of conventionally managed and large crop fields might not only increase pest abundance but can also threaten pollination and biological control services at a landscape scale (Holzschuh et al. 2010).

For example, colorful flower strips in the vicinity of cereal or vegetable fields not only make the landscape more beautiful but can also promote numerous beneficial insects such as parasitic wasps, ground beetles, birds, and spiders, which help to decimate pests feeding on crops (Geiger et al. 2009). Crop rotation and intercropping, i.e., the cultivation of several crops on the same field, also contribute to soil health and ensures that soil-borne plant diseases and pests do not become rampant, that weed populations are suppressed, and that the humus content in the soil is built up. A high humus content also improves ecosystem resilience against extreme weather events such as heavy rainfall or drought periods. Experiments in organic cotton cultivation in Egypt have shown, for example, that the cultivation of basil (*Ocimum basilicum*) between cotton rows reduced cotton pest species (Schader et al. 2005). Basil contains many essential oils and has a repellent effect on many pest insects. Additionally, the cotton farmers had a new source of income with the sale of basil. This mixed cultivation of two or more crops called intercropping or polyculture was more often practiced in the past especially in tropical regions (Vandermeer 1989), but due to increasing specialization of the farms it was abandoned.

Natural pest control, as it goes unnoticed in the landscape, is a free ecosystem service of great value. For the USA, this service has been estimated at approximately US\$ 12 billion a year (Losey and Vaughan 2006). In general, a richly structured landscape with flowering and meadow strips or wooded areas can increase natural pest control in the fields, as beneficial insects find shelter in these ecosystem elements. In the meantime, there are also special seed mixtures on the market that promote beneficial insects. These flower strips can even keep pests in neighboring cereals below damage thresholds (Tschumi et al. 2015). It also works very well in apple plantations as a cross-European study with sown perennial flower strips within apple plantations demonstrated (Cahenzli et al. 2019). As a result of the flower strips, natural enemies were attracted and the most important insect pests the rosy apple aphid (*Dysaphis plantaginea*) and the codling moth (*Cydia pomonella*) decreased and fruit damage due to pest infections was reduced.

With this promotion of natural enemies also the overall plant diversity in this agroecosystem increases and benefits insect pollinators (Pfiffner et al. 2019). Additionally, providing habitats or nesting boxes can encourage owls, which in turn control rodents. The promotion of bats in the agricultural landscape can decimate harmful moths, such as the codling moth. One could easily fill a complete book on these examples.

Diversification is not only necessary in a given landscape but also in a time dimension, by a diverse crop rotation that means the sequence of different crop species on the same field from year to year. Cover crops or intermediate crops such as alfalfa or clover not only provide a protein-rich feed for insect pollinators or dairy cows. These leguminous plants also draw nitrogen directly from the air, where there is plenty of it, into the soil with the help of so-called rhizobacteria forming a symbiosis with plant roots of legumes. These legumes are among the few plant families that can do this. And this is but one way for organic farming to maintain their agroecosystem without mineral nitrogen inputs. Nobody says that conventional farmers cannot adopt some of these tricks from organic farmers.

Instead of fighting pests or diseases when there is already fire on the roof, organic farmers try to avoid such problems beforehand, for example, by growing grain varieties that form long stalks. Then the leaves are far away from the ground, where there is less danger of infection with harmful fungal spores. Interesting are also other experiences of organic farmers who have noticed that when the rows of potatoes are aligned in the prevailing wind direction, less potato beetle infestation occurs than when they are cultivated at right angles to the wind (Grossarth 2016).

Most crop production systems in industrialized countries and increasingly also in developing countries are characterized by low crop diversity, decreased crop rotation, and high use of fossil energy and pesticides. Hence, the challenge of modern agriculture would be to balance productivity, profitability, and environmental health. A study conducted over 9 years varied the diversification of the cropping system in order to promote biological control while maintaining crop productivity (Davis et al. 2012). The results were quite promising as grain yields and profits in the more diverse systems (3-year rotation with maize-soybean-small grain + red clover and a 4-year rotation with maize-soybean-small grain + alfalfa-alfalfa) were similar to or even greater than those in the conventional system, despite reductions of herbicides and fertilizers.

Aside from switching to organic farming, also conventional farms could use substantially less pesticides. In a large study from France, 946 conventional farms were examined using a wide variety of production methods (Lechenet et al. 2017). The result was astounding, as about 60% of the farms were able to reduce their pesticide use by 42%, without any loss in yields or income. Once again, the question arises as to how it can be that farmers are advised to use so many pesticides? This study has also shown that the more diverse the cultivation methods were, the less pesticides had to be used.

My own experiences fit in here well: At a training event for farmers, a representative of the food industry openly admitted that 10 years ago they recommended twice the amount of applications for their contract farmers. However, due to groundwater pollution with nitrate and public pressure, these fertilizer recommendations have now been halved without affecting crop yields. This is only a symptom of the general situation where industry representatives are advising farmers. Governments cut independent, state-run extension services for farmers in the last few decades and industry gladly filled this gap. It can be assumed that an advisor from the agrochemical industry recommends more pesticide use—that is after all their job. It would be unrealistic to think that agriculture itself will come forward with proposals for substantial pesticide reduction. This would look as if they advised for too high pesticide uses in the past which would also counteract the so-called good agricultural practice. Outgoing from a quotation of Alfred Einstein who said that "We cannot solve our problems with the same thinking we used when we created them", we could translate this to our topic by stating: "Don't expect solutions to problems from people who created them and made a lot of profits with them".

In this context, stories of winegrowers who are experimenting with reduced pesticide applications are also very promising. In viticulture, many pesticides are applied without exactly knowing that a pathogen or pest will make problems at all. During dry summers with little rainfall, the majority of fungicide applications in eastern Austrian vineyards against downy mildew (*Plasmopara viticola*) were actually not necessary. These on-farm experiments showed that grapevines treated with pesticides yielded as much as those not treated with fungicides. This also points to the need to better adapt plant protection strategies to actual weather conditions also in the wake of climate change with prognosticated longer periods of dry weather.

Also, there are many biofungicides besides synthetic fungicides. Biofungicides are naturally based microbial or biochemical pesticides that can directly affect disease-causing organisms or may stimulate host plant resistance against the fungal pathogen. Biopesticides generally are narrow spectrum, have low toxicity, decompose quickly, and thus are considered to have low potential for negative impacts on the environment and are approved for organic crop production (OMRI 2019).

I have already mentioned earlier that herbicides for weed control under grapevines have only become a fashion in viticulture since about 15 years (at least in Austrian and German vineyards). Before that, weeds were kept in check mechanically. The justification for fighting weeds is competition for water and nutrients. However, it is difficult to understand why weeds with a root depth of just a few centimeters are a major competitor for vines that grow their roots several meters deep. In fact, there are very few studies that actually demonstrate that competition by weeds is a problem for vines. In discussions with winegrowers, it becomes clear that many of these weeding activities in the vineyard are performed because the vineyards should look tidy and perfectly managed. In the opinion of many winegrowers, this includes green inter-rows and bare soil under the grapevines. Perhaps science could provide data ensuring that winegrowers and consumers also value biodiversity and less pesticide contamination.

In reality, the situation is more complex than outlined here. For example, there is a grapevine trunk disease complex called Esca (formed by fungi of the species *Phaeoacremonium aleophilum*, *Phaeomoniella chlamydospora*, and *Fomitiporia mediterranea*), which can cause the grapevine to die. The fungi penetrate the wood through wounds in the wood from pruning. An increased occurrence of Esca in recent years has also been associated with the increasing use of pruning shears. With these pneumatic pruning shears, it is possible to cut much deeper into the wood than with manual shears, causing greater damage to the wood, which makes it easier for the pathogenic fungus to enter the vine. Such observations by individual, smart winegrowers should more often be taken up by science. Concepts for integrating ideas from the nonscientists and practitioners into scientific research projects are becoming increasingly important in ecology and agriculture via so-called participatory or citizen science. Perhaps, social media could be better used to communicate with farmers to more quickly hear from urgent problems faced by practitioners.

Although various policies to limit the use of agrochemicals have recently been implemented in the European Union, we have seen that the use of herbicides (and fertilizers) has remained fairly constant. The narrative of intensive agriculture is that it currently works at its optimum and every reduction in pesticides or fertilizers would also lead to reductions in yields. Coupling field surveys with experimental data where herbicide use and nitrogen input were manipulated, it could be shown that the high use of nitrogen fertilizer or intense herbicide use only slightly increased yields, but that this increase is not enough to offset the additional costs incurred by their use (Catarino et al. 2019b). Cereal yields were slightly lower when reducing herbicide use;

however, gross margins after herbicide reduction were actually higher because the costs for herbicides were also reduced.

There is also high potential of reducing pesticide inputs in rice production. With 770 million tons produced in 2017, rice was the third most important cereal crop after maize (1.1 billion tons) and wheat (772 million tons) and every reduction in the pesticide input in this system is very significant (FAOSTAT 2020). In Sri Lanka, there exists the traditional so-called maavee rice production scheme producing a high-value rice product without chemical inputs but instead relying on alluvial deposits and natural pest and disease resistance of traditional varieties (Horgan et al. 2017). The maavee rice can attain three times the selling price of rice from conventional farms making it also more economically viable.

A reduced pesticide use of course also benefits insect pollinators and overall biodiversity (Samways 2019). A large study including 294 farms quantified the effects of pesticide use, insect pollination, and soil quality on oilseed rape (*Brassica napus*) yield and gross margin (Catarino et al. 2019a). Interestingly, data revealed that greater yields may be achieved by either increasing pesticide use or increasing bee abundance. However, crop economic returns were only increased when bee abundance was high, because pesticides did not increase yields while their costs reduced gross margins. Given the many negative effects of pesticides mentioned in the former chapters, it seems economically sensible to reduce the pesticide use wherever possible.

In Europe, a panel for the future of science and technology answered the question whether farming without using herbicides, fungicides, and insecticides is possible (EPRS 2019). The panel compiled state-of-the-art information regarding the role of pesticides in securing global food production, preserving biodiversity, and supporting farmers' income in order to stimulate a further reduction of pesticide use in European agriculture. However, the line of argumentation is rather mainstream identifiable because pesticides are termed "plant protection products" throughout this report. But the argumentation is nevertheless interesting. The report states that food security and healthy food for 11 billion people by 2100 is one of the biggest challenges of this century. It is one of the most important, if not the most important, human rights, and any agricultural system has to fulfill this requirement within the planetary sustainability boundaries. This implies that no further land increase for agriculture is acceptable, since this is the most important driver for biodiversity loss, greenhouse gas increase, and environmental distruction. The panel estimates that without pesticides, yields will be reduced between 19% (wheat) and 42% (potato). It is concluded that pesticide reductions seem feasible in the case of high actual pesticide use, but perhaps not in the case of low use.

Reduction of pesticides seems possible based on sophisticated warning and decision support systems, but such a reduction is only realistic when the risk of yield or food quality reduction is acceptable for the farmer. Precision farming, including remote sensing with unmanned aerial vehicles, can also contribute to more targeted application and reduction of pesticide use. An important contribution will also come from the breeding of more resistant varieties. To what extent new breeding techniques such as genetic engineering is the way to go will be elaborated in a following chapter.

Interestingly, even the agrochemical industry assumes that at least in Europe a drastic decline in pesticides is inevitable and that from over 1000 pesticides about 200–300 will remain (Strassheim 2019).

3.1.3 Can the Growing World Population Be Fed?

Proponents of industrial agriculture and its heavy reliance on new chemicals argue that, although it would be ideal to reduce the use of pesticides, there are just too many mouths to feed on our planet. However, in this discussion it is often ignored that pesticides can also interfere with other sectors of food production. For instance, we have seen that arable farmers using neonicotinoids can seriously affect fish yields in Japan (Yamamuro et al. 2019). Other examples are anoxic zones in many river mouths of the globe such as the Yangtze in China, the Rhine in Germany, or the Mississippi in the Gulf of Mexico which bring loads of nutrients and pesticides from intensive agricultural areas causing serious economic losses for the fish industries (Smith et al. 2017). Clearly, a more comprehensive comparison of the environmental impacts of the world's food systems would be needed and must also include interactions between different food production systems (Jensen 2019).

The advocates of industrial agriculture keep repeating the mantra that organic farming would only lead to more hunger in the world because of decreased yields. As a consequence, pesticide-intensive agriculture is presented as the only way to feed the world's growing population. Much of it is a myth well-maintained from the marketing departments of the big agrochemical companies rather than backed by scientific evidence (Leu 2014).

The world population reached seven billion in 2011, and will reach nine billion people by 2050, according to a report by the Food and Agriculture Organization (FAO). Projections are that more meat and dairy products will be consumed in future. Additionally, more biofuel will be used which is also produced on agricultural land. It is expected that the global demand for food will rise by 70% by the middle of the century. The question is how an increased population will be fed when even today hundreds of millions are already starving? By 2050, the yields of rice, corn, wheat, and soybeans would have to be twice as high as at present (Ray et al. 2013).

In order to increase harvests, there are basically two options: either the yield per plant or the area under cultivation needs to be increased. The former can be achieved through more efficient cultivation techniques, improved varieties, more efficient fertilization, and more efficient control over pests and diseases. However, the increase in cultivated areas is mainly at the expense of non-crop and semi-natural areas such as forests, fens, bogs, marshland, and grasslands. This would be bad for biodiversity because these non-crop ecosystems often inhabit a great diversity of species which will affect the self-regulation of agro-ecosystems and affect our climate, as these areas represent huge reservoirs for carbon. Indeed, in a recent quantitative synthesis, landscape simplification due to a reduction in semi-natural area was shown to reduce biological pest control in crops on average by 46% (Rusch et al. 2016). It has been shown that non-crop field margins benefitted pest control of cotton bollworms mainly in unsprayed cotton fields suggesting that pesticide management should be planned by including the influence of non-crop elements in the landscape (Gagic et al. 2019).

When talking about the future nutrition of the world population, it should be clear that this cannot be achieved by maintaining or even globalizing the current Western lifestyle for 9–10 billion people. This refers not only to consumption patterns of industrial nations, the waste of energy, and natural resources but also regarding agriculture. The problem lies mainly in the excessive consumption of meat, which already accounts for most of the world's arable land. Industrialized meat production also causes severe harm to farm animals. For the production of animal feed crops, vast quantities of grain end up in the feed troughs of animals instead of on the plates of people. On top of that it uses a lot of water and energy and pollutes our water bodies. In addition, too much agricultural land is used for the production of so-called agro-fuels or biofuels (Erb et al. 2012).

It has already been stated that more than three-quarters of the world's food is produced by small family farms with less than one hectare land, not by industrialized farms (FAO 2014). Yes, industrialized agriculture feeds people in industrialized countries and produces agrofuels, but the majority of the world's population is fed by small farmers. Since industrialized agriculture cannot simply be established in remote areas or in developing countries, this also means that the situation of small farmers must be improved if more people are to be fed in the future.

While mechanization reduces costs for large farms, it does not automatically mean that more food will be produced. Even in the USA, 88% of all farms are still family farms (USDA 2015). They also produce more food per hectare than large farms. Large farms produce a great deal of certain crops that are used in industrial processes, such as corn or wheat. But small, diversified farms generally produce larger quantities and greater diversity per unit of land (Larson et al. 2016).

The before-mentioned World Agricultural Report also dispels the myth of the superiority of industrial agriculture from an economic, social, and ecological point of view (IAASTD 2009). As a new paradigm of twenty-first century agriculture, the report advocates for small-scale, labor-intensive, and diversity-based structures. This also guarantees a socially, economically, and ecologically sustainable food supply through resilient cultivation and distribution systems (ZS-L 2013). However, this does not mean that the real existing smallholder and traditional agriculture needs to be romanticized or that this is a call for a return to pre-industrial conditions. Many farmers in developing countries often simply lack qualified knowledge of sustainable farming methods. A combination of traditional and modern knowledge could trigger a huge boost to innovation for the benefit of the environment and health.

The ban on pesticides can also lead to a boost in innovation in the field of agricultural technology—digitization and precision farming are the keywords. Just as lawnmower robots have now found their way into domestic gardens, self-propelled weeding robots with special tasks could also bustle around in agricultural fields. These small, electrically-driven weeding machines compact the soil only slightly and do not emit climate-damaging CO_2 (Menzler and Griepentrog 2017). Autonomous weeding machines could even be equipped with sensors that remove only weed species that really do reduce crop yields while leaving others that do not interfere with the crop plants, all for the benefit of keeping biodiversity in agricultural fields high, and the countryside and landscape more aesthetic. This robot technology also has the advantage that it can run day and night. Unfortunately, research for alternatives to pesticide treatments was not economically attractive, because pesticides are so cheap. I have the feeling that the public sensitivity to look for alternatives has increased and that this might perhaps trigger more research initiatives.

There is definitely enough food produced to feed the world at least in terms of the produced calories. The main issue is the unequal production and distribution that prevent the needy from having access to food. A food systems model study addressed the agronomic characteristics of organic agriculture to specifically test whether it would be feasible to feed the growing world population (Müller et al. 2017). It is clear from this study that a 100% conversion to organic agriculture needs more land than conventional agriculture but reduces nitrogen pollution and pesticide use. A transformation to fully organic agriculture is only possible in combination with reductions of food wastage and food-competing feed from arable land (used for meat production), and with correspondingly reduced production and consumption of animal products. Other indicators such as greenhouse gas emissions also improve, but adequate nitrogen supply would be challenging but manageable. Therefore, according to this study it is possible to globally shift to organic farming if we reduce the high consumption of animal products in industrialized countries, use less concentrated feed (soybeans, cereals) in animal husbandry, and avoid food waste. Consequently, organic farming would secure the food supply of the world's population at over 9 billion in 2050, land consumption would not increase, greenhouse gas emissions would be reduced, and the negative effects of today's intensive food system such as large nitrogen surpluses or high pesticide loads would be greatly reduced.

Another study quantified the extent to which reductions in the amount of human-edible crops fed to animals and reductions in food waste could increase food supply (Berners-Lee et al. 2018). The study comes to a similar conclusion that the current production of crops is sufficient to provide enough food for the projected global population of about 10 billion in 2050, although very significant changes to the socioeconomic conditions and radical changes to the dietary choices of many people would be required. Importantly, full access to the global food supply needs to be guaranteed. Also, most meat and dairy would need to be replaced with plant-based alternatives, and a greater acceptance of human-edible crops currently fed to animals, especially maize, as directly consumed human food is required. Under these scenarios, the scope for biofuel production is limited. If society continues on a "business-as-usual" dietary trajectory, a 119% increase in edible crops grown will be required by 2050.

Food security and world hunger should not be used as pretexts for the continuation of a system that brought us into the environmental mess we face today all over the globe.

When talking about the effects of agriculture and climate change the carbon emission for the production, packaging, storage, and distribution of pesticides is rarely considered. The production of pesticides is very energy intensive and uses fossil fuel. It is estimated that per kilogram of herbicide between 1.7–12.6 kg of carbon is emitted; for insecticides this is 1.2–8.1 kg carbon and for fungicides 1.2–8.0 kg carbon (Lal 2004). In comparison, for each kilogram of nitrogen fertilizer only 0.9–1.8 kg carbon is emitted; for each kilogram phosphorus 0.1–0.3 kg carbon and for potassium 0.1–0.2 kg carbon is emitted. For a calculation on a land area of course, it makes a great difference for the carbon budget when about 100 kg of nitrogen fertilizer is applied versus only about 4 kg of herbicides.

The carbon footprint of agriculture can also be influenced by our diet selection. If one in five people in richer countries adopted low-meat diets, and threw away a third less food than they currently do, while poor countries were assisted to preserve their forests and restore degraded land, the world's agricultural systems could be absorbing carbon dioxide by 2050 instead of adding massively to global heating as they do at present (Roe et al. 2019).

Nutrient, herbicide, and sediment loading from agricultural fields causes environmental and economic damage. Nutrient leaching and runoff pollution can lead to eutrophication and impaired drinking water resources, while soil erosion reduces water quality and agronomic productivity. Often the use of herbicides is justified with erosion protection via no-till systems. However, long-term experiences with reduced tillage indicate that weed populations shift to perennial and grass species and the diversity and abundance of broad-leaf plants may decrease further making weeds more difficult to control while reducing biodiversity (Swanton et al. 1993). Important is also a greater diversification in the cropping system itself. A change of the cropping system from a two-year to a 4-year system has been shown to reduce erosion losses up to 60% (Hunt et al. 2019). The associated reductions in herbicide use intensity generally did neither affect nutrient and sediment losses nor crop yields and profitability for instance in the corn-soybean rotation that dominates the central USA.

3.1.4 Green Genetic Engineering, the Solution?

The so-called green genetic engineering comprises a variety of techniques to create genetically modified organisms (GMOs) for agriculture (both crops and livestock). Genetic engineering started in the 1980s, and we are now in the third generation. In first-generation transgenic crop plants, one or two plant genes are modified by adding genetic material from another species to the host plant in order to improve the agronomic properties of the plant, but usually not the quality of its product: resistance to insects, herbicides, drought, and cold. Oilseed rape, maize, soybean, cotton, and rice are important GMO-crops that are widely cultivated. Second-generation crops contain modifications that change existing metabolic pathways to obtain new food properties: composition of oils, allergens, or vitamin content; the so-called golden rice, which is engineered to produce beta carotene (pro-vitamin A), is a prominent example here. Third-generation crops produce non-plant products such as pharmaceutical substances and antibodies in plants. A new era of developing GMOs came with new technologies, e.g., CRISPR gene editing. CRISPR stands for clustered regularly interspaced short palindromic repeats and is manipulating an organism's genome without introducing alien genes. This technology is already applied in the field for instance on mosquitoes in order to fight mosquito-borne viruses and diseases.

In purely scientific terms, it is fascinating to see that we understand the function of organisms so well that we can genetically manipulate it in such a way that it develops desired properties. But when we see that the GM plants have to be acquired by farmers through high license fees and that they have to commit themselves to buying seeds and pesticides from the same company, doubts arise whether this concept is driven by science or rather by financial interests of multinational companies. At least the classical GMOs seem like a simple business model: the patent holder earns money by selling herbicide-resistant seeds (e.g., Roundup-ready corn) and the associated herbicide (glyphosate-based Roundup) as an all-round carefree package for farmers. Genetically modified seeds are widely used, particularly in the USA, Brazil, and Argentina. In Europe, people are still predominantly skeptical, although genetically modified crops are already being cultivated in Spain, Portugal, the Czech Republic, Slovakia, and Romania.

Generally, the European Union has declared its political intention to support the coexistence of three farming types: conventional and organic farming without GMOs and farming including GMOs. We already saw that the coexistence of pesticide-intensive farming and organic farming is almost impossible because of pesticide drift. It makes it even more complicated when GMOs are cultivated in the same landscape as contaminations of conventional and organic fields and products with GMOs are very likely. There is also a legal dimension to this as licensed GMOs can unintentionally develop in fields where they have not been planted.

In 2017, GM crops grew on around 190 million hectares worldwide. On 47% of this area, crops were tolerant to herbicides such as glyphosate and glufosinate. Crops that produce an insecticide (mainly *Bacillus thuringiensis*) make up 12% and 41% of GMO crops have both properties (ISAAA 2017). Less than 1% of the area concerned plants with other characteristics, such as flowers with colors that do not occur naturally. Around 92% of all GMOs are grown in just five countries: USA, Brazil, Argentina, India, Canada, China. These countries mainly grow just four GM crops: soy, maize, oilseed rape (canola), and cotton (Robinson et al. 2015). Still, 88% of arable land across the globe remains GM free.

Advocates of GMOs claim increased productivity of GMO crops, while conserving biodiversity, precluding deforestation, mitigating climate change, and improving economic returns, health, and social benefits. How can anyone be critical about GMOs? The main point is that after 25 years of GMO cropping promises that GMOs will bring higher yields while reducing pesticide use are still not fulfilled (Umweltinstitut München 2019).

Environmental authorities in Germany (Federal Agency for Nature Conservation), Austria (Environment Agency Austria), and Switzerland (Federal Office for the Environment) concluded in a report that effects of herbicide-resistant GM crops do not yield higher harvests but rather lead to a decline in biodiversity (Tappeser et al. 2014). The following paragraphs highlight findings from this report. The decline in biodiversity was seen in cropping systems with transgenic crops resistant to the herbicides glyphosate and glufosinate and the further intensification of farming and increased pressure on biodiversity. GMOs will have various impacts on the agricultural practice, including weed control, soil tillage, planting, crop rotation, yield, and net income. However, studies on yields of herbicide-resistant crops compared to conventional crops do not consistently show higher yields of GM crops compared to conventional crops without GMOs.

Herbicide usage in GM crop systems and their impacts are difficult to compare with conventional crop management because different herbicides are applied at different rates and the specific environmental impacts may vary. As herbicide-tolerant GM crops gained market share in 1996–2000 in the USA, agricultural applications of glyphosate rose by 186% from 12.5 million kilograms application in 1995 to 35.7 million kilograms by 2000; until 2014, there was an additional 217% increase in the amount of glyphosate used in the USA (Benbrook 2016). Additionally, in regions where herbicide-resistant crops are widely adopted, less crop rotations and crop diversification are seen and can enhance disease and pest pressure.

Reasons for farmers to adopt herbicide-resistant systems are, besides a simplification of weed control, reduced production risks, the currently low herbicide prices, and expected lower costs (e.g., in combination with conservation tillage and other production factors such as less labor and fuel consumption). It seems that mainly the higher flexibility rather than the crop yield and the final economic success (costs vs. returns) are the decisive factors for adopting herbicide-resistant systems (Tappeser et al. 2014).

Although recommended for many years, many farmers did not address weed resistance by integrated weed management, but often continued to rely on herbicides as a single measure (Schütte et al. 2017). Despite occurrence of widespread resistance in weeds to other herbicides, industry rather develops transgenic crops with additional herbicide resistance genes, the so-called stacked herbicide-resistant traits, which combine glyphosate resistance with resistance to glufosinate and/or resistance to other herbicides such as synthetic auxins like 2,4-D or ALS (acetolactate synthase)-inhibiting herbicides.

Effects of GMOs on biodiversity can be explained by a more efficient removal of weeds leading to a further reduction of biodiversity of various organisms in farmland. This includes direct effects, such as depletion of the weed seedbank and low weed density and diversity, as well as indirect effects, e.g., impacts on animals feeding on weeds and on predators of these animals. Thus, farmland birds may be particularly affected. Further, the dramatic reduction in monarch butterfly (*Danaus plexippus*) populations in the USA has been linked to the widespread cultivation of herbicide-resistant crops in the Midwest which drastically reduced the population of milkweed, the feeding plant of monarch larvae (Pleasants and Oberhauser 2013). According to the experience in countries adopting herbicide-resistant crops, herbicide use was increased instead of reduced. Therefore, it is highly questionable whether herbicide-resistant systems comply with aims to stop the loss of biodiversity on farmland. From a nature protection perspective, herbicide-resistant crops seem to be not an option for sustainable agriculture aiming also to protect biodiversity.

Also, the Intergovernmental Science-Policy Platform on Biodiversity and Ecosystem Services (IPBES) who assesses the worldwide status of biodiversity and ecosystem service concluded in their report that GMOs can have indirect effects on pollinators by reducing the availability of weeds, or sublethal and direct effects on insect pollinators that are usually accounted for in risk assessments (IPBES 2017). The few studies dealing with the evaluation of the effects of commercial GM crops on soil microarthropods have generally reported a lack of significant deleterious effect of for instance GM herbicide-resistant soybean on the springtail community in the soil. However, the scarcity of data on the effect of GM crops on soil microarthropods and on soil biodiversity in general underlines the need for further, independent studies (Jeffery et al. 2010).

Another criticism of GM crops is the transfer of new genes from GM organisms to wild or domesticated non-GM populations which could happen through various ways (Pascher et al. 2017). Moreover, there are also concerns that the introduction of GM genes into nontarget species could have negative consequences for the nutritional composition of food and affect both human and environmental health.

A rather new method is to reduce mosquito populations by releasing into the wild of GMO mosquitoes that express a lethal gene into the wild. These male GM mosquitoes of the species *Aedes aegypti* have a self-limiting gene which is inherited by offspring of parents of normal mosquitoes and GM mosquitos and is supposed to prevent them surviving into adulthood. In theory, when these mosquitoes are released in high numbers, a dramatic reduction in the mosquito population should follow. However, serious concerns have been raised that these GMO mosquitoes carry unknown risks for native mosquito populations and eventually perhaps also for humans (Kofler et al. 2019).

Field trials involving recurring releases of GM mosquitoes demonstrated a reduction of nearly 95% of target populations in Brazil (Carvalho et al. 2015). Another study showed that some of the offspring of GM mosquitoes survived to adulthood and mated with native mosquitoes and thereby introduced some of their genes into the wild mosquito population (Evans et al. 2019). The impact of mosquitoes carrying these new genes remains largely unknown, but it is feared that a new breed of mosquito might emerge that is more difficult to control or that virus host–insect interactions are changed. Thus, like GM soybean or corn, there is legitimate concern about the propagation of new genetic material in wild populations with as yet unknown consequences.

Despite these concerns, field trials have not been stopped but will rather be extended to other countries. Critics state that these genetic modification technologies need to be more transparent and their environmental risks assessed in more detail—at least as it is requested for pesticides.

The large-scale cultivation of GMOs, the associated pesticide use, and its severe impacts on rural workers and their families in Paraguay were the reason for a statement of the UN Human Rights Committee to prosecute those responsible for the damage (UNHRC 2019b). The large-scale use of pesticides in a soy-growing region in Paraguay (Colonia Yerutí) has had severe impacts on rural people's living conditions, health, livelihoods, contaminating water resources and aquifers, preventing the use of streams, and causing the loss of fruit trees, the death of various farm animals, and severe crop damage. Paraguay was made responsible for human rights violations in the context of massive pesticide applications and must now undertake an effective and thorough investigation into this and the subsequent poisoning of people (UNHRC 2019a). The rural people have experienced a range of physical symptoms, including nausea, dizziness, headaches, fever, stomach pains, vomiting, diarrhea, coughing, and skin lesions. According to this UN statement the contamination has so far resulted in the death of one person and the poisoning of 22 other inhabitants of this community.

Since 2012, new and powerful technologies have been developed that have multiplied the possibilities for altering the genome of organisms. Describing these methods, their potentials and dangers would be beyond the scope of this book. Briefly, genome editing (e.g., through the method CRISPR/Cas) is one of these methods that changes the genome via supposedly precisely identifiable locations in the DNA. Another method is called gene drive (also called mutagenic chain reaction) and is a method with which the spread of genetic modifications can no longer be inherited according to Mendelian inheritance rules with a 50% probability, but rather dominantly, so that entire populations are rapidly altered. The European Court of Justice classified the new processes such as CRISPR/Cas as genetic engineering for which the strict EU guidelines for genetically modified organisms apply (Callaway 2018). This means that all products made through gene editing should be regulated, assessed for their health and environmental impacts, and labeled. Many plant breeders and scientists find that gene editing should be considered a mutagenesis, just like it happens in nature or through other breeding techniques such as irradiation. Thus, it should be exempt from the directive, because they can involve changes to DNA and not the insertion of foreign genes. Meanwhile, there is increasing pressure on the EU to change this decision.

Especially in the USA there is a heavy debate about the labeling of GMO products. Biotech companies are strictly against this labeling because they fear consumers would turn away from these products. They also state that their products had "substantial equivalence" to other products anyway. The best-selling author Michael Pollan very aptly observes in this regard (Pollan 2002): "The new plants are novel enough to be patented, yet not so novel as to warrant a label telling us what it is we're eating. It would seem they are chimeras: 'revolutionary' in the patent office and on the farm, 'nothing new' in the supermarket and the environment."

Meanwhile studies have shown that when manipulating the genome of crop plants using CRISPR/Cas, changes always occur in places that have not been targeted, but in most cases the participating scientists do not search for such "off-target effects" (Modrzejewski et al. 2019). Undesirable effects can also occur at the target site of genetic manipulation, for example when the genome of bacteria, which serve as aids in technology, is incorporated into the target organism (Norris et al. 2019). Undesirable side effects can lead to altered nutrient compositions, an increased content of natural toxins, or the formation of allergens (ENSSER 2017).

There is a widely spread assertion that these GM products from gene editing are not technically distinguishable from classical breeding and consequently cannot be regulated is incorrect (Kawall 2019). In the meantime methods for tracking manipulations via gene editing have been developed (Chhalliyil et al. 2020). When genetically modified organisms are released, they can spread and multiply unplanned. In plants, these properties created by humans or incorporated by other living organisms can also spread through crossbreeding in related species. So far we cannot fully estimate the consequences of their spread. However, we can say that it is practically impossible to reverse changes imposed in wild populations of plants, animals, fungi, or bacteria. Plants multiply by pollen, which is spread by wind or insects (and more rarely also vertebrates). Bees and bumblebees fly for miles; the wind even brings dust from the Sahara over the Alps. The use of genetically manipulated plants in agriculture therefore inevitably leads over time to the spread of new genes in the genetic material of plants that have not been genetically engineered. A coexistence between farming with GM crops, conventional farming without GM crops, and organic agriculture seems impossible (Umweltinstitut München 2019).

Another aspect is that genetic engineering concentrates the power over seeds and animal breeding in a few corporations. The multinational companies Bayer/Monsanto, DowChemical/DuPont, and ChemChina/Syngenta are not only the most important suppliers of genetically modified seeds, but also the largest players in the market for pesticides. Thus, they have an enormous economic interest in selling both seeds and pesticides.

Scientific research and real-world farming experience show that GM crops have not yet delivered their promises. Above all, corporate manufacturers launched GMO crops based on two main promises: better yields and less pesticide use—neither was fulfilled so far (Mart 2015). In reality, they have presented farmers with the new challenges of controlling herbicide-resistant superweeds and Bt toxin-resistant super-pests. GM crops are no less dependent on fertilizers than any other chemically grown crop. Critiques also state that the GMO approach is based on an outdated understanding of genetics and is bound to fail (Robinson et al. 2015). Our modern understanding of genetics tells us that we need to take a holistic "systems biology" approach in crop development that preserves gene organization and regulation, rather than disrupting it, as GM does. The way to safely and effectively generate crops with complex desirable properties such as higher yield, drought tolerance, and disease resistance is through natural breeding, augmented where useful by marker assisted selection. Also, it is important to note that there is no scientific consensus over the safety of GMOs (Hilbeck et al. 2015); as long as this is the case, the precautionary principle should apply.

Meanwhile, the European Parliament has called for a global moratorium on the release of gene drives to prevent the new technologies from being released prematurely (EP 2020). Such a moratorium would respect the precautionary principle of the European Union and the UN Convention on Biological Diversity.

3.1.5 Let Us Get Rid of the Most Toxic Pesticides

When criticizing the current pesticide use, journalists often ask if I do not think that more harmful substances would be used if, for example, glyphosate were banned? The formulation of this question is already a result of the successful propaganda machinery of the agrochemical industry during the last decades. Why should more dangerous substances be used when they have been banned for good reason? Quite often there are simply nonchemical alternatives around. After all, there is organic farming and similar approaches suche as permaculture or regenerative farming, which successfully demonstrates for decades that it is possible to produce food in highest quality without synthetic pesticides.

The strategy of replacing one pesticide with another one, where it then turns out that the replacement product is more dangerous than the old product, is nothing unusual. In fact, in the past many pesticides were replaced by newer, supposedly safer pesticides. For example, Roundup was a replacement for the herbicide 2,4,5-trichlorophenoxyacetic acid (also known as 2,4,5-T), an herbicidal ingredient in Agent Orange, a defoliant used by the British in the Malayan Emergency and the USA in the Vietnam War. When the herbicide atrazine was banned in the EU due to its cancerogeneity, a replacement product called terbuthylazine was launched. Terbuthylazine is chemically similar to atrazine and has similar effects on the environment and health. *Bacillus thuringiensis* pesticides produced by genetically modified plants are regarded as safe substitutes for organochlorines, carbamates, and organophosphorus were substitutes for DDT; DDT itself replaced lead arsenate and so on.

Another example are neonicotinoids. Neonicotinoids were also in part designed to replace pyrethroid insecticides due to concerns over their impacts on the environment. The contention that older "nastier" pesticides are replaced by newer and thus safer chemicals is often not true. More often are newer pesticides added to the pesticide loads. In the UK, the use of pyrethroids in oilseeds increased from 10,051 kg in the year 2000 to 20,335 kg in 2016 (increase 102%), while at the same time also the use of super bee-toxic neonicotinoids increased by 9700% from 27 kg in the year 2000 to 2646 kg in 2016 (PAN UK 2017b). Many other such examples of substitute products could be listed here.

Pesticide bans are usually celebrated as successes by NGOs or pesticide opponents. However, it should also be noted that manufacturers know that the regular replacement of pesticides is part of their business. One of the main reasons is that weeds and pests develop resistance to pesticides so that they become ineffective over time. Another reason is that patents expire and other companies can manufacture these products, as is currently the case with glyphosate-based herbicides. For manufacturers, this also means that they are constantly on the lookout for replacement products, even independently of campaigns by environmental groups. To what extent the big companies to some extent actually "control" these NGO successes is hard to tell.

Manufacturers also know that product registration is based on their own research and testing. This means that problems with the safety or toxicity of an approved pesticide are only revealed years later through independent studies or practical experience. Until then, good business can be made with these pesticides. Looking at this system of current authorization of pesticides without any long-term monitoring of their nontarget effects, it is clear that old, widespread pesticides automatically have a worse reputation than newer products, simply because we know less about them.

I hope that after these remarks it has also become clear that the statement that pesticides are the best tested substances are simply wrong. This statement is also contradicted by regular withdrawals and bans of widely used, allegedly super-safe pesticides. If the approval procedure would be so rigorous, potential problems would have been identified prior to a market release.

3.2   Does the Use of Pesticides Pay Off at All?

We already addressed economic issues a couple of times, but let us consider this in more detail. It is estimated the global pesticide expenditures at the producer level totaled nearly US$ 56 billion in 2012 (EPA 2017). Herbicides (and plant growth regulators) account for about 44%, insecticides for 29%, fungicides for 26%, and fumigants for 1% of the global market share. US farmers use on average 5% for pesticides, but this depends heavily on the farm sector, with fruit farmers having a much higher expenditure for pesticides than dairy farmers. Cotton farms in Australia for instance apply on average 11 insecticide and three herbicide sprayings per season, which accounts for 21% of the total crop production expenses, not including the costs for professional consultants (Kennedy et al. 2002). Nevertheless, many advocates of pesticides argue about the cost efficiency of pesticide use over other methods of weed and pest control.

If we take a closer look, it becomes apparent that most cost calculations simply ignore the environmental and social consequences of pesticide use. Table 3.2 provides an overview of such a calculation conducted for the USA (Pimentel 2005). According to this calculation, the largest economic and

Table 3.2 Estimated economic costs of pesticide use

Cost category	Type	Total costs (million US $)
Public health effects	Hospitalized patients, outpatient-treated poisoning, lost work because of poisoning, pesticide cancers, fatalities	1228.5
	Pesticide residues in food	No data
Environmental costs	Domestic animal poisoning and deaths	30.8
	Destruction of beneficial natural enemies (predators, parasites)	257.4
	Pesticide resistance in pests	1500.0
	Losses of honey bees and wild bees and their services	283.6
	Crop losses through excess use; contamination, testing	1391.0
	Ground and surface water contamination	2000.0
	Fishery losses	100.0
	Wild birds and mammals	2100.0
	Microbes and invertebrates	No data
Pesticide pollution control	State and federal regulatory actions, pesticide monitoring	470.0
Ethical and moral issues		No data
Total costs per year		9361.3

Data according to Pimentel et al. (2005)

environmental losses associated with pesticide use in the USA are the loss of wild birds and mammals through nontarget effects. Then follows ground and surface water contamination with pesticides, costs for pesticide resistance in pests and the need for additional applications. Also big contributions are crop losses through contamination via spray drift, wrong pesticide usage, and the costs for pesticide testing. Ethically tricky are economical calculations valuing nonlethal and especially lethal pesticide poisonings (Richter 2002). This issue has become worldwide attention through legal cases in the USA with glyphosate applicators who developed cancer.

Despite the widespread use of pesticides in the USA, it is estimated that around 37% of the crop is still destroyed by insects, plant diseases, or weeds (Pimentel 1997). Although the use of insecticides increased in the USA, crop losses due to insect feeding almost doubled from 7 to 13% (Pimentel et al. 1991). This increase in insect damage is partly due to changes in farming practices associated with pesticide-intensive agriculture, such as the shift from crop rotation, where corn was one of several crops, to pure corn monocultures over many years. Cost-benefit calculations are necessary to come up with a fair scientific assessment of the benefits of pesticides.

One argument put forward by the pesticide industry for the continued use of neonics is that it brings a huge economic benefit to agriculture and that alternative products are worse. There is always the warning that the growing world population cannot be fed and that no better products are available. In an overview study, for which 19 studies on neonics were combined, eight studies showed no yield advantage and 11 studies unclear yield effects (Stevens and Jenkins 2014). The use of neonics did not show economic benefits for farmers compared to other methods or no use of pesticides at all when pest pressure was low. Another report concludes that the economic and environmental losses associated with the use of seed dressing do not outweigh the benefits (CFS 2016). Further, cost analyses across soybean-growing regions in the USA showed only negligible benefits of about 0.01–0.22 tons per hectare with inconsistent evidence of a break-even cost of neonicotinoid use (Mourtzinis et al. 2019).

After the EU's moratorium on neonicotinoids in 2013 the agrochemical company Syngenta has applied for an exemption for Great Britain because a danger was assumed to production. In 2014, an industry representative was asked by the British Environment Committee to submit a study that can scientifically prove that neonicotinoids actually increase yields in arable crops. Apparently, the company representative could not present a simple study to prove this (Goulson 2014). It seems that neonics have been used for more than 20 years now only for the benefit of the manufacturers. Such findings do not really strengthen confidence in the agrochemical industry and in political authorities that approve these products despite a lack in critical scrutiny.

It has already been mentioned that the assessment of economic consequences of pesticide use can get more complicated if for example neonicotinoid use in one food producing sector such as arable farming leads to significant reductions in yields in another food producing sector such as fish farming (Yamamuro et al. 2019). However, this aspect is rarely considered in cost-benefit calculations for pesticides.

Additionally, the impairment of natural pest control and nutrient conditions is usually not included in cost–benefit analyses for pesticides. We already mentioned the diverse effects of pesticides on birds. The economic impact of pesticide poisoning on birds alone is estimated at approximately US$ 2 billion annually in the USA (Pimentel 2005). Insects have been estimated to generate a value of US$ 3 billion for pollination of agricultural crops (Pimentel 2005). In addition, predatory and parasitic beneficial insects and other arthropods provide a value of US$ 4.5 billion per year through natural pest control. Indirect influences of pesticides are the impoverishment of natural ecosystems caused by a decrease of pollinator insects and their reduced water purification and water retention capacity. The indirect influences, together with reduced diversity of plant communities, ultimately also have an impact on landscape aesthetics, their recreational value, and their tourist attractiveness. Newer data value the ecosystem services provided by wild insects in the USA with about US$ 60 billion per year (Losey and Vaughan 2006). However, it is also important to note that effects such as a decline in biodiversity, or a decline in soil fertility and yields, human or animal diseases, cannot be attributed solely to the use of pesticides. Thus, coming up with proper estimates on the economic contribution of pesticides is the great challenge.

Fungicides can also lead to pest outbreaks when their use decimates fungal pathogens that normally infest other pest organisms (Pimentel 2005). If pest outbreaks occur because their natural counterparts have been destroyed by pesticides, this may also mean that other pesticides have to be used to secure yields.

Such detailed economic assessments of pesticide use are only available for a few countries. A very comprehensive study reviewed 61 papers and 30 independent datasets between 1980 and 2014 (Bourguet and Guillemaud 2016). They distinguished between regulatory costs, human health costs, environmental costs, and defensive expenditures. Those costs are either internal to the market, but hidden to the users, or external to the market and most often paid by a third party. Study authors estimate the regulatory costs in the USA in the 2000s at a minimum of US$ 4 billion, if all regulations were considered even at US$ 22 billion. Across the reviewed studies, health costs generally did not consider fatalities due to chronic exposure (e.g., fatal outcomes of cancers). Including these costs would amount to health costs of US$ 15 billion in the USA in 2005. Environmental costs of pesticide use are mainly ignored. Additionally, defensive expenditures have rarely been considered; they would include at least the extra cost of the part of organic food consumption due to aversive behavior linked to pesticide use. According to a retrospective calculation, the hidden and external costs of pesticide use in the USA were in the range of US$ 39.5 billion per year in the late 1980s (Bourguet and Guillemaud 2016). These authors also calculated that benefit-cost ratio of pesticide use and found that in the USA the cost of pesticide use has most likely outreach its benefits, especially if the costs for chronic exposures and evoked illnesses and deaths are included in the calculations.

For Switzerland, the economic costs of pesticide use for the year 2012 amount to approximately 47–94 million € per year (Zandonella et al. 2014). Regulatory costs account for around 18.7 million €; health costs strongly depend on the underlying methodology and show a range between 23–70 million €. The costs of ecosystem damage were estimated to be around 9 million €; however, they were difficult to monetize. As a contrast the annual expenditure of Swiss agriculture on pesticides amounts to about 117 million € per year and is thus about in the same range as the external costs.

We already mentioned that European law actually requires the implementation of Integrated Pest Management (IPM), according to which pesticides may only be used in the absence of alternatives and technical necessity (Lefebvre et al. 2015). It also calls for the minimization of pesticide inputs into waters bodies. Compliance with the regulatory provisions on the safe use of pesticides requires monitoring, but given the number of users (around 600,000 farms in Germany, Austria, and Switzerland alone) and the area to be monitored, this can only be done on a random basis. Accordingly, the inspection rate for agricultural, horticultural, and forestry holdings in Austria in 2013 was 1.8%. This is the situation in a wealthy country. Everyone can imagine what the actual situation is in other not so wealthy ones.

Another economical aspect hardly addressed in these cost-benefit calculations is that billions of taxpayers' money is flowing into conventional farming systems, pesticide research and development in Europe and the USA because agricultural universities there are mainly working on mainstream, pesticide-intensive projects. The share for the development and application of ecological agricultural practices for the benefit of the environment and health, on the other hand, is negligible.

And what would the pesticide phase-out cost? The annual environmental cost of pesticide damage is approximately US$ 50 per hectare, based on an estimated cost of nine billion $ per year of environmental damage caused by pesticides worldwide (Pimentel et al. 1993). If the costs of fertilizer loss, soil erosion, and drinking water pollution are added, the environmental damage costs are at least US$ 300 per hectare, for example, for intensive corn production. When these environmental costs are added to the production costs of corn, the total cost of intensive corn production is US$ 850 per hectare. With average yields of about 11 tons and an average corn price of US$ 150 per ton in the USA, environmental damage costs would account for at least 18% of the proceeds from the sale of the crop.

Often forgotten when calculating the economic impact of pesticides are the costs incurred by the authorities for the necessary control of application and residues in soils, food, water bodies and drinking water. These controls are necessary because compliance with legal limits is necessary to protect the environment and human health. But why does the society need to pay for these costs caused by pesticide applicators? In addition, costs are incurred for training courses for pesticide applications, which are usually financed and organized by the chambers of agriculture or similar bodies financed by taxpayers. For the USA, these costs are estimated at around US$ 500 million per year (Pimentel 2005). Generally the question arises as to whether the polluter-pays principle should not be applied more strictly and actually making the pesticide manufacturers pay for these costs.

Not really expressible in money are ethical and moral aspects when it comes to cost–benefit considerations for the use of pesticides. Do the increased agricultural yields have a negative impact on public health? Does this also apply to pesticides suspected of causing cancer or other chronic diseases? Studies show that more than 10,000 cases of cancer a year in the USA are caused by pesticides, but that pesticides contribute to an increase in agricultural income of US$ 32 billion a year (Pimentel 2005). Conversely, does that mean that every cancer case is worth US$ 3.2 million? The evaluation of human life is unethical but still necessary, the US EPA for instance calculates the value of a human life at US$ 3.7 million (Kaiser 2003). According to estimates, pesticide-induced poisonings and diseases against humans cost the USA more than US$ one billion per year (Pimentel and Greiner 1997).

Most of these cost-benefit calculations are made on a country scale. However, we have seen that pesticide effects are indeed global. It is practically impossible to include all aspects affected by pesticides because of the complexity of the interactions of the modern agricultural business and food industry. In any case, even with this rough calculation it becomes clear that the pesticide applications may be economically viable for one farm, but make little sense for the whole society from an macroeconomic point of view.

A fair assessment of the benefits of pesticides would need to include all these costs mentioned above that are currently externalized to the society. The question is who pays for the loss of biodiversity, water pollution, the consequences for climate change, and human health? Sure is that current food prices do not reflect the true costs including the costs for adverse effects on human health and environmental damages. As far as I know, no study included all externalized costs of pesticides use. However, as an example a study calculated the real costs of different farm products by including costs for energy, greenhouse gases, and nitrogen pollution associated with food production in Germany (Gaugler and Michalke 2018). Fig. 3.1 summarizes the findings separated for conventional vs. organic agriculture.

By including environmental follow-up costs, the prices for meat products from conventional agriculture would have to be about three times more expensive (+196%) and those for organic meat products 82% higher. Price surcharges of plant products (vegetables, cereals) would then be 28% higher for conventionally and 6% higher for organically farmed products. The high

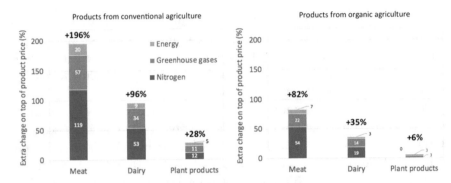

Fig. 3.1 Price surcharges through environmental costs based on of costs for energy, greenhouse gases, and nitrogen during production of meat, dairy, and plant products when commonly externalized costs were included to product prices. Note that costs for pesticide applications and effects were not included due to a lack of data. Data according to Gaugler and Michalke (2018)

external costs and price increases of meat products are explained by the energy-intensive rearing of livestock. These include the cultivation of animal feed, heating, and ventilation of stables and the metabolism of the animals. These factors lead, among other things, to a significantly higher release of reactive nitrogen and greenhouse gases as well as a higher energy requirement than for plant products. Of course, these surcharges based on energy input for heating would be lower in warmer climates.

Dairy products would be about 96% more expensive when coming from conventional farms and about 35% more expensive when organically produced. For organically produced products, the abandonment of mineral nitrogen fertilizers in fodder production and the reduced use of concentrated feed in livestock farming lead to lower external costs in all food categories examined.

These calculations are only an estimate and surcharges are likely to be much higher as the study authors could not include the costs of pesticide use because no data were available. This general lack in concrete pesticide usage data is a common problem and unacceptable as these substances are applied in our environment in thousands of tons and farms are legally required to write down exact dates and dosages of pesticide applications. Moreover, the calculations do not include effects of antibiotic resistance in livestock farming. Hence, it is very likely that if all these costs of the ecological damage caused by food production were included, the prices for organic food would perhaps be lower than of conventionally produced food.

This discussion of higher food prices is perhaps a topic in rich industrialized countries and is therefore a socially sensitive matter. Also, increased, ecologically more realistic prices for the consumer do not guarantee that farmers will actually benefit from this. The best option would be that agricultural policy includes these ecological criteria and transparently informs the consumer about the true costs of food production. Having a label on the product informing about the incurred environmental costs including the intensity of fertiliser, pesticide and antibiotic input would be a great help for more sustainable consumer choices.

3.3 Food Waste Reduction and Municipalities Without Pesticides

Food waste is food that is lost during any of the four stages of the food supply chain: producers, processors, retailers, and consumers. Global food loss and waste amount to about one-third of all food produced (FAO 2019). In low-income countries, most loss occurs during production, while in industrialized countries much food—about 100 kg per person per year—is wasted at the consumption stage.

Wasting food has been called the "world's dumbest environmental problem" (Outrider 2019). Every year, the average family of four in the USA tosses roughly US$ 2000 worth of food; therefore, 30–40% of food produced in the USA ends up discarded. It is estimated that in the USA across the food system, more than 180 kilograms of food per person per year is wasted. To produce this amount of food about 19% of all US cropland would be necessary. And when that food is wasted, so are the resources that go into producing it, including 21% of freshwater used by the US agriculture and climate change pollution equivalent to 37 million cars per year. For the wasted food 18% of all farming fertilizers are also wasted. In the USA alone in 2010 about 31% of the food produced, worth a total of US$ 160 billion, was lost after harvesting and was not available as food (Buzby et al. 2014). This corresponds to about 1250 calories per capita and day that were wasted. In contrast, the average yield differences between conventional and organic farming of about 25% seem comparatively small.

In Germany alone, 11 million tons of food worth around 20 billion € ends up in the garbage every year. This corresponds to an amount that would require an incredible 275,000 trucks to transport (VBZ NRW 2016).

No data on the wasted pesticides are available, but it is probably in the same range as for fertilizers. If we could redirect just one-third of the food that we now toss to people in need, it would more than cover unmet food needs across the countries.

Food waste also includes that entire fields are left unharvested and plowed in (e.g., onions or carrots) because market prices are too low, or produce is rejected for cosmetic reasons. Also consumers throw out anything past or even close to its sell-by date, or forget food in the back of their fridges, or restaurants or private people cook too large portions and the leftovers must be disposed of (NRDC 2017). Other foods such as strawberries often spoil during transport and in storage when conditions are unfavorable.

The situation is most likely similar in all industrialized countries but also in developing countries where more than 40% of food is spoiled postharvest or during processing because of inadequate storage and transport conditions (Foley et al. 2011). Industrialized countries have lower losses at the side of the producers, but more on the retail or consumer level.

Fortunately, food waste and loss have become an issue of great public concern. The 2030 Agenda for Sustainable Development reflects the increased global awareness of the problem. Under Goal 12—*Ensure sustainable consumption and production patterns*, there is target 12.3 calling for halving per capita global food waste at retail and consumer levels by 2030, as well as reducing food losses along the production and supply chains (UN 2019a). However, reducing food loss and waste is also critical to creating a Zero Hunger world and reaching the world's Sustainable Development Goal 2 (*End Hunger*).

Governments are slowly recognizing this problem, for example the USA set the goal to cutting food waste in half by 2030 (EPA 2019). France has become the first country in the world to ban supermarkets from throwing away or destroying unsold food, forcing them instead to donate it to charities and food banks (Gore-Langton 2017). The law follows a grassroots campaign in France by shoppers, anti-poverty campaigners, and those opposed to food waste. In the UK, the government has a voluntary agreement with the grocery and retail sector to cut food and packaging waste in the supply chain and does not have mandatory targets. In Austria, some supermarket chains sell imperfect-looking products at a lower price: crooked cucumbers, not-perfectly looking apples, or carrots that are just as delicious and nutritious as the perfect looking ones. Reducing food waste and rethinking our dietary could substantially improve the delivery of calories and nutrition with no accompanying environmental harm.

In industrialized countries, one reason for this disregard for food products is probably also its low price. In Germany, in the 1950s each household spent about 50% of its income on food and non-alcoholic beverages; nowadays, it is about 11%; in Austria, it is even below 10% (EC 2019). There are great variations in Europe regarding this. The highest share of household expenditures for food and non-alcoholic beverages in Europe is of Romanian citizens with almost 28%, while the lowest share is of UK citizens with only 7.8% of their household incomes. It would be interesting to know if citizens who have to spend more money for food also waste less.

With every discarded food, not only a product with a certain monetary value is destroyed. The production costs of working time, use of energy, water, pesticides, fertilizers, agricultural area and soil are also wasted. It is estimated that more than half of all the food waste in the household can be avoided. In the public debate, consumers are often accused of wasting food, because they allegedly misunderstand the labeling on best before date and throw away food that could still be consumed without smelling or tasting if it is actually spoiled.

Reducing food losses and waste is seen as a promising way to improve food supply in the coming decades (Kummu et al. 2012). In agriculture, even in Central Europe, also losses in the grain harvest can amount to 10%. Losses occur before the harvest due to plant diseases or mechanical damage during threshing and storage (aid.de 2016). But in countries where food is most urgently needed, an even higher amount is wasted in the field. The developing countries still have to cope with numerous logistical storage problems that have been solved in wealthy countries for decades. It is hard to understand that so little thought is given to postharvest losses. In Asia, postharvest losses for rice average around 13%; for Brazil and Bangladesh they are estimated at around 20%.

On a global scale based on publicly available data, it has been calculated that around a quarter of the food produced worldwide (around 614 kilocalories per capita per day) is lost in the food chain (Kummu et al. 2012). The production of these lost and wasted food crops accounts for 24% of total freshwater resources used in food crop production, 23% of total global cropland area, and 23% of total global fertilizer use. The per capita use of resources for food losses is largest in North Africa and West-Central Asia (freshwater and cropland) and North America and Oceania (fertilizers). The smallest per capita use of resources for food losses is found in Sub-Saharan Africa (freshwater and fertilizers) and in industrialized Asia (cropland). If the lowest loss and waste percentages achieved in any region in each step of the food supply chain could be reached globally, food supply losses could be halved. By doing this, there would be enough food for approximately one billion extra people.

Besides the moral and economic dimension of food waste, there is also the negative impact on the climate. It is estimated that at least 10% of the world's greenhouse gases from agriculture could be caused by food waste alone by 2050 (Hiç et al. 2016). Agriculture as a whole is one of the biggest drivers of climate change, accounting for about 20% of global greenhouse gas emissions in 2010 (Pradhan et al. 2013).

In addition to reducing food waste, a change in diet with less meat consumption would also lead to a reduction in pesticide use. A summary of 100 specialist publications has shown that a change in diet would even prolong life by up to 10 years and prevent massive environmental damages (Tilman and Clark 2014). In this way, greenhouse gas emissions could also be reduced by an amount corresponding to the current emissions of all cars, trucks, aircraft, and ships.

In cities and municipalities, pesticides are used both on cultivated land (horticulturally used areas such as beds, green areas, parks, lawns) and on noncultivated land (roads, paths, car parks, sports fields, and playgrounds without lawns). No information can be given about the amount of pesticides used, because this data is not collected and made available to the public. The most common agent is glyphosate, which removes wild herbs from traffic areas and squares. But insecticides and fungicides are also used, for example in public rose gardens or parks against fungal diseases and aphids.

Because of public debates on pesticide effects, many cities and local authorities have taken an interest in reducing the use of pesticides on public land have pledged to avoid no pesticides altogether. As municipalities have a special role to set examples for environmentally friendly and sustainable behavior, this is very important.

Moreover, municipal land space is becoming increasingly important because the quality of agricultural land as habitat for many species of animals and plants (e.g., wild pollinating insects, birds, mammals, and typical field weeds) has been decimated dramatically (Reichholf 2007). Early planning in management and the appropriate selection of plant types of green areas can help to reduce the use of pesticides. When combined with the use of alternative weed control methods, municipal land areas can make a meaningful contribution to the conservation of biodiversity. There is also a great potential in the use of smart trap systems and trap protection stations to monitor and control rat infestation in sewer systems, or to use brushes or thermic weed control systems.

Municipalities can act immediately and change the management and maintenance of their municipal land (BUND 2019). There are already more than 50 cities and municipalities in Germany that are completely or partially pesticide free; some cities have even been pesticide free for over 20 years. On their way to pesticide reduction, some cities have decided not to use glyphosate in the first step. Many cities still have exceptions such as the use of fungicides in the historic rose garden or individual pesticides in the cultivation of plants in municipal plant nurseries. As long as the exception is not the rule, these communities still make a valuable contribution to the protection of biodiversity and human health.

The measures for the pesticide-free communities are manifold. Import is a different sense of what is really beautiful. Do all areas have to look tidy? Are daisies in a lawn or dandelions in pavement scratches really such a nuisance?

Having bee-friendly, site-adapted perennial shrubs with all-year flowering instead of exotic alternating planting is also less labor intensive and helps to save taxpayer money. Many alternatives are available for weed control instead of herbicide use: there are all sorts of sweepers, mowers, brush cutters, weed brushes, joint scrapers, hand weeders, thermal devices with infrared, hot steam, hot foam, electricity or water. There is more to come in this area as members of the European Parliament even signed a declaration in order to call the European Commission for a total ban on chemical pesticides in public areas particularly to demonstrate responsibility toward European citizens (PAN Europe 2019).

In the meantime, some do-it-yourself chains have also decided to no longer offer glyphosate-containing herbicides or neonicotinoid-containing insecticides in their product ranges. Some garden centers even offered their customers exchange campaigns: chemical-synthetic pesticides returned by the consumers were exchanged for new harmless biological products. The pressure from civil society is now so great that over a hundred supermarket chains, including the largest in the world, have decided to remove bee-endangering neonicotinoids from their assortments (FOE 2017).

3.4   Agricultural Subsidies to Curb the Use of Pesticides

Because of the many side effects of pesticide use on humans and the environment, one would assume that at least agricultural subsidies would provide some steering in the right direction. Unfortunately, this is not really the case. The European Union has enacted various laws and directives to protect water bodies and the environment; none of them has such a great influence on ecosystems as the so-called Common Agricultural Policy, CAP (Pe'er et al. 2014). Despite or because of the CAP payments, the intensification of agriculture is increasing strongly both in Western Europe and especially in the new EU member states.

Almost 40% of the EU budget, is spent on these agricultural subsidies, about 59 billion €. The greatest social acceptance of public and financial resources for the agricultural sector is for small farmers and farmers in less-favored and mountain areas. However, the real distribution of public funds shows a different picture because subsidies depend on the farm size: the larger the farm, the more money they get. Thus, the current agricultural subsidies mainly benefit large industrialized farms and the food industry (Weiss 2010). A closer look at the receiver of farm subsidies shows that also big agrochemical companies received millions of subsidies (farmsubsidy.org). For instance, BASF received 6.2 million € on subsidies between 2001 and 2013, Syngenta received a total of 1.7 million €, and Monsanto 662.805 €. Nestlé, the worlds biggest food company based in Switzerland even received an unbelievable amount of 625.9 million € between 1999 and 2013. This is all taxpayer money for the support of private companies in a so-called free market economy. By the way, the Royal Farms Windsor in the UK also received several million € on public support. Therefore, 80% of the European money for agriculture goes to the 20% largest farmers (Factcheck.eu 2019). Only about 6% of public funds flow actually to small farmers. No wonder that the number of farms throughout Europe is steadily declining, and the big ones are swallowing the small ones.

Despite this large sum of funding, success in the environmental sector remains quite moderate. In some areas in Europe, the groundwater is not drinkable due to contamination with agrochemicals and has to be filtered at high costs. Besides pesticide-intensive and fertiliser-intensive maize production is promoted for energy from biogas or ethanol, although it is clear that this type of energy production is inefficient. Agricultural subsidies also promote seed dressings and other pesticides that endanger biodiversity. It is perhaps naïve to expect that agricultural policy should also help to avoid environmental problems and not contribute to their reinforcement.

In so-called agro-environmental programs, measures to reduce pesticides and protect water have been promoted for decades. Nevertheless, pesticide use is not decreasing substantially. Around 17,000 agricultural enterprises in Germany took part in the funding measure "Abandoning fungicides in arable farming," at a cost of around 5 million € per year (Jahn et al. 2014). A glance at the fungicide statistics, however, shows that the use of fungicides has risen from 773 tons to 835 tons per year in a couple of years. The fact that pesticide-intensive agriculture is also allowed in about one-third of Europe's nature reserves (so-called Natura 2000 areas) indicates how half-hearted this approach is.

Critical scrutiny of agricultural subsidies seems legitimate, since about 20% of the value of agricultural production in the EU is subsidized by the taxpayer. On the other hand, less than 1% are invested to mitigate the negative impact on the environment. Ecological follow-up costs are passed on to the general public, while tax revenues are used to promote them. We are not even talking about the problems of shifting environmental costs, because national decisions can have consequences in other regions of the world. Imported feedstuffs such as soybeans for our livestock farms generate environmental costs in the countries of production, but these are not passed on to consumers in the import countries in the form of higher meat prices.

A consequence to inefficient support policies could be to leave agriculture completely to the market and not to allow any intervention or subsidy possibilities. New Zealand, for example, has left agriculture completely to market regulation since the mid-1980s and no longer provides subsidies. However, this has led to an even greater intensification of agriculture. Another example is China, which massively subsidizes input factors such as fertilizers and pesticides, leading to high yields but also to high greenhouse gas emissions and environmental pollution. On the other hand, China also uses its state control for radical measures that transform arable land into nature conservation areas.

The most logical solution would be to tax the environmental impact in such a way that products from harmful production are more expensive than those from more environmentally friendly production. Agriculture deserves the support of society—but only if it provides clearly defined services that produce food without negative effects on the environment and human health. With a fair payment for nature and the environment, agriculture would also lose its reputation as a pure recipient of subsidies. To be fair, there is already support available through agro-environmental programs to reduce fertilizer use, actively protect the climate, or increase soil fertility, but their effectiveness is quite limited.

Another point is how the costs of pesticide applications are distributed. It has already been explained that pesticide manufacturers contribute little to the damage caused by their products, despite their profits running into billions. But even the agricultural sector, which spends 100s of millions € every year on the use of pesticides, does very little to clean up pollution. The environmental or public health costs outlined before are instead passed on to the general public.

Agri-environmental programs in Europe also allocate millions of € each year for the spraying of glyphosate herbicides to promote conservation tillage. The aim is to promote soil and climate protection, since plowing allegedly increases soil erosion and releases greenhouse gases from the soil. In the German federal state of Saxony alone, conservation tillage without plowing was subsidized in 2008 on 76,620 hectares and the associated use of glyphosate with 3.6 million € (Sächsischer Landtag 2008). Of course, the large agrochemical companies are vehement supporters of this practice for soil conservation.

An article in the *New York Times* even made a case that the EU is actually financing the very environmental problems it is trying to solve because the most heavily subsidized areas also showed the worst pollution (Apuzzo et al. 2019). The farm subsidies have had serious environmental consequences from algae blooming in the Baltic Sea due to excess nitrogen and phosphorus input, declining biodiversity, and high greenhouse gas emissions. Well-intentioned policies to improve the environmental impacts of agriculture were amended under the pressure of mighty lobby groups. Additionally, often agricultural subsidies are administered by politicians and officials who benefit themselves from these payments (Apuzzo and Gebrekidan 2019).

3.5   Policy Makers Should Decide and Act on the Basis of Facts

It seems undisputed that policy measures are needed to reduce the risks associated with pesticides' application in agriculture. Taxes on pesticides are often discussed as a proper tool to make pesticides economically less attractive and ultimately lead to a reduced pesticide use. Some European countries have already introduced pesticide taxes, for instance France, Denmark, Norway, and Sweden. An analysis of the effectiveness of these pesticide taxes found mixed results (Böcker and Finger 2016). The main result was that if a tax on a specific pesticide is high enough, the application and the associated risks will be reduced significantly.

Although the EU has committed itself to sustainable agriculture in various international agreements to protect biodiversity and the climate, this is not reflected in reform proposals for the future Common Agricultural Policy (Pe'er et al. 2019). According to the planned financial framework for 2021–2027, the funds for rural development that are important for a "greener" EU agricultural policy, e.g., for organic farming, are to be further reduced from their already low level and subsequent shifts of funds toward direct payments are to be permitted. It is suggested that direct payments for farmers should be phased out gradually and replaced by a system that balances all the objectives of the common agricultural policy and above all supports farmers who operate sustainably and environmentally friendly. For a sustainable reform, the entire process would also have to be improved by reducing the influence of lobbying associations and involving experts from science and society alongside agricultural committees, ministries, and agencies. Also, it has to be emphasized that despite a clear increase in pesticide use during the last 40 years, crop losses have not significantly decreased (Oerke 2006).

Perhaps undecided policy makers will be encouraged to act more decisively against the excessive use of pesticides when they hear that even in the highest circles of agriculture there is criticism of pesticide-intensive agriculture. Even at the German Agricultural Society, the renowned specialist organization of the agricultural and food industries, it is admitted that too few crop rotations and a "gigantic expenditure of chemicals" lead to more and more resistance and even a decline in crop yields. Such insights give cause for optimism. However, there was immediate opposition from the so-called representatives of the farmers, who vehemently defend themselves against these "system debates" (Grossarth 2017). It is also clear from this reaction that agricultural policy, international trading agreements, and the influence of multinational companies on policy must be critically questioned if farmers should be freed from the pesticide treadmill.

Similar to international climate agreements, it would be important to also reach agreement to reduce the use of pesticides. Therefore, it would also be necessary to hold the manufacturers of pesticides responsible for any damage caused by their products, instead of always passing everything on to the general public.

Their remains the initiative of individual consumers. The most effective and long-term way to prevent people from being exposed to even more toxic chemicals is to move away from pesticide-intensive, industrial agriculture. These movements would support small-scale agriculture and free it from the dependence of the agro-industry. In conversations with farmers, I often see that many farmers have a bad feeling about using so many pesticides. Most farmers blindly believe what advisors tell them without questioning that an advisor that comes from an agrochemical company has a clear conflict of interest and will most certainly recommend more pesticide use. We heard before that German agricultural experts estimate that every second use of pesticides occurs at the wrong time and would therefore be unnecessary.

A recent United Nations report even proposes a human right to adequate food and health without pesticide exposure (UNHRC 2017). In this context, the agrochemical industry is also reminded of its responsibility to produce and sell pesticides that do not violate these general rights. The report also criticizes the lack of awareness of the risks posed by some pesticides. This lack of awareness is often exacerbated by industry lobbying, who play down the risks of pesticides, and complacent governments, which pretend that existing laws, and controls are sufficient to ensure protection against dangerous pesticides anyway.

A political and ethical aim should be to eliminate double standards for pesticides. It is not acceptable and unethical that a pesticide that is banned in a country because of health concerns, can still be produced and exported to countries with less stringent legislation. In the interests of human health, ecology, and the sustainable development of our planet, a global policy framework should be developed to reduce the use of pesticides. Politicians should do much more to hold the producers of pesticides responsible for the pollution of drinking water and health damage caused by their products.

Recently, a pesticide manufacturer was pleaded guilty to illegally spraying banned pesticides in Hawaii (USAO 2019). Agrochemicals company Monsanto, has agreed to pay US$ 10.2 million in fines and pleaded guilty to spraying a banned pesticide (Penncap-M) on research crops in Maui, Hawaii. Monsanto's parent company Bayer apologized in a statement, saying "we did not live up to our standards or the law," but justified that Monsanto had reported the spraying to the Hawaii Department of Agriculture.

We have heard of many cover-ups and lobby interventions in pesticide registration. This is absolutely unacceptable given the scale of the problem. Politicians must work to ensure that the environmental risk analysis for pesticides is carried out by independent agencies and that all studies that are carried out on the subject are disclosed. The whole approval process should be based on the precautionary principle, as agreed by all EU countries: As soon as there is a suspicion of a health impairment, a pesticide must be withdrawn from the market and manufacturers must be held liable for the subsequent health costs.

In the context of the authorization of pesticides, manufacturers should be required to demonstrate transparently that there are no pesticide alternatives around. This means that pesticides should not receive approval in any country until their safety and necessity has been demonstrated. Why use herbicides and contaminate the soils when mechanical weeding is just as effective? The prophylactic use of pesticides such as in seed dressings should be radically restricted because it leads to excessive pesticide use and is not scientifically backed. We also do not constantly wash our hair with an anti-louse shampoo just because we might catch head lice!

Politicians must ensure more rigorous residue analyses of authorized pesticides in foodstuffs and publish them on a regular basis. These analyses should, of course, be financed by a fund fed by the pesticide manufacturers.

For farmers, more support for agro-ecological practices should be provided in order to increase biodiversity in the agricultural landscape and thus promote biological pest control. This also includes proven plant cultivation methods such as crop rotation designs, site-specific variety selections, and in improvements of soil fertility and soil health. Greater support for organic farming and alike systems should lead to the abolition of subsidies for pesticides and instead to the introduction of a comprehensive pesticide tax.

If the world stays on its current path, the state of biodiversity will continue to decline. This is due to projected further increases in pressures, most prominently habitat loss, land use intensity and climate change. The agricultural and forestry sectors together have globally caused almost 60% of the total reduction in terrestrial biodiversity until 2010 (Kok et al. 2018). By technological improvements and increased productivity, an increased use of ecological methods in agriculture and forestry, and consumption changes, this biodiversity loss could be lowered. The changes needed in the agricultural and forestry sector to achieve this go well beyond current efforts to reduce their impact on biodiversity, as for instance the EU Pollinators Initiative (EC 2018). It will also provide valuable information on EU progress toward the relevant UN Sustainable Development Goals.

Environmental risk assessment of pesticides also needs to better consider the agricultural practices of long-term use of multiple pesticides including all additional stressors (land use, light pollution, microplastic pollution, overall chemical pollution) affecting nontarget effects of pesticides in a changing world (Brühl and Zaller 2019).

The concentration of the agrochemical industry on a few global corporations is also worrying, according to a UN report (UNHRC 2017). The two US corporations DuPont and Dow Chemical merged to Corteva, ChemChina bought Syngenta from Switzerland, and the German Bayer group took over the US company Monsanto. In the end, three companies dominate more than 60% of the markets for commercial seeds and agrochemicals (Heinrich-Böll-Stiftung et al. 2017). These three conglomerates also account for the majority of applications for plant ownership at the European Patent Office. With this economic power there comes political influence. For Europe, it is to be feared that these global corporations and the politicians who are willing to support it will target the fundamental achievements of EU legislation. The agrochemical companies have already filed claims for damages against the EU Commission, after some bee-damaging neonicotinoids have been banned (Erickson 2013). The situation continues and, as it sounds in diplomatic language, legal options are still assessed by the companies while restriction on neonicotinoids continues in Europe. After appeals of the manufacturers, the European Court of Justice (ECJ) said the EU had correctly applied its "precautionary principle," which allows restrictions on chemicals even when conclusive evidence of harm is lacking. In a separate ruling, the ECJ backed chemicals giant BASF in its complaint against EU restrictions on fipronil, a different class of insecticides. So, the current EU regulation that restricts fipronil will have to be amended.

The courts must still examine whether the European ban of neonicotinoids is legally sound: After all, the manufacturers had previously received approval for their pesticides and invested in their production. The more powerful the agrochemical companies become, the more frequently such legal interventions will occur. The agrochemical industry is now so powerful that it does not shy away from intervening with governments for pesticide bans and threatening them with lawsuits (UNHRC 2017). According to an UN report there was even the threat of suing individual officials who were involved in drawing up the official document on the harmfulness of neonicotinoids.

As we have seen, this has also resulted in manufacturers generally questioning scientific studies that demonstrate the health hazards of their products and trying to discern the scientists involved. Also, examples have been published that the expertise of supposedly independent scientists were bought by these companies in order to provide expert opinions on the company's products. Other practices include the infiltration of authorities and ministries with former industry representatives or the promise of lucrative positions in industry after leaving political office, the so-called revolving doors between industry and politics. For example, the US Food and Drug Administration (FDA), which is responsible for monitoring foods and their health effects, was led by former executives of the largest herbicide manufacturer (MultiWatch 2016).

Strategic partnerships between industry and research institutions or professorships at universities, which serve to increase the credibility of the companies, should also be critically observed. When pesticide studies are carried out with company funds, these supposedly independent institutions become dependent on the donors. Very often there is a "proofreading" by the donor before the results are published.

In politics, the economic arguments usually are regarded stronger than the ones regarding public health or even ecological issues. In this context, the economic evaluation of the use of pesticides should be convincing. Undoctored calculations clearly show that pesticides are actually a loss-making business.

Ultimately, the dispute over the health risk posed by pesticides is not a scientific one, but a political one. Society seems to accept the current level of risk and uncertainty for the benefit of pesticides. However, in my opinion there is insufficient transparency about the extent of pesticide use in the appropriate industries to make an informed decision.

Since 2009 in the European Union a directive was adopted to achieve a sustainable use of pesticides by reducing the risks and impacts of pesticide use on human health and the environment and promoting the use of Integrated Pest Management (IPM) and of alternative approaches or techniques. All 28 member states had adopted national action plans; however, only five actually set high-level measurable targets, of which four relate to risk reduction and one to use reduction (EC 2017). Integrated pest management would be a cornerstone of this directive, but compliance with the principles of IPM at individual grower level is not being systematically monitored. The level of compliance with the establishment of inspection systems for spraying equipment, as well as training and certification systems, is difficult because of unreliable data provided by the member states.

In the last few years, we have seen a change from pesticide reduction targets to risk reduction. Experience in Sweden and Denmark showed that reduction in use was not the most effective means to reduce the risks associated with pesticides. The rationale behind this is that pesticides have different intrinsic properties and rates of use, and using a pesticide with a relatively high dose rate, but a relatively benign environmental and/or toxicological profile, may be a lower risk than a product used at a lower rate, but with a high toxicity (EC 2017). Accordingly, in Sweden for example, the health risks decreased by 69% and environmental risks associated with pesticides decreased by 31% in 2015 compared to 1988. In Germany, the environmental risks associated with pesticide use were reduced by over 50% during the 1987–2007 period. In both cases, these significant reductions in risk occurred against a backdrop of static, or even increased pesticide use. In the Netherlands, the national measure of pesticide impact on surface water was 85% lower, and in drinking water was 75% lower in 2010 compared to 1998.

Pesticide taxes can be an important tool of policy makers and are of increasing importance in European countries. We have already seen that the effectiveness of pesticide taxes in France, Denmark, Norway, and Sweden (Böcker and Finger 2016) is high only when the tax is high enough and differentiated regarding the toxicity of pesticides. Interestingly, pesticide hoarding activities of farmers were observed in various countries especially before an introduction or increase of taxes.

When the costs of pesticides that are paid for by the general public are passed on to the price of pesticides this can be an important tax instrument to reduce the use of pesticides. Some European countries have already done this. Pesticide consumption fell by 21% in Denmark, 43% in the Netherlands, 46% in Finland, and 79% in Sweden compared to the long-term average (Böcker and Finger 2016). Politics could facilitate the choice of healthier food by rigorously taxing the use of pesticides. Denmark has implemented a new pesticide taxation system in 2013 intended to reducing the negative effects of pesticide use on humans and the environment and promoting plant protection measures that have comparatively fewer negative effects on the environment and health. According to the Danish system pesticide products that are particularly problematic for humans or the environment are taxed at a high rate, while less harmful plant protection products are subject to relatively lower taxes (MEF DK 2019). If the farmer chooses a cheaper pesticide product, he/she automatically chooses the more environmentally friendly and healthier alternative. The revenue from the tax is invested in agricultural projects in Denmark and is used to implement and cover the administrative costs of the national pesticide action plan. So far, pesticides have for several years been subject to ad valorem tax of 25% for herbicides and fungicides and 35% for insecticides. This differentiated taxing on pesticides will ensure that farmers have a financial incentive to use the pesticide with the least load on the environment and/or human health.

The situation in most other countries is quite different. Governments have no serious intentions in involving the pesticide manufacturers in the external costs of their products. The pesticide tax system currently used in Denmark, France, and Sweden has also been suggested for Germany (Möckel et al. 2015). The concept provides for a basic levy rate of 20 €/ha per year for the maximum permissible application rate per pesticide. On average, the researchers calculated for 66 pesticide active substances that there would be a price increase of about 40%. And in agriculture this would be a burden above all for farmers applying expensive pesticides and hardly any burden for organic farmers. According to the study, such an increase in costs would not only be permissible under constitutional and EU law, but would also make sense as a steering measure. The possible revenues from the tax are estimated at 1 billion Euros per year in Germany alone that could be used for environmental measures.

An important but often neglected issue are so-called emergency approvals for officially banned pesticides. These exemptions allow pesticide manufacturers to bypass the established pesticide-approval process intended to protect people, wildlife, and the environment and are regularly granted from authorities in Europe and the USA and perhaps other countries. The US EPA for instance issued emergency approvals to spray neonicotinoids — pesticides the agency itself recognizes as "very highly toxic" to bees—onto more than 16 million acres of crops known to attract bees (Donley 2020). Five of the nine approvals granted for the neonicotinoids in 2019 have been handed out for nine consecutive years for the same emergency. By the way, most alleged "emergencies" were often foreseeable occurrences, not emergencies. A critical report of the US EPA on this matter states that routine misuse of these exemptions for neonicotinoids (e.g., sulfoxaflor, thiamethoxam, clothianidin, and dinotefuran) poses significant risks not only to pollinators such as bees, small birds, and butterflies. Study authors clearly conclude that if a pesticide cannot gain approval under the standard pesticide-approval process, then agricultural practices must change to reflect that reality.

3.6   What Must Be Done Concretely?

What is needed is a radical transformation of the pesticide-intensive agriculture to a more sustainable agriculture. Radical literally means from the roots coming, thus it includes a change of our rather careless attitude toward pesticides, their authorization and control. One would think that the countless issues with pesticides across the globe are too serious to be dealt with by small corrections here and there. However, the bottom line is that although the costs of pesticide-intensive agriculture seemed to increase with increased resistances, a decline in soil health due to overuse of pesticides, a decline in soil organic matter, contaminated ground and drinking water, a decline in overall biodiversity, and health concerns, the commitment to industrialized agriculture has not diminished so far (Mart 2015). Also, the increasing cultivation of GMOs has not led to new ways of farming, but only solidified large units, monocultures, and heavy agrochemical inputs with a loss in biodiversity.

The following is a wrap-up of what was mentioned earlier in this book.

3.6.1   Transform Pesticide-Intensive Agriculture and Change Farm Subsidies

Scientific studies and on-farm experiences show that a rapid success in reducing pesticide exposure is possible. However, a future without pesticides requires other costly but feasible adaptations of the currently structured, conventional, and industrialized agroecosystems (FiBL 2019). A rather conservative measure with a great pesticide reduction potential lies in the development of new application methods that bring pesticides more efficiently to the target organisms.

- Practice in organic farming and other forms of agriculture free of synthetic pesticides shows that herbicides can be completely replaced with state-of-the-art mechanical weed control methods, mixed crops, and soil coverings. For the replacement of other pesticides, the range of possible solutions such as natural antagonists (insects, viruses, nematodes), plant extracts, or natural materials (clay minerals, milk extracts, etc.) is enormous. However, the development of standardized organic plant protection products is extremely expensive. It is hard to understand why similar standards as for synthetic chemicals should also apply for substances that are clearly harmless. Things are developing in this direction and already about half of all applications for approval of new active ingredients in the EU now involve biological plant protection products. Just to get an idea of what is possible, there is for instance the method that bumblebees get coated with a biofungicide that gets directly carried from crop to crop by bumblebees. The biofungicide is actually just another fungi (*Clonostachys rosea*) that feeds, for instance on gray mold on strawberries and reduce damage there (Paynter 2019).

- Important to be successful in this matter is thinking in terms of systems rather than only on a case-by-case basis. Preventive crop protection is not practical without interactive effects of multi-unit crop rotations, mixed crops, wildflower strips and hedges, and the tolerating of yield-neutral residual weeds in the field. This approach needs to be developed in a dialogue between farmers, plant protectionists, cultivation technicians, and ecologists.
- More complex disease and pest problems, especially in specialty crops, cannot be tackled without the breeding ot new varieties and selecting crop varieties that are appropriate for a given location. However, as long as environmental costs for pesticides are externalized to the public, such a transformation away from pesticides appear overly costly and less competitive compared with standard pesticide use.
- Spurred by wide public concern about a decline in biodiversity the German government announced in the fall of 2019 a 100 million € action plan for insect protection, research and monitoring of insect populations (Vogel 2019). Besides the promotion of more habitats suitable for insects, it is also planned to reduce pesticide use. In this process, farmers must be assured that these activities are not against them and that a reduced pesticide use not only automatically hampers yields but can as well be economically profitable (Catarino et al. 2019b).
- Decreasing exposure of nontarget organisms, private gardens and public land to pesticides can be achieved by reducing their usage, seeking alternative forms of pest control, and improving current application techniques to reduce pesticide drift.
- Small fields and many types of different crops promote biodiversity in agricultural landscapes (Sirami et al. 2019). In particular, a reduction in the size of arable land to less than 6 hectares has been shown to greatly increase diversity of insects, birds, and plant species. Smaller field sizes and more crop diversity can have similar effects on biodiversity than non-crop habitats in the landscape.

- Public subsidies should only go to producers who do not endanger the environment or our health.
- Future farming needs to be more focused on agro-ecological solutions that improve soil quality, crop diversity, crop health, and overall biodiversity in agroecosystems. So far the avenue including agricultural genetic engineering has been shown to mainly serve the interests of multinational companies and their governmental allies as a route to patented ownership of the food supply (Robinson et al. 2015).
- The more away we move from pesticide-, oil-, and capital-intensive agriculture, the more we will support an agriculture that is characterized by high biodiversity, ecological cycles, and traditional and innovative knowledge in a society and also considers humanistic, ethical, and sociological aspects.

3.6.2 Reform Pesticide Environmental Risk Assessment and Establish Pesticide Monitoring

- Health for people and the environment first! As soon as there is any doubt that pesticides are detrimental to our environment and health, the precautionary principle must be adopted and approval be withdrawn.
- The current environmental risk assessment of pesticides underestimates the real-world situation in the field and is inappropriate to halt the massive decline in biodiversity (Brühl and Zaller 2019).
- The approval of pesticides must be completely revised. The authorization must test both the side effects of the active substances and of the co-formulants, cross-activities of pesticide cocktails, and endocrine effects. These tests must be financed by the pesticide manufacturers, but should be carried out by independent scientists with an obligation to disclose all data and analytical methods.
- Pesticide taxes have been shown to accelerate the process toward their reduction but only when they are high enough and specific. Revenues from these taxes should be used to support pesticide-relevant environmental issues.
- The environmental risk assessments should consider temporal and spatial dimensions of pesticide exposures and chronic, long-term, and multigenerational effects that are currently ignored (Sgolastra et al. 2020).
- Current safety assessments of pesticides rely heavily on studies conducted over 30 years ago (Vandenberg et al. 2017). Outdated safety assessment for pesticides could be improved by regular and widespread monitoring of pesticides and their metabolites in the environment and in human bodies.

- Pesticides should be increasingly discussed in conjunction with global change aspects and environmental stressors prevalent in the field.
- It is necessary to disclose all pesticide consumption data, preferably for each agricultural field. Those data should actually be available, as farmers are obliged to keep records. To protect privacy, aggregated, and of course anonymized, data could be published to the public, but for scientific purposes all details should be made accessible.
- Pesticides should no longer be allowed to be sold to private individuals. There are no plausible reasons to voluntarily poison the own garden or house. Also, private users have no proper training and are not aware of the dangers of the pesticides they are applying.
- Above all, pesticide manufacturers shall provide unstamped, publicly available scientific studies demonstrating the real need for the use of a given pesticides. Evidence should be provided that there is no pesticide-free alternative feasible.
- The use of pesticides in public spaces should be banned as many municipalities already did. If pesticides have to be used at all, then only those approved for organic farming should be allowed.
- Foodstuffs, fruits, and vegetables should be regularly tested for residues of all pesticides authorized in the country and not just for a few selected ones. Of course should pesticide manufacturers be required to bear the costs for this.
- The widespread overuse of pesticides must be tackled. A renaissance of integrated pest management approaches, where nonchemical alternatives to pest control should be prioritized, and pesticides approved in organic agriculture are allowed as a last resort.

3.6.3 Educate and Train Pesticide Users, Consumers and Researchers

- The complex and integrated development challenges we face today demand that decision making be based on sound science and consider indigenous and local knowledge. Especially in countries where industrialized agriculture is not yet established there is a great knowledge on traditional pesticide-free practices including multispecies cropping systems.
- A transition will only be successful if ecologically sound farming methods are more comprehensively taught in agricultural schools, technical colleges, and universities.

- Improving education and exchange of knowledge among farmers, scientists, communities, and the general public.
- Advice and training for farmers on plant protection and pesticide use should be provided by independent experts and not by industry representatives as it happens nowadays. Studies have also shown that ecological illiteracy can deepen farmers' dependency on pesticides (Wyckhuys et al. 2019). An analysis spanning more than hundred years (1910–2016) shows that tribal and indigenous people possess sophisticated knowledge of insects that occur within farm settings, a knowledge that many contemporary farmers do not have any more.
- Just as there is a human right to physical integrity, religious freedom, freedom of expression, clean water, and adequate food, there should also be a universal human right to be nonpoisoned (Cribb 2016). The protection of nature and human health should be as important as the protection of human rights.
- Consumers need to be informed that food products differ considerably in their intensity of pesticide applications. Ideally this could be stated on the product label. By choosing products that have fewer pesticides, it is less likely to have pesticide residues in the food.
- Farming methods have a huge impact on the use of pesticides. Organic produce almost never has pesticide residues present. Buying food from ecologically sound farms can help to curb global insect collapse (Carrington 2019).
- If possible, consumers should be encouraged to grow some of their own vegetables and not only cultivate lawn monocultures.
- Climate change denial is widespread; however, there is also a pesticide denial (Karlsson 2019). The characteristics of pesticide or chemical denial show similarity with those of climate denial, including reliance on fake experts, cherry-picked facts, and attacks on critical scientists.

3.6.4   Homework for Policy Makers

- The optimistic aspect of pesticide reduction is that its achievement would be much easier than that of tackling climate change aspects, for which a complete change of our lifestyles is necessary. This should be encouraging for policy makers as well as the fact that political measures would immediately show results.
- Encouraging might also be that humans have already demonstrated that they can do something for the improvement of our environment. The suc-

cessful phaseout of chlorofluorocarbons (CFCs) that caused stratospheric ozone depletion and increased harmful ultraviolet-B radiation on earth surface is a prominent example. At least three things made the phaseout of CFCs feasible (Jackson 2002). First, compelling scientific evidence warned of a potential ozone problem a decade before the ozone hole was ever detected. Second, there was a potential threat to human health due to ozone depletion and increases in cataracts and carcinomas. Third, there was a technological feasible solution; CFCs could be substituted by other less ozone-damaging chemicals. The first two aspects are analogous for pesticides; however, instead of only substituting harmful pesticides by less harmful ones, I would rather prefer the promotion of all nonchemical measures before—and there are plenty of them available. Otherwise we would continue curing symptoms rather than treating the underlying problem. Similarly to climate change one has to consider that even if we would phase out pesticide completely, problems with pesticide residues will continue for decades. That our grandchildren and their children will be dealing with the problems that we create is a questionable environmental legacy (Jackson 2002).

- With all environmental problems we face, it is no coincidence that the United Nations General Assembly itself has declared that the decade between 2021 and 2030 must be dedicated to ecosystem restoration because this is fundamental to achieving several Sustainable Development Goals (SDGs), mainly those on climate change, poverty eradication, food security, water, and biodiversity conservation (UN 2019b).
- Ultimately government subsidies should only support agriculture and food systems that deliver on the SDGs (in line with "public funds for public goods"). However, it is important to consider that powerful vested interests, including global and national agribusiness corporations, food companies, and commodity groups, command ever-greater market power and heavily influence policies (Eyhorn et al. 2019). It will take a critical mass of scientists, farmers, policy makers, businesses, and above all civil society organizations to align on a transformation agenda and pull these powerful players along to achieve the SDGs.

3.7   Afterword

Perhaps, after reading this book, you have gained the impression that we expose us, our environment and fellow co-organisms on this earth with thousands of inadequately tested pesticides. Well, unfortunately, this corresponds to the sad reality. For the first time in the history of the world, a single species (we human beings) has managed to poison the entire planet. Some scientists consider the Earth's exposure to pesticides and other chemicals an even greater threat to humankind than climate change because the chemical pollution is still underestimated while climate change finally gets the attention it deserves.

However, when you drift into depression after reading the book or even turn into resignation, then I have unfortunately missed my goal. Actually, I wanted to sensitize people to the topic and encourage them to critically question established practices. And the solution to solve the pesticide problem is much easier than tackling climate change because we do not need to change our lifestyles in order to get rid of pesticides.

When the French diplomat Stéphane Hessel published his two small books *Indignez-vous!* [*Time for Outrage!*, Hessel (2011) and *Engagez-Vous!* (*Get Involved!*, Hessel and Vanderpooten (2011)], he motivated me to speak up against our pesticide use. These booklets argue that the French need to again become outraged about societal problems. Hessel's reasons for personal outrage include the growing gap between the very rich and the very poor, France's treatment of its illegal immigrants, the need to reestablish a free press, and the need to protect the environment. The books were a success in France and became bestsellers and were translated into several languages. Hessel, who died at the age of 95 in 2013, was a personality of great moral integrity who participated in the drafting of the Charter of Human Rights, was in the French Resistance, and survived the Nazi concentration camp Buchenwald. In his books, he particularly calls on young people to make a new start.

The realization that things cannot go on like this should also be the motto for the use of pesticides. There are optimistic signs that there is some movement regarding environmental matters. Meanwhile, we see the *Fridays for Future* movement all over the globe initiated by the 16-year-old Greta Thunberg in Sweden. Many imitators emerged since then from *Scientists for Future* to *Farmers for Future*; we also see public protests against the agroindustry, but also counter-protests of farmers against too harsh environmental laws forcing them to use less fertilizers and pesticides. In the German state of Bavaria in 2019, 1.7 million people (18% of the eligible voters) signed a petition for the protection of biodiversity and to establish organic farming on 30% of Bavarian agricultural area by 2030 (Kunzig 2019). Also in 2019, a European Citizens' Initiative was launched aiming to cut the use of synthetic pesticides by 80% by 2030 and completely ban them by 2035. The initiative calls for smaller, diverse, and more sustainable farms, an expansion of organic farming, and more research to pesticide-free and GMO-free agriculture. Such initiatives and protests from the civil society are very encouraging and important as long as policy makers tend to persist in the status quo. It is important that the protesting parties keep respecting each other despite different opinions, keeping in mind the famous quote of Daniel Patrick Moynihan: "You are entitled to your opinion. But you are not entitled to your own facts." So far, political players only very slowly realize that they have to offer a plan and clear commitment toward building on more sustainable agriculture and societies.

We should commit ourselves to a human right to nontoxic food production, even if this seems utopian in the complex world situation. We should be indignant about the widespread use of pesticides without the benefits being scientifically proven. And we should be outraged that pesticide residues can be found in the body of newborn babies. Or that adverse health effects for pesticide-using farmers or other people are treated as collateral damage.

We should also be outraged by the connectedness between agrochemical industry and politics. We know the patterns of averting the dangers from the harmfulness of tobacco consumption and from the denial of the human contribution to climate change. Accordingly industry representatives will always try to dispel skepticism and doubt, claiming that we still do not have enough scientific evidence on the harmful effects of pesticides. The profit margins generated with these products are just tantalizingly high to perpetue the current system.

Although most pesticides are used in agriculture, farmers are both affected most and responsible at the same time. The main culprits are profit-oriented agricultural lobbyists who advise on senseless pesticide use and do not care about the health of the environment and their customers. Research institutions who use scientific blinkers to bite into the details and forget the big picture also play an important role here. Farmers magazines are usually full of advertisements for agrochemicals and create a demand for their use. Is this necessary? When talking about a ban on tobacco advertising and restrictions of tobacco consumption, the end of the world has also been written. Tobacco consumption decreased and who really misses these ads today?

One thing is important to emphasize. It does not help us to fall into an apocalyptic doomsday mood. Rather, it is important to take a critical look at developments, point out grievances, and call our elected representatives to act. Excuses that the facts have not yet been conclusively clarified do not count, because the facts are actually already overwhelming and above all we should act according to the precautionary principle. It gives reason for optimism that there are already enough examples to demonstrate that food production is successful without using pesticides. There are also tentative examples where consumer pressure was so strong that pesticides had to be taken off the market.

We must understand that pesticides are used so widely only because they are heavily subsidized and because we, as consumers, demand food products that are as cheaply as possible without realizing that this behavior favors pesticide use and damage to the environment and societies. In the rich industrialized countries, we are proud that we only spend about a tenth of our monthly income for food while ignoring the damage that we are causing all over the world.

People are tired of hearing politicians' ritual Sunday speeches about sustainability, but finally want to see credible activities in this direction! It remains to be hoped that the sensitization to the topic and the pressure of civil society movements will be so great that a rethinking process will finally be initiated.

My lectures at the university, too, often deal with depressing aspects of environmental science and human ecology. In order not to discourage the students, at the end of the term I usually give them a statement of the American psychologist Margaret Mead on their way: "Never doubt that a small group of thoughtful, committed, citizens can change the world. Indeed, it is the only thing that ever has."

I think this statement would also be appropriate for the topic of this book. If the pressure of an enlightened civil society becomes great enough, and if the political and fiscal framework conditions are created in such a way that our consumer behavior is directed toward more ecologically friendly products, then those who refuse to make progress will finally also have to move in order to keep up.

Acknowledgments In the German version of the book I thanked numerous people for supporting me in this endeavor. Some of them were criticized for providing information for such an outrageous book, so I refrain from mentioning names here to not subject people to such criticism.

I have collected much of the content of this book during the preparation of lectures at the University of Natural Resources and Life Sciences Vienna. In this context, I also would like to thank many students for their critical questions and suggestions, which helped to clarify my arguments. I am grateful to my superiors for always supporting me with this project and also to technical staff, fellow colleagues, and bachelor's, master's, and doctoral students for productive discussions. When presenting my German book in front of different audiences I sometimes had to master delicate situations when people accused me of bashing farmers or ignoring modern science, but nevertheless I always learned my lesson.

I am grateful for the input from numerous people I spoke during the writing process; fellow scientists gave valuable tips, some provided experiences from agricultural practices in different parts of the world, some shared insights into official procedures, and some told me pesticide-relevant experiences and accidents. I am also very grateful for the many suggestions on this topic I got from reading very knowledgeable books, blogs, newsletters, websites of NGOs, and articles in independent newspapers and magazines. Listening and watching pesticide-critical and investigative documentaries in public radio and TV stations was also very thought-provoking. I found it very encouraging to see so many dedicated, critical and very knowledgeable journalists for the pesticide topic.

Although I carefully checked all statements in the text and tried to properly reference them, there still might be errors that have crept in—in this case I apologize for it.

Big thanks go to my family for putting up with me rather pesticide-minded in the last couple of months of writing. Nevertheless, they always fully supported me with this book project. Last but not least, I thank the team of the Zsolnay-Deuticke Verlag in Vienna for supporting the German book and the team of Springer Nature for believing that the content of this book would also be worthwhile for the English-speaking world.

References

Ackerman-Leist P (2017) A precautionary tale. How one small town banned pesticides, preserved its food heritage, and inspired a movement. Chelsea Green Publishing, White River Junction, VT

aid.de (2016) In: Ernährung Bf (ed), Video: Verlustquellen bei der Getreideernte: Gesamternteverluste senken. https://www.bzfe.de/inhalt/videoclip-verlustquellen-bei-der-getreideernte-2633.html. Accessed 23 Nov 2019

Altieri MA (1995) Agroecology: the science of sustainable agriculture, 2nd edn. Westview Press, Boulder, CO

Apuzzo M, Gebrekidan S (2019) Who keeps Europe's farm billions flowing? Often, those who benefit. The New York Times. https://www.nytimes.com/2019/12/11/world/europe/eu-farm-subsidy-lobbying.html. Accessed 12 Jan 2020

Apuzzo M, Gebrekidan S, Armendariz A, Wu J (2019) Killer slime, dead birds, an expunged map: the dirty secrets of European farm subsidies. The New York Times. https://www.nytimes.com/interactive/2019/12/25/world/europe/farms-environment.html. Accessed 12 Jan 2020

Badenhausser I, Gross N, Mornet V, Roncoroni M, Saintilan A, Rusch A (2020) Increasing amount and quality of green infrastructures at different scales promotes biological control in agricultural landscapes. Agric Ecosyst Env 290:106735

Barański M et al (2014) Higher antioxidant and lower cadmium concentrations and lower incidence of pesticide residues in organically grown crops: a systematic literature review and meta-analyses. Brit J Nutr 112:794–811

Benbrook CM (2016) Trends in glyphosate herbicide use in the United States and globally. Environ Sci Eur 28:3

Berners-Lee M, Kennelly C, Watson R, Hewitt CN (2018) Current global food production is sufficient to meet human nutritional needs in 2050 provided there is radical societal adaptation. Elem Sci Anth 6:52

BMEL (2019) Organic farming in Germany. https://www.bmel.de/SharedDocs/Downloads/EN/Agriculture/OrganicFarming/Organic-Farming-in-Germany.pdf?__blob=publicationFile. Accessed 10 Nov 2019

Böcker T, Finger R (2016) European pesticide tax schemes in comparison: an analysis of experiences and developments. Sustainability 8:378

Bourguet D, Guillemaud T (2016) The hidden and external costs of pesticide use. In: Lichtfouse E (ed) Sustainable agriculture reviews, vol 19. Springer International Publishing, Cham, pp 35–120

Brühl CA, Zaller JG (2019) Biodiversity decline as a consequence of an inappropriate environmental risk assessment of pesticides. Front Environ Sci 7:177

BUND (2019) Pestizidfreie Kommunen: Es tut sich was. https://www.bund.net/themen/umweltgifte/pestizide/pestizidfreie-kommune/. Accessed 10 Nov 2019

Buzby JC, Wells HF, Hyman J (2014) The estimated amount, value, and calories of postharvest food losses at the retail and consumer levels in the United States. EIB-121. U.S. Department of Agriculture, Economic Research Service. www.ers.usda.gov/webdocs/publications/43833/43680_eib121.pdf. Accessed 17 Oct 2019

Cahenzli F et al (2019) Perennial flower strips for pest control in organic apple orchards—a pan-European study. Agric Ecosyst Environ 278:43–53

Callaway E (2018) CRISPR plants now subject to tough GM laws in European Union. Nature 560:16. https://doi.org/10.1038/d41586-018-05814-6

Carrington D (2019) Buy organic food to help curb global insect collapse, say scientists. The Guardian. https://www.theguardian.com/environment/2019/feb/13/buy-organic-food-to-help-curb-global-insect-collapse-say-scientists?CMP=Share_iOSApp_Other. Accessed 1 Feb 2020

Carvalho DO et al (2015) Suppression of a field population of Aedes aegypti in Brazil by sustained release of transgenic male mosquitoes. PLoS Negl Trop Dis 9:e0003864

Catarino R, Bretagnolle V, Perrot T, Vialloux F, Gaba S (2019a) Bee pollination outperforms pesticides for oilseed crop production and profitability. Proc R Soc B 286:20191550

Catarino R, Gaba S, Bretagnolle V (2019b) Experimental and empirical evidence shows that reducing weed control in winter cereal fields is a viable strategy for farmers. Sci Rep 9:9004

Chhalliyil P, Ilves H, Kazakov SA, Howard SJ, Johnston BH, Fagan J (2020) A Real-Time Quantitative PCR Method Specific for Detection and Quantification of the First Commercialized Genome-Edited Plant. Foods 9:1245

CFS (2016) Net loss: economic efficacy and costs of neonicotinoid insecticides used as seed coatings: updates from the United States and Europe. Center for Food Safety, 23pp. https://www.centerforfoodsafety.org/reports/4591/net-losseconomic-efficacy-and-costs-of-neonicotinoid-insecticides-used-as-seed-coatings-updates-from-the-united-states-and-europe. Accessed 2 Nov 2019

Conway G (1997) The doubly green revolution. Penguin Books, London

Cribb J (2016) Surviving the 21st century: Humanity's ten great challenges and how we can overcome them. Springer, Switzerland

Davis AS, Hill JD, Chase CA, Johanns AM, Liebman M (2012) Increasing cropping system diversity balances productivity, profitability and environmental health. PLoS One 7:e47149

Donley N (2020) Trump EPA used 'Emergency' loophole to approve pesticides toxic to bees on 16 million acres in 2019. https://biologicaldiversity.org/w/news/press-releases/trump-epa-used-emergency-loophole-to-approve-pesticides-toxic-to-bees-on-16-million-acres-in-2019-2020-01-02/. Accessed 12 Jan 2020

EC (2017) Overview report. Sustainable use of pesticides. DG Health and Food Safety. http://www.europarl.europa.eu/cmsdata/149097/EC%20overview%20report%20on%20sust%20pesticides%202017.pdf. Accessed 11 Jan 2020

EC (2018) Communication from the commission to the European Parliament, The Council, The European economic and social committee and the committee of the regions. Commission E. https://eur-lex.europa.eu/legal-content/EN/TXT/PDF/?uri=CELEX:52018DC0395&from=PL. Accessed 25 Oct 2019

EC (2019) Eurostat. How much are households spending on food? ec.europa.eu/eurostat/web/products-eurostat-news/produ...LIN_viewMode=print&_101_INSTANCE_VWJkHuaYvLIN_languageId=en_GB. Accessed 30 Jan 2020

ECFR (2019) US National Organic Program. The National list of allowed and prohibited substances. https://www.ecfr.gov/cgi-bin/text-idx?c=ecfr&SID=9874504b6f1025eb0e6b67cadf9d3b40&rgn=div6&view=text&node=7:3.1.1.9.32.7&idno=7#se7.3.205_1601. Accessed 4 Jan 2020

EFSA (2019a) The 2017 European Union report on pesticide residues in food. EFSA J 17:e05743

EFSA (2019b) Special eurobarometer "Food safety in the EU". Wave EB91.3, 104pp. European Food Safety Authority. https://www.efsa.europa.eu/sites/default/files/corporate_publications/files/Eurobarometer2019_Food-safety-in-the-EU_Full-report.pdf. Accessed 9 Nov 2019

ENSSER (2017) Products of new genetic modification techniques should be strictly regulated as GMOs. https://ensser.org/wp-content/uploads/2017/09/ENSSER-NGMT-Statement-v27-9-2017.pdf. Accessed 6 Jan 2020

EP (2020) European Parliament resolution of 16 January 2020 on the 15th meeting of the Conference of Parties (COP15) to the Convention on Biological Diversity (2019/2824(RSP)). P9_TA-PROV(2020)0015 COP15 to the Convention on Biological Diversity (Kunming 2020). http://www.europarl.europa.eu/doceo/document/TA-9-2020-0015_EN.pdf. Accessed 29 Jan 2020

EPA (2017) Pesticides Industry Sales and Usage. 2008–2012 Market estimates. Agency USEP. https://www.epa.gov/pesticides/pesticides-industry-sales-and-usage-2008-2012-market-estimates, https://www.epa.gov/sites/production/files/2017-01/documents/pesticides-industry-sales-usage-2016_0.pdf. Accessed 29 Sept 2019

EPA (2019) United States 2030 food loss and waste reduction goal. https://www.epa.gov/sustainable-management-food/united-states-2030-food-loss-and-waste-reduction-goal. Accessed 6 Jan 2020

EPRS (2019) Farming without plant protection products. Can we grow without using herbicides, fungicides and insecticides? In: Service EEPR (ed) IN-DEPTH ANALYSIS. Panel for the Future of Science and Technology, PE 634.416—March 2019; http://www.europarl.europa.eu/RegData/etudes/IDAN/2019/634416/EPRS_IDA(2019)634416_EN.pdf. Accessed 25 Oct 2019

Erb K-H, Haberl H, Plutzar C (2012) Dependency of global primary bioenergy crop potentials in 2050 on food systems, yields, biodiversity conservation and political stability. Energy Policy 47:260–269

Erickson B (2013) LAWSUIT Bayer, Syngenta challenge European Commission's ban on neonicotinoid pesticides. Chem Eng News Arch 91:15

Evans BR et al (2019) Transgenic Aedes aegypti mosquitoes transfer genes into a natural population. Sci Rep 9:13047

Eyhorn F et al (2019) Sustainability in global agriculture driven by organic farming. Nat Sustain 2:253–255

Factcheck.eu (2019) True: "80% of the European money for agriculture goes to the 20% largest farmers". https://eufactcheck.eu/factcheck/true-80-percent-of-the-european-money-for-agriculture-goes-to-the-20-percent-largest-farmers/. Accessed 30 Jan 2020

FAO (2014) Family farmers: feeding the world, caring for the earth, www.fao.org/resources/infographics/infographics-details/en/c/270462/. Accessed 22 Dec 2017

FAO (2019) Food loss and food waste, www.fao.org/food-loss-and-food-waste/en/. Accessed 23 Nov 2019

FAOSTAT (2020) Crops—Cereals, www.fao.org/faostat/en/#data/QC. Accessed 5 Jan 2020

FiBL (2019) Pestizide: Nur Systemlösungen bringen den Ausweg, www.fibl.org/de/infothek/meldung/pestizide-nur-systemloesungen-bringen-den-ausweg.html. Accessed 20 Jan 2020

FOE (2017) Walmart and True Value to phase out bee-killing pesticides while Ace Hardware lags behind, foe.org/news/2017-05-walmart-and-true-value-to-phase-out-bee-killing-pest/. Accessed 11 Jan 2020

Foley JA et al (2011) Solutions for a cultivated planet. Nature 478:337–342

Freyer B (ed) (2016) Ökologischer Landbau: Grundlagen, Wissensstand und Herausforderungen. Haupt Verlag, UTB, Bern, Schweiz

Gagic V, Hulthen A, Marcora A, Wang X, Jones L, Schellhorn N (2019) Biocontrol in insecticide sprayed crops does not benefit from semi-natural habitats and recovers slowly after spraying. J Appl Ecol 56:2176–2185

Gaugler T, Michalke A (2018) "How much is the dish?"—Was kosten uns Lebensmittel wirklich? www.tollwood.de/wp-content/uploads/2018/09/20180914_how_much_is_the_dish_-_was_kosten_uns_lebensmittel_langfassungfinal-2.pdf. Accessed 03 Oct 2018

Geiger F, Wäckers FL, Bianchi F (2009) Hibernation of predatory arthropods in semi-natural habitats. BioControl 54:529–535

GlobalOrganics (2018) Are pesticides allowed in organic farming? https://www.global-organics.com/post.php?s=2018-02-02-are-pesticides-allowed-in-organic-farming. Accessed 4 Jan 2020

Gore-Langton L (2017) France's food waste ban: One year on. https://www.food-navigator.com/Article/2017/03/24/France-s-food-waste-ban-One-year-on. Accessed 6 Jan 2020

Gosnell H, Gill N, Voyer M (2019) Transformational adaptation on the farm: processes of change and persistence in transitions to 'climate-smart' regenerative agriculture. Glob Environ Change Hum Policy Dimens 59:13

Goulson D (2014) David's blog posts for June 2014. Launch of the Worldwide Integrated Assessment (WIA) on the environmental impacts of systemic pesticides. http://www.sussex.ac.uk/lifesci/goulsonlab/blog/wia. Accessed 19 Oct 2019

Grossarth J (2016) Vom Land in den Mund. Warum sich die Nahrungsindustrie neu erfinden muss. Nagel & Kimche, Hamburg

Grossarth J (2017) Zukunft der Landwirtschaft. Große Sorge ums Essen: Frankfurter Allgemeine. http://www.faz.net/aktuell/wirtschaft/landwirtschaft-fordert-wegen-ernterueckgang-einen-systemwechsel-14690355.html. Accessed 23 Nov 2019

Heinrich-Böll-Stiftung, Rosa-Luxemburg-Stiftung, Bund für Umwelt und Naturschutz Deutschland, Oxfam Deutschland, Germanwatch, Le Monde diplomatique (2017) Konzernatlas. Daten und Fakten über die Agrar- und Lebensmittelindustrie 2017, 52pp, www.boell.de/de/konzernatlas. Accessed 23 Dec 2017

Hessel S (2011) Time of outrage! Quartet Books, London

Hessel S, Vanderpooten G (2011) Engagez-vous! Editions De L'aube. La Tour d'Aigues, France

Hiç C, Pradhan P, Rybski D, Kropp JP (2016) Food surplus and its climate burdens. Environ Sci Technol 50:4269–4277

Hilbeck A et al (2015) No scientific consensus on GMO safety. Environ Sci Eur 27:4

Holzschuh A, Steffan-Dewenter I, Tscharntke T (2010) How do landscape composition and configuration, organic farming and fallow strips affect the diversity of bees, wasps and their parasitoids? J Anim Ecol 79:491–500

Horgan FG, Kudavidanage EP, Weragodaarachchi A, Ramp D (2017) Traditional 'maavee' rice production in Sri Lanka: environmental, economic and social pressures revealed through stakeholder interviews. Paddy Water Environ 16:225.241

Hunt ND, Hill JD, Liebman M (2019) Cropping system diversity effects on nutrient discharge, soil Erosion, and agronomic performance. Environ Sci Technol 53:1344–1352

IAASTD (2009) Agriculture at a crossroads. IAASTD—International Assessment of Agricultural Knowledge, Science and Technology for Development, Washington, DC

IPBES (2017) The assessment report of the intergovernmental science-policy platform on biodiversity and ecosystem services on pollinators, pollination and food production. In: Potts SG, Imperatriz-Fonseca VL, Ngo HT (eds) Secretariat of the Intergovernmental Science-Policy Platform on Biodiversity and Ecosystem Services, Bonn, Germany. https://doi.org/10.5281/zenodo.3402856, 552pp. Accessed 24 Dec 2019

ISAAA (2017) Global status of commercialized biotech/GM crops in 2017: Biotech Crop Adoption Surges as Economic Benefits Accumulate in 22 Years. In: Applications ISftAoA-B (ed) Brief 53. http://www.isaaa.org/resources/publications/briefs/53/download/isaaa-brief-53-2017.pdf. Accessed 23 Nov 2019

Jackson R (2002) The earth remains forever. Generations at a crossroads. University of Texas Press, Austin, TX

Jahn T, Hötker H, Oppermann R, Bleil R, Vele L (2014) Protection of biodiversity of free living birds and mammals in respect of the effects of pesticides. In: NABU, vol 30/2014. https://www.umweltbundesamt.de/publikationen/protection-of-biodiversity-of-free-living-birds. Accessed 20 Oct 2019

Jeffery S et al (2010) European atlas of soil biodiversity. European Commission, Publications Office of the European Union, Luxembourg

Jensen OP (2019) Pesticide impacts through aquatic food webs. Science 366:566–567

Kaiser J (2003) How much are human lives and health worth? Science 299:1836–1837

Karlsson M (2019) Chemicals denial—a challenge to science and policy. Sustainability 11:4785

Kawall K (2019) New possibilities on the horizon: genome editing makes the whole genome accessible for changes. Front Plant Sci 10:525

Kennedy I, Sánchez-Bayo F, Caldwell R (2002) Cotton pesticides in perspective. Department of Agricultural Chemistry and Soil Science, The University of Sydney, Australia

Kofler N, Baylis F, Dellaire G, Getz LJ (2019) Genetically modifying mosquitoes to control the spread of disease carries unknown risks. The Conversation Academic Rigour, Journalistic Flair, theconversation.com/genetically-modifying-mosquitoes-to-control-the-spread-of-disease-carries-unknown-risks-123862. Accessed 05 Oct 2019

Kok MTJ et al (2018) Pathways for agriculture and forestry to contribute to terrestrial biodiversity conservation: a global scenario-study. Biol Conserv 221:137–150

Kummu M, de Moel H, Porkka M, Siebert S, Varis O, Ward PJ (2012) Lost food, wasted resources: global food supply chain losses and their impacts on freshwater, cropland, and fertiliser use. Sci Total Environ 438:477–489

Kunzig R (2019) Bavarians vote to save bugs and birds—and change farming. National Geographic. https://www.nationalgeographic.com/environment/2019/02/bavarians-vote-save-bugs-birds-change-farming/. Accessed 12 Jan 2020

Lal R (2004) Carbon emission from farm operations. Environ Int 30:981–990

Larson DF, Muraoka R, Otsuka K (2016) On the central role of small farms in African rural development strategies. Policy Research working paper; no. WPS 7710. World Bank Group, Washington, DC

Lechenet M, Dessaint F, Py G, Makowski D, Munier-Jolain N (2017) Reducing pesticide use while preserving crop productivity and profitability on arable farms. Nat Plants 3:17008

Lefebvre M, Langrell SRH, Gomez-y-Paloma S (2015) Incentives and policies for integrated pest management in Europe: a review. Agron Sustain Develop 35:27–45

Leu A (2014) The myths of safe pesticides. Acres, Austin, TX

Losey JE, Vaughan M (2006) The economic value of ecological services provided by insects. Bioscience 56:311–323

Mäder P, Fliessbach A, Dubois D, Gunst L, Fried P, Niggli U (2002) Soil fertility and biodiversity in organic farming. Science 296:1694–1697

Mart M (2015) Pesticides, a love story. America's enduring embrace of dangerous chemicals. University Press of Kansas, Lawrence, KS

McIntyre BD, Herren HR, Wakhungu J, Watson RT (2009) International assessment of agricultural knowledge, science and technology for development (IAASTD): global report. Washington, DC

MEF DK (2019) Pesticide tax. Denmark MoEaFo (ed) https://eng.mst.dk/chemicals/pesticides/reducing-the-impact-on-the-environment/initiatives-under-the-green-growth-action-plan/pesticide-tax/. Accessed 23 Nov 2019

Menzler M, Griepentrog HW (2017) Digitalisiere oder weiche? Ökologie and Landbau 2:25–26

Möckel S, Gawel E, Kästner M, Knillmann S, Liess M, Bretschneider W (2015) Einführung einer Abgabe auf Pflanzenschutzmittel in Deutschland. Duncker & Humblot, Berlin

Modrzejewski D, Hartung F, Sprink T, Krause D, Kohl C, Wilhelm R (2019) What is the available evidence for the range of applications of genome-editing as a new tool for plant trait modification and the potential occurrence of associated off-target effects: a systematic map. Environ Evid 8:27

Mollison B (1993) Permaculture: a designer's manual. Permaculture-Resources, Califon, NJ

Mourtzinis S et al (2019) Neonicotinoid seed treatments of soybean provide negligible benefits to US farmers. Sci Rep 9:11207

Müller A et al (2017) Strategies for feeding the world more sustainably with organic agriculture. Nat Commun 8:1290

MultiWatch (ed) (2016) Schwarzbuch Syngenta. Dem Basler Agromulti auf der Spur. edition 8, Liebefeld, Schweiz

Mumelter G (2017) Giftiger Kampf um Pestizide in Südtirol. Der Standard. https://www.derstandard.at/story/2000065829421/giftiger-kampf-um-pestizide-in-suedtirol. Accessed 10 Nov 2019

Navarro-Roldán MA, Amat C, Bau J, Gemeno C (2019) Extremely low neonicotinoid doses alter navigation of pest insects along pheromone plumes. Sci Rep 9:8150

Niggli U (2014) Sustainability of organic food production: challenges and innovations. Proc Nutr Soc 74:83–88

Norris AL, Lee SS, Greenlees KJ, Tadesse DA, Miller MF, Lombardi H (2019) Template plasmid integration in germline genome-edited cattle. bioRxiv:715482

NRDC (2017) Wasted: how America is losing up to 40% of its food from farm to fork to landfill. https://www.nrdc.org/sites/default/files/wasted-2017-report.pdf. Accessed 6 Jan 2020

Oates L, Cohen M, Braun L, Schembri A, Taskova R (2014) Reduction in urinary organophosphate pesticide metabolites in adults after a week-long organic diet. Environ Res 132:105–111

Oerke EC (2006) Crop losses due to pests. J Agric Sci Cambr 144:31–43

OMRI (2019) Organic materials review institute. https://www.omri.org. Accessed 7 Dec 2019

Outrider (2019) Opinion - Food waste is the world's dumbest environmental problem, therising.co/2019/09/16/food-waste-is-the-worlds-dumbest-environmental-problem/. Accessed 23 Nov 2019

PAN Europe (2019) Pesticide-free towns. https://www.pesticide-free-towns.info. Accessed 10 Nov 2019

PAN UK (2017a) Is organic better? https://www.pan-uk.org/organic/. Accessed 4 Jan 2020

PAN UK (2017b) The truth about pesticide use in the UK. https://www.pan-uk.org/pesticides-agriculture-uk/. Accessed 11 Jan 2020

Pascher K, Hainz-Renetzeder C, Gollmann G, Schneeweiss GM (2017) Spillage of viable seeds of oilseed rape along transportation routes: ecological risk assessment and perspectives on management efforts. Front Ecol Evol 5:104

Paynter B (2019) There's a new group of workers spreading organic pesticide on crops: bees. Fast Company 09-07-19, www.fastcompany.com/90399623/theres-a-new-group-of-workers-spreading-organic-pesticide-on-crops-bees. Accessed 1 Feb 2020

Pe'er G et al (2014) EU agricultural reform fails on biodiversity. Science 344:1090–1092

Pe'er G et al (2019) A greener path for the EU common agricultural policy. Science 365:449–451

Pfiffner L et al (2019) Design, implementation and management of perennial flower strips to promote functional agrobiodiversity in organic apple orchards: a pan-European study. Agric Ecosyst Environ 278:61–71

Pimentel D (1997) Pest management in agriculture. In: Pimentel D (ed) Techniques for reducing pesticide use: environmental and economic benefits. Wiley, Chichester, pp 1–11

Pimentel D (2005) Environmental and economic costs of the application of pesticides primarily in the United States. Environ Dev Sustain 7:229–252

Pimentel D, Greiner A (1997) Environmental and socio-economic costs of pesticide use. In: Pimentel D (ed) Techniques for reducing pesticide use: environmental and economic benefits. Wiley, Chichester, pp 51–78

Pimentel D et al (1991) Environmental and economic impacts of reducing US agricultural pesticide use. In: Pimentel D (ed) Handbook on pest management in agriculture. CRC Press, Boca Raton, FL, pp 679–718

Pimentel D et al (1993) Environmental and economic effects of reducing pesticide use in agriculture. Agric Ecosyst Environ 46:273–288

Pimentel D, Hepperly P, Hanson J, Douds D, Seidel R (2005) Environmental, energetic, and economic comparisons of organic and conventional farming systems. Bioscience 55:573–582

Pleasants JM, Oberhauser KS (2013) Milkweed loss in agricultural fields because of herbicide use: effect on the monarch butterfly population. Insect Conserv Div 6:135–144

Pollan M (2002) The botany of desire: a plant's-eye view of the world. Random House Trade Paperbacks, New York

Pradhan P, Reusser DE, Kropp JP (2013) Embodied greenhouse gas emissions in diets. PLoS One 8:e62228

Ray DK, Mueller ND, West PC, Foley JA (2013) Yield trends are insufficient to double global crop production by 2050. PLoS One 8:e66428

Reganold JP, Wachter JM (2016) Organic agriculture in the twenty-first century. Nature Plants 2:15221

Reichholf JH (2007) Stadtnatur. Eine neue Heimat für Tiere und Pflanzen. oekom, Munich, Germany

Richter ED (2002) Acute human pesticide poisonings. In: Pimentel D (ed) Encyclopedia of pest management. Dekker, New York, pp 3–6

Robinson CH, Antoniou M, Fagan J (2015) GMO myths and truths. Condensed and updated. a citizen's guide to the evidence on the safety and efficacy of genetically modified crops and foods. Earth Open Source, London

Roe S et al (2019) Contribution of the land sector to a 1.5°C world. Nat Clim Chang 9:817–828

Rusch A et al (2016) Agricultural landscape simplification reduces natural pest control: a quantitative synthesis. Agric Ecosyst Environ 221:198–204

Sächsischer Landtag (2008) Kleine Anfrage: Einsatz von Herbiziden auf sächsischen Äckern. edas.landtag.sachsen.de/viewer.aspx?dok_nr=14834&dok_art=Drs&leg_per=5&pos_dok=202. Accessed 05 Nov 2017

Samways MJ (2019) Insect conservation. A global synthesis. CABI Publishing, Wallingford

Schader C, Zaller JG, Köpke U (2005) Cotton-basil intercropping: effects on pests, yields and economical parameters in an organic field in Fayoum, Egypt. Biol Agric Hortic 23:59–72

Schiebel A (2017) Das Wunder von Mals: Wie ein Dorf der Agrarindustrie die Stirn bietet. Oekom Verlag, München

Schmidt H (2018) Das neue Unionsrecht der Biolebensmittel: Krieg in den Dörfern und Konformität statt Gleichwertigkeit. Z ges Lebensmittelrecht 45:434–485

Schmitz J, Hahn M, Brühl CA (2014) Agrochemicals in field margins—an experimental field study to assess the impacts of pesticides and fertilizers on a natural plant community. Agric Ecosyst Environ 193:60–69

Schütte G et al (2017) Herbicide resistance and biodiversity: agronomic and environmental aspects of genetically modified herbicide-resistant plants. Environ Sci Eur 29:5

Sgolastra F et al (2020) Bees and pesticide regulation: lessons from the neonicotinoid experience. Biol Conserv 241:108356

Sirami C et al (2019) Increasing crop heterogeneity enhances multitrophic diversity across agricultural regions. Proc Natl Acad Sci 116:16442–16447

Smith MD et al (2017) Seafood prices reveal impacts of a major ecological disturbance. Proc Natl Acad Sci 114:1512–1517

Stevens S, Jenkins P (2014) Heavy costs. Weighing the value of neonicotinoid insecticides in agriculture. Center for Food Safety. https://www.centerforfoodsafety. org/files/neonic-efficacy_digital_29226.pdf. Accessed 19 Oct 2019

Strassheim I (2019) Selbst Syngenta rechnet mit Verbot der meisten Pestizide. Tagesanzeiger. https://www.tagesanzeiger.ch/wirtschaft/selbst-syngenta-rechnet-mit-verbot-der-meisten-pestizide/story/26294606. Accessed 3 Jan 2020

Suslow T (2000) Chlorination in the production and postharvest handling of fresh fruits and vegetables. Use of chlorine-based sanitizers and disinfectants in the food manufacturing industry, Chap. 6, 15pp. www.siphidaho.org/env/pdf/ Chlorination_of_fruits_and_veggies.PDF. Accessed 05 Nov 2017

Swanton CJ, Clements DR, Derksen DA (1993) Weed succession under conservation tillage: a hierarchical framework for research and management. Weed Technol 7:286–297

Tappeser B, Reichenbecher W, Teichmann H (eds) (2014) Agronomic and environmental aspects of the cultivation of genetically modified herbicide-resistant plants. A joint paper of BfN (Germany), FOEN (Switzerland) and EAA (Austria). http:// www.bfn.de/fileadmin/MDB/documents/service/skript362.pdf, 77pp. Accessed 05 Nov 2017

Tilman D, Clark M (2014) Global diets link environmental sustainability and human health. Nature 515:518–522

Tschumi M, Albrecht M, Entling MH, Jacot K (2015) High effectiveness of tailored flower strips in reducing pests and crop plant damage. Proc R Soc B 282:20151369

Umweltinstitut München (2019) Gentechnik 2.0—Argumentationshilfe für die neue Gentechnikdebatte in Europa. Umweltinstitut München. http://www. umweltinstitut.org/fileadmin/Mediapool/Druckprodukte/Gentechnik/PDF/ Argumentationshilfe_Gentechnik_Download.pdf. Accessed 23 Nov 2019

UN (2019a) Goal 12: ensure sustainable consumption and production patterns. https://www.un.org/sustainabledevelopment/sustainable-consumption-production/. Accessed 6 Jan 2020

UN (2019b) The United Nations General Assembly declare 2021–2030 the UN decade on ecosystem restoration. https://www.unwater.org/the-united-nations-general-assembly-declare-2021-2030-the-un-decade-on-ecosystem-restoration/. Accessed 1 Feb 2020

UNCTAD (2013) Trade and environment review 2013. Wake up before it is too late. Make agriculture truly sustainable now for food security in a changing climate. United Nations Conference on Trade and Development. unctad.org/en/PublicationsLibrary/ditcted2012d3_en.pdf. Accessed 05 Nov 2017

UNHRC (2017) Report of the Special Rapporteur on the right to food, vol. A/HRC/34/48. United Nations Human Rights Council, documents-dds-ny.un.org/doc/UNDOC/GEN/G17/017/85/PDF/G1701785.pdf?OpenElement. Accessed 29 Sept 2019

UNHRC (2019a) International Covenant on Civil and Political Rights, CCPR/C/126/D/2751/2016. United Nations Human Rights. Office of the High Commissioner (ed) UN Treaty Body Database. https://tbinternet.ohchr.org/_layouts/15/treatybodyexternal/Download.aspx?symbolno=CCPR/C/126/D/2751/2016&Lang=en. Accessed 23 Nov 2019

UNHRC (2019b) Paraguay responsible for human rights violations in context of massive agrochemical fumigations. United Nations Human Rights. Office of the High Commissioner, www.ohchr.org/EN/NewsEvents/Pages/DisplayNews.aspx?NewsID=24890&LangID=E. Accessed 23 Nov 2019

USAO (2019) Monsanto agrees to plead guilty to illegally spraying banned pesticide at Maui Facility. The United States Attorney Office, www.justice.gov/usao-cdca/pr/monsanto-agrees-plead-guilty-illegally-spraying-banned-pesticide-maui-facility. Accessed 21 Dec 2019

USDA (2015) Family farms are the focus of new agriculture census data. https://www.usda.gov/media/press-releases/2015/03/17/family-farms-are-focus-new-agriculture-census-data. www.usda.gov/media/press-releases/2015/03/17/family-farms-are-focus-new-agriculture-census-data. Accessed 15 Nov 2017

Vandenberg LN et al (2017) Is it time to reassess current safety standards for glyphosate-based herbicides? J Epidemol Comm Health 71:613–618

Vandermeer J (1989) The ecology of intercropping. Cambridge University Press, Cambridge

VBZ NRW (2016) Lebensmittel: Zwischen Wertschätzung und Verschwendung. https://www.verbraucherzentrale.nrw/lebensmittelverschwendung.   Accessed 10 Nov 2019

Vitousek PM, Mooney HA, Lubchenco J, Melillo JM (1997) Human domination of Earth's ecosystem. Science 277:494–499

Vogel G (2019) €100 million German insect protection plan will protect habitats, restrict weed killers, and boost research. Science. https://doi.org/10.1126/science.aaz4151

Wagner SL (2000) Fatal asthma in a child after use of an animal shampoo containing pyrethrin. West J Med 173:86–87

Weiss H (2010) Schwarzbuch Landwirtschaft: Die Machenschaften der Agrarpolitik. Deuticke Verlag, Wien

Willer H, Lernoud J (eds) (2016) The world of organic agriculture. Statistics and emerging trends 2016. Research Institute of Organic Agriculture (FiBL) IFOAM—Organics International, Frick, Switzerland and Bonn, Germany

Willer H, Lernoud J (eds) (2019) The world of organic agriculture. Statistics and emerging trends 2019. Research Institute of Organic Agriculture (FiBL) IFOAM—Organics International, Frick, Switzerland and Bonn, Germany

Wyckhuys KAG, Heong KL, Sanchez-Bayo F, Bianchi FJJA, Lundgren JG, Bentley JW (2019) Ecological illiteracy can deepen farmers' pesticide dependency. Environ Res Lett 14:093004

Yamamuro M, Komuro T, Kamiya H, Kato T, Hasegawa H, Kameda Y (2019) Neonicotinoids disrupt aquatic food webs and decrease fishery yields. Science 366:620–623

Yang T, Doherty J, Zhao B, Kinchla AJ, Clark JM, He L (2017) Effectiveness of commercial and homemade washing agents in removing pesticide residues on and in apples. J Agric Food Chem 65:9744–9752

Yu HL, Zhang Y, Wu K, Wyckhuys K, Guo Y (2009) Flight potential of Microplitis mediator, a parasitoid of various lepidopteran pests. BioControl 54:183–193

Zaller JG, Moser D, Drapela T, Schmöger C, Frank T (2008) Effect of within-field and landscape factors on insect damage in winter oilseed rape. Agric Ecosyst Environ 123:233–238

Zandonella R, Sutter D, Liechti R, Stokar Tv (2014) Volkswirtschaftliche Kosten des Pestizideinsatzes in der Schweiz. Pilotrechnung. https://heidismist.files.wordpress.com/2016/03/ch-kosten-pestizideinsatz-zusammenfassung.pdf. Accessed 1 Feb 2020

Zhang W (2018) Global pesticide use: profile, trend, cost /benefit and more. Proc Int Acad Ecol Environ Sci 8:1–27

ZS-L (2013) Wege aus der Hungerkrise. Die Erkenntnisse und Folgen des Weltagrarberichts: Vorschläge für eine Landwirtschaft von morgen. Zukunftsstiftung Landwirtschaft, Berlin, 52pp

Index

© Springer Nature Switzerland AG 2020
J. G. Zaller, *Daily Poison*, https://doi.org/10.1007/978-3-030-50530-1

Printed in the United States
By Bookmasters